SCALING

SCALING: A Sourcebook for Behavioral Scientists

Edited by GARY M. MARANELL

Routledge
Taylor & Francis Group

LONDON AND NEW YORK

First published 1974 by Transaction Publishers

Published 2017 by Routledge
2 Park Square, Milton Park, Abingdon, Oxon OX14 4RN
711 Third Avenue, New York, NY 10017, USA

Routledge is an imprint of the Taylor & Francis Group, an informa business

Library of Congress Catalog Number: 2007025981

Library of Congress Cataloging-in-Publication Data

Scaling : a sourcebook for behavioral scientists / Gary M. Maranell, editor.
   p. cm.
   Originally published: Scaling / edited by Gary M. Maranell. Chicago : Aldine Pub. Co., [1974].
   Includes bibliographical references and index.
   ISBN 978-0-202-36175-8
   1. Scaling (Social sciences) 2. Attitude (Psychology)--Testing. 3. Psychometrics. I. Maranell, Gary Michael, 1932-

H61.M423 2007
300.1'519535--dc22

2007025981

ISBN 13: 978-0-202-36175-8 (pbk)

Contents

Preface   ix

Introduction   xi

PART I. MEASUREMENT AND SCALING   1

  1. Ernest Nagel, Measurement   3
  2. S. Smith Stevens, Measurement   22
  3. Cletus J. Burke, Measurement Scales and
     Statistical Models   42

PART II. THURSTONIAN METHODS   57

  4. Louis L. Thurstone, Psychophysical Analysis   59
  5. Louis L. Thurstone, A Law of Comparative Judgment 81
  6. Bert Green, Paired Comparison Scaling Procedure   93
  7. Allen L. Edwards, Circular Triads, the Coefficient
     of Consistence, and the Coefficient of Agreement   98
  8. Robert T. Ross, Optimal Orders in the Method
     of Paired Comparisons   106
  9. Allen L. Edwards, Paired Comparison Attitude Scales 110
 10. Allen L. Edwards, The Method of Equal-Appearing Intervals 113
 11. Bert Green, The Method of Successive Intervals   122

PART III. SCALOGRAM ANALYSIS   129

12. Samuel A. Stouffer, An Overview of the
    Contributions to Scaling and Scale Theory   131
13. Louis L. Guttman, The Basis for Scalogram Analysis   142
14. Benjamin W. White and Eli Saltz, Measurement
    of Reproducibility   172
15. Karl F. Schuessler, A Note on Statistical
    Significance of Scalogram   197
16. Roland J. Chilton, Computer Generated Data and
    the Statistical Significance of Scalogram   205
17. Roland J. Chilton, A Review and Comparison of Simple
    Statistical Tests for Scalogram Analysis   212
18. Carmi Schooler, A Note of Extreme Caution on the
    Use of Guttman Scales   223

PART IV. SUMMATED RATING METHODS   231

19. Rensis Likert, The Method of Constructing an
    Attitude Scale   233
20. John P. Robinson, Jerrold G. Rush and Kendra B. Head
    Criteria for an Attitude Scale   244
21. H. J. Butcher, A Note on the Scale Product and
    Related Methods of Scoring Attitude Scales   258
22. P. K. Poppleton and G. W. Pilkington, A Comparison of
    Four Methods of Scoring an Attitude Scale in
    Relation to its Reliability and Validity   267

PART V. UNFOLDING THEORY   273

23. C. H. Coombs, Some Aspects of the Metatheory
    of Measurement   275
24. C. H. Coombs, Psychological Scaling without a
    Unit of Measurement   281
25. C. H. Coombs, A Real Example of Unfolding   300
26. David Goldberg and C. H. Coombs, Some Applications of
    Unfolding Theory to Fertility Analysis   311

PART VI. RELATED MATERIALS   333

27. Allen L. Edwards and F. P. Kilpatrick, A Technique for the Construction of Attitude Scales   336
28. Samuel A. Stouffer, Edgar F. Borgatta, David G. Hays and Andrew F. Henry, A Technique for Improving Cumulative Scales   347
29. Robert K. Leik, and M. Matthews, A Scale for Developmental Processes   365
30. Sanford Labovitz, The Assignment of Numbers to Rank Order Categories   388
31. Frederic M. Lord, On the Statistical Treatment of Football Numbers   402

Selected Bibliography   407

Appendix: Tables   419

Index   429

Preface

This sourcebook is an attempt to cover the generally recognized scaling methods. However, to keep it from becoming overly large, it was necessary to exclude certain types of material. The decisions regarding what should be excluded are never easy to make, and some readers may disagree with those made here. I can only hope that a substantial number of teachers will find merit in the content that has resulted from these difficult decisions. Because of size and cost considerations, this volume does not include highly mathematical articles. The materials that have been included should be quite understandable to any behavioral science graduate student and to most undergraduate students as well.

Ideally, sourcebooks should draw together related and useful material from a wide range of diverse sources that are otherwise difficult to assemble. There is little justification for bringing together papers that appeared in only the major journals of a particular discipline. The material here has been drawn from a wide spectrum of journals and books, both sociological and psychological. Further, the materials have been drawn from a time spectrum covering more than 40 years.

It has been my impression that the average student in the behavioral sciences does not have an opportunity to read a sufficient number of primary sources for himself, and this is especially true in the area of methodology. I have, however, hesitated to send a dozen or more students to journals when I knew that the journals were in tatters already. Yet I felt that they were missing some of the excitement of discovery and continuity this literature can provide. This is one reason I have prepared this book.

The reader should understand that this volume is intended for students who are engaged in learning scaling, and much of the material may be familiar to experienced professionals. The book is intended to be mathematically within the reach of everyone who is willing to read with some care. It is not the exhaustive tome originally planned but I believe that it is a more useful volume in its present form.

Introduction

This book focuses upon the process of scaling as a problem of empirical measurement and, even more importantly, as the process of translating theoretical ideas and concepts into variables. These variables or systems of observations, which are coded in a sufficiently reliable and precise manner, are necessary to make it possible for scientists to perceive, record, and study differences, correlations, and change in the phenomena they are interested in understanding. Without the scaling procedures that record and measure variables, the process of studying social and individual behavior would be seriously limited, and research in many areas of study would be made impossible. Behavioral science may be said to have advanced in many respects with the developments that have occurred in the level of sophistication of scaling and measurement procedures. Such continued advances, coupled with an increasing sophistication in theory and model construction, are prerequisites for its subsequent development.

Discussions of measurement often advance the accurate but rather proasaic assertion that "measurement involves the assignment of numbers, or sometimes numerals, to objects according to rules." This volume attempts to provide a variety of such sets of rules. Sections II through V each describe a different procedure, or in some cases a set of procedures, for the assignment of numbers to objects. These procedures are the operations needed to construct the measuring devices that behavioral scientists use to do the research and theory testing necessary to develop and advance their field of study. In a fundamental sense, the procedures of scaling are consonant with and subtlely and consistently encouraging of more systematic and precise problem formation and theory construction. This is because scaling and measurement procedures strongly encourage and, in most instances, demand an attention to and a focusing upon variables that are quantifiable rather than upon less precise pieces of observation of social and psychological phenomena. In an important sense, one fundamental prerequisite for the continued development of the behavioral sciences is the increasing quantification of our basic variables that, in turn, demands some theory and technique of measurement.

This volume is also an attempt to provide a glimpse of the growth and development of scaling as it has been employed in the behavioral sciences. The development of scaling methods has been somewhat cumulative, each approach building in some ways upon antecedent attacks upon the measurement problem. In the instances when this development has not been cumulative, the innovation was often stimulated by a shortcoming, an unrealistic or unacceptable assumption, or a weakness in the available methods. In some of these cases, the innovation has been based upon alternative assumptions that seem more realistic or are easier to accept. The development within scaling also involves an elaboration and improvement of techniques and procedures and a needed attention to oversights and to the dangers involved in the misapplication and the naive use of particular scaling methods.

Many behavioral scientists have been somewhat indifferent to the problems and issues involved in measurement and scaling. This frequently has led them to allocate much less thought to this aspect of their research than is necessary for careful inquiry. When measurement problems are ignored, the effects and the results are not unlike those that would be encountered by astronomers who are forced to use cracked lens and no calibration or surveyors who are forced to use rubber yardsticks or no yardsticks at all, or physicists with watches which run randomly fast and slow. In view of this one can only be dismayed at the consistently cavalier approach to measurement which has frequently typified behavioral science research. Superficial and naive measurements are occasionally subjected to inappropriately careful analysis that, of course, is not better or worse than the instances when the grossness of the analysis rivals the casualness of the measurement. This casualness in regard to measurement is often in marked contrast to the fact that such gross measures are used to do research that has been otherwise methodically designed and has grown out of extremely subtle, careful, lucid, and sophisticated theoretical consideration. It is safe to say that an analysis is nearly meaningless when the phenomena being analyzed have been carelessly measured.

The indifference to measurement, or its lack of a position of importance in many of the behavioral sciences (particularly Sociology, Political Science, and Anthropology), suggests that it is seen as peripheral rather than centrally important to the field. This lack of integration of something as important as measurement into the body of the behavioral sciences is a matter of some seriousness. The fact of this lack of integration is revealed through an examination of course offerings in many behavioral science departments. These listings of courses reveal that there exists an apparent plethora of courses on all facets of, for example, the Sociology of——, and that independent from all these courses there is a set of courses on how to do the science—in case anyone might be interested. These separate and peripheral courses include those devoted to scaling and measurement, design and analysis and are taught by a species, unique to the less established sciences, called methodologists. Apparently few are concerned about how a science can develop when it separates training in the ways of doing the science from knowledge of the content of the science.

There are many reasons why scaling and measurement are important to the behavioral sciences. First, *science is nomothetic,* or put another way, nomothetic measurement is necessary to do scientific research and is needed to develop a science. This means that science finds it necessary to measure or scale the same property in all the units, subjects, or phenomena being studied. The relationship between two variables can be discovered only when all subjects or units of study have been scaled or measured on the same two variables. Obviously the nomothetic nature of science strongly encourages the use of scales of measurement among behavioral scientists. The construction of a scale designed to measure, for example, anxiety does, when it is applied or administered to all the members of a group of subjects, provide nomothetic measurement. Similarly the employment of a scale of urban integration to measure this property in a sample of cities provides an assessment of the same property in all the cities being studied.

A second reason scaling and measurement are important to the development of science is that they provide *standardization.* For a science to grow other researchers must be able to replicate one's research, and to do this others must be able to perform the same measurements one has previously performed. This means that the procedures of measurement must be clear, open or public, and repeatable, and the easiest way to accomplish this is through the development, use, and publication of the scale or measurement device. When one has conducted a study and has employed a scale of, for example, social integration or cohesion, other researchers can by employing the same scales secure comparable measurement in other samples, for other studies and purposes. A related advantage is that by being standardized a scale increases objectivity since the biases of each researcher are controlled and minimized by using the standardized measurement. Furthermore even the biases of the scientist that devised the scale are, by virtue of the openness and availablity of the scale, subject to greater scrutiny and consequently greater control.

Another advantage of scaling is that it provides for increased *precision*. Instead of simply classifying groups in terms of being formalized or not, a scale that identifies the characteristics and indicators of the variable "formalization" allows the scientist to make more extensive and subtle discriminations between groups on this variable and permits a more precise description of the ways the groups vary in formalization.

A fourth advantage of scaling and measurement is that they allow for and increase *conciseness*. Quantitative measurement and scaling permit the condensation and conciseness of mathematics to be employed by scientist. This conciseness occurs in the description of the quantitative measurement, the reporting of the results of the measurement, as well as in the description of the theoretical and empirical relationships discovered or predicted. This advantage of scaling and measurement involves more explicitly an increased economy. The economy is involved in the time, effort, and words gained or saved when one expresses theories, variables, relationships, and findings in general, as well as in the processes involved in quantitative and mathematical analysis. It is for reasons such as these that some have said, "If you haven't measured it, you do not know what you are talking about."

The present limited concern for scaling and scale construction grew out of a need to secure useful measurement of phenomena that were of interest to behavioral scientists and did not lend themselves easily to quantification—that is, the development of scaling was stimulated by an interest in securing something Thurstone referred to as "subjective measurement." Behavioral scientists were and still are interested in obtaining measurements of phenomena such as values, attitudes, and opinions, in addition to other phenomena. In the 1920's, Thurstone, having these interests, considered the issues involved in measurement seriously and provided the early and still substantial developments associated with his name. His substantive empirical research illustrated the usefulness of the techniques he developed. Thurstone studied food preferences; social values; views of criminal offenses; attitudes toward war, the church, and movies among other things. The continued use of these methods documents the importance of his accomplishments.

Summated rating scales have also focused upon measuring subjective phenomena, explicitly attitudes and opinions. Because of Rensis Likert's advocacy of the method, as well as his contributions to the method, such scales are sometimes referred to as Likert scales. He viewed the process as an indirect method of measuring dispositions and values that are most easily expressed verbally. Scales of this type, which involve the summing of weighted responses, have several advantages. The advantages include the facts that such scales are easy to construct, have been found to be highly reliable, and have consistently provided useful and meaningful results in research. Therefore, it is not surprising to find that Likert scaling is still very commonly used.

Louis Guttman's interest was in the area of attitude measurement as well. Guttman scaling theory developed from the basic concept of a scalable or unidimensional universe of attributes. One of the main purposes of the technique is to ascertain whether the attitudes or characteristics (universe of content or attributes) being studied actually involve only a single dimension, or form a unidimensional scale. Guttman's association with the research group that produced *The American Soldier* (Stouffer, et al., 1950) provided an opportunity for an appreciation of the importance of his contribution to disseminate quite widely. The publication also explained the theoretical background, procedures, and modes of evaluation of the cumulative or Guttman scales in a comprehensive manner and reported the results of a program of research that constructed and employed hundreds of such scales, measuring attitude areas such as soldiers' adjustment to military life, the effects of combat, and postwar plans. The interest in cumulative scales has remained high, and it is still one of the most commonly employed methods of scaling many sociological, psychological, and political variables.

A more recent innovative contribution is that of C. H. Coombs. This method, called unfolding, is also used for attitude scaling although it is useful for scaling other objects and subject preferences as well. The procedure derives information on the unidimensionality and relative spacing of items from individual preference orders. The simplicity of the assumptions make this one of the most elegant and soundly based methods of those available. The process is only now beginning to be used in research to any appreciable extent. It is included in this volume in the hope that this will increase its use to some extent.

The scientists mentioned above were, for the most part, concerned with subjective measurement and perceived its importance to the behavioral sciences. Subsequent examination reveals, however, that the scaling procedures are more generally useful than anticipated. The Thurstonian, summated rating, and unfolding methods are principally oriented to the measurement of attitudes and opinions, which is, of course, an important task. Perhaps because of its extensive use by sociologists and political scientists, scalogram analysis has been more widely applied. Laws, behaviors, voting patterns, cultural and national characteristics, among other phenomena, have been measured with scalogram procedures.

These methods of scaling have achieved reliability that is greater than that associated with measurement based upon a single question or observation. Scaling also can reveal an order present in our observations that might otherwise elude us. We discover, for example, that a scaling of congressional votes can reveal a clear order of both men and issues. Scales can provide a structure, a fixed order, and stable measurement that less systematic approaches may leave undetected.

Scaling is not without some problems. Some of the Thurstonian methods, for example, demand a great amount of effort on the part of the subjects, which is a serious shortcoming. Other approaches, such as the summated rating method and certain Thurstonian methods, attend to sums of responses and means of responses that disregard individual differences in response variation and pattern. This inattention can result in a tendency to ignore the fact that identical scores can be obtained in startlingly different ways.

One larger problem in the field of scaling is a result of the strong data orientation which has marked the work of those using such techniques. Given the fact that most scales have been and still are devised for a particular study or program of research, the attention to the problem of validity (or that a scale actually measures what it proports to measure) has been typically somewhat inadequate. Face validity (the circumstance in which a scale simply looks like it is measuring what it is supposed to measure) has been grossly overworked, and more sophisticated approaches that also demand more time and effort have often been avoided. Validation is an onerous but necessary task. It is a task that social scientists frequently avoid.

Some of the articles that follow do not reveal a concern with certain subtleties of scaling, while others are quite attentive to them. The former articles tend to suggest that the ordering of objects is sufficient and that scaling beyond the ordinal level may be a waste of time. This position is consistent with the use of scaling methods that serve to determine only the order of a set of objects. This position is somewhat inconsistent with other articles that treat techniques designed to make appraisals of the intervals between ordered objects. However, both orientations are relevant to current conceptions of the problem of scaling, and it would be a serious oversight to ignore this issue in a sourcebook devoted to scaling.

Part I of the book presents a series of general articles dealing philosophically with the problem of measurement. This section examines what is meant by measurement and scaling as well as the notions underlying the process of measuring. The articles present positions ranging from highly restrictive interpretations of scaling to more permissive definitions of the problem. The former position, taken by Nagel and Stevens, demands that the numerals assigned to the observations reflect only the characteristics of the observations and that the statistics used to analyze such data reflect these same characteristics. Burke takes a more permissive position which encourages the use of relatively elaborate statistical analysis on observations that are assigned numbers reflecting only the ordered nature of the observations. A paper by Labovitz, included in Part VI of this volume also reflects this more permissive position.

Part II deals with the scaling methods developed and advanced by L. L. Thurstone. These include paired-comparison scaling, equal-appearing interval scaling, and successive-interval scaling. The paired-comparison method, scales stimuli and objects through an analysis of judgments that subjects have made regarding the relative positions of pairs of stimuli. The subject's or judge's task is made as simple as possible by focusing attention upon only two stimuli or objects at a time. The subject is asked to compare these two in regard to some aspect or property, decide which has more of the property, and then proceed to another pair, make the same comparison on the same property, and so on until all the paired-comparison judgments requested have been made. Judgments acquired in this manner can be employed to provide estimates of the relative positions of the entire set of stimuli or objects. The method of equal-appearing intervals requires the subjects or respondents to place each of the items or stimuli being scaled into one of a set of ordered categories. The scale values of the items are typically scores (such as medians, or means) summarizing where a group of respondents have placed the items in the ordered categories. The extent of agreement on the placement is described with a measure of the dispersion of the judgments. The third approach to scaling associated with L. L. Thurstone is the method of successive-intervals. The task performed by the respondents is similar to that required in the equal-appearing interval method, however the analysis of the responses differs and it provides through a more elaborate procedure a somewhat improved and more stable set of scores. Part II of this book also describes the application of Thurstone methods to attitude scaling and presents statistical tests that can be used to appraise the significance of agreement between subjects as well as the consistency of subjects.

Part III focuses on scalogram analysis, or Guttman scaling, and presents the background, rationale, and procedures involved in this method. In addition to the assumption of the existence of a scalable or unidimensional universe of content or attributes, scalogram analysis is based further on a notion of ordered indicators or successive barriers or hurdles. For example, once an individual has acquired or expressed a difficult or extreme position or has passed a difficult hurdle or indicator he should express or pass all simpler and less demanding ones as well. Guttman scaling can therefore be said to order cases or respondents along a single dimension in terms of some property; it also creates a variable or measurement by combining characteristics, items, or indicators of the property, and at the same time it ascertains or tests the extent to which these characteristics or items "hang together" or are themselves systematically ordered. Unless both indicators and subjects can be ordered with regard to a single attribute, we have no grounds, according to Guttman, for assuming that we have produced an ordinal scale of a unitary concept. The simultaneous ordering of indicators and respondents provides the essential feature of a true Guttman scale—if each respondent or object is assigned a rank position according to their total responses or characteristics the process can be reversed and the responses can be deduced from the rank position. This quality of being able to deduce or reproduce the responses to each item, knowing only the total score or ranking, is called "reproducibility" and is one test of whether a set of items constitutes a scale in the Guttman sense. Cumulative or Guttman scales are evaluated in terms of the extent to which the pattern of response approximates these ideal forms and is not simply accounted for by the operation of chance. The measures that indicate the degree of approximation to the ideal forms are referred to as coefficients of reproducibility or scalability. Part III of the book describes a number of such coefficients and provides ways of appraising the role of probability in scalogram analysis.

Part IV is concerned with summated rating or Likert scaling. This type of scaling is used in the measurement of attitudes, opinions, beliefs, preferences, information, and other phenomena that employ multiple response items. This method also focuses upon the problem of unidimensionality; however, in this case it typically involves only an attempt to insure that all the items are measuring the same attitude or opinion area and is accomplished through an item analysis. The scaling procedure involves the collection of many statements or items pertaining to the attitude or belief in question. These items are then administered to a trial group of subjects who indicate whether they strongly agree, agree, are undecided about, disagree, or strongly disagree with each statement. These response categories are weighted in a manner consistent with the content of the attitude being scaled. Each item is then examined to see how well the responses to it agree with the total scores based on all items. Items are eliminated if the agreement is low. The items that remain after this analysis make up the final scale and the set is assumed to measure the same or a similar attitude or belief. This process increases the homogeniety or consistency of the items but does not guarantee that only one property is being measured nor insure that the resulting scale is unidimensional. The problems of concern to someone developing a summated rating scale—problems of item analysis and scale evaluation—are dealt with in this section of the book.

Part V is an introduction to unfolding theory and methods. This method is also a technique used for attitude scaling although it is useful for measuring other subject preferences as well. Unfolding is a recent and innovative contribution to scaling by C. H. Coombs. It is based on the assumption that the judging process involves an appraisal of the relative distance between a stimuli and the judge's ideal position. The method views the ordering of stimuli as a consequence of the existence of ordered distances between each stimuli or item and the judge's or subject's ideal position. Therefore, the technique locates stimuli and subjects on the same scale and with sufficient data it can determine the ordering of the intervals between the stimuli as well as the ordering of the stimuli themselves. The method is in less general use than most of the other procedures described in this volume. Most of the research that has used unfolding techniques has been done by scientists associated with the originator of the method. An increase in familiarity with the method among behavioral scientists generally will almost certainly lead to a greater appreciation for its elegance and increase its use.

Part VI of the book is made up of a selection of rather loosely related papers. These articles focus upon various special cases and problems relevant to scaling. They include discussions of some matters that are developments of certain of the previous discussed methods as well as other material which is not intimately associated with any of the five scaling methods discussed earlier in the book.

The six parts of this book constitute a rather complete introduction to the major forms of scaling. In total they provide an overview of the development of scaling in the behavioral sciences.

PART I

Measurement and Scaling

Part I is concerned with the phenomena of measurement and scaling in general. This includes a consideration of the postulates upon which measurement is based as well as a discussion of the processes of measurement in science generally. An examination of the nature of scales and the relationship between measurement and analysis procedures also are included. Measurement is both an operation employed by scientists in the pursuit of their research and an area of concern of philosophers of science. This part of the volume focuses upon the general philosophic considerations and the implications of these for scientific activity.

In the first paper, Ernest Nagel presents a treatment of measurement that cuts across science in general. He emphasizes the point that quantitative measurement serves various important goals in science. For example, measurement rules eliminate ambiguity in classification and serve to achieve observational agreement. Measurement accomplishes this by providing objective procedures that are employable by any scientist. Another important result of measurement is that, by providing quantification and metric, the formulation of scientific theories and equations is facilitated. The development of theory and equations is a matter of great importance to science: It is indeed a goal of science. Nagel considers the condition of measurement and presents the axioms of quantity or measurement; the primitive and most simple relations upon which a measurement system can be built. His paper also contains an extensive discussion of the importance of addition in measurement. Some scholars contend that measurement is possible only when the scale employed allows clear and appropriate addition. Nagel points out that most qualities are incapable of addition, and therefore, the general measurement process cannot be so restricted without recognizing that this makes even such properties as density and acceleration unmeasureable.

In the second paper S. S. Stevens presents the classical view of measurement as well as the development of his own position. Stevens defines the levels of measurement—nominal, ordinal, interval, and ratio. He also discusses the relationships between the scales and the operations necessary to establish each scale. Examples of variables of each scale type are presented. Stevens contends that the basic classification of the scales of measurement serves as a device that dictates the selection of appropriate statistical methods. His work, among others, has precipitated an interest in non-parametric statistics. Non-parametric statistics include analytic procedures that do not assume the existence of interval or ratio scales or a particular distribution of variables. They require only nominal and ordinal scales of measurement and make no specific assumption about the distribution of the variables being examined. Therefore, when an investigator has only order in his measurements, only knowing that one person exceeds another and is exceeded in turn by a third in some property, and when he knows nothing regarding the amount or interval that separates them, he is encouraged by Stevens to employ statistics that are based on the assumption that only the order he knows is resident in the measurement. The relationship between measurement scale and statistics is of immense importance to the position Stevens advances. However, we will find later that this is not a position shared by everyone. Stevens discusses some additional scales of measurement developed by C. H. Coombs. (These techniques are more fully discussed in Part IV of this book.)

The final paper in this section is a discussion of the relationship between measurement scales and statistics. Two opposing orientations are presented. The first of these is the measurement-directed position that is the one presented by Stevens in the previous article; this position is described and analyzed. The second position is a measurement-independent position that insists that the statistician must be free to employ methods without the constraints imposed by scale considerations. This paper, by Cletus J. Burke, identifies the issues central to the debate between those occupying these two positions. The issues raised in this article and in this first section in general will come to our attention again at the end of this book.

Chapter 1

MEASUREMENT

ERNEST NAGEL
Columbia University

I. The Occasion and Conditions for Measurement

Measurement has been defined as the correlation with numbers of entities which are not numbers.[1] As practiced in the developed sciences this is a sufficiently comprehensive, though cryptic, statement of the object of measurement. But in a larger sense, in a sense to include most of those acts of identification, delimitation, comparison, present in every day thought and practice, numerical measurement is only infrequently used. "*This* is the missing book," or "He had a *good* sleep," or "The cake is *too* sweet," are judgments making no explicit reference to number. From this larger point of view, measurement can be regarded as the delimitation and fixation of our ideas of things, so that the determination of what it is to be a man or to be a circle is a case of measurement. The problems of measurement merge, at one end, with the problems of predication.

There are indeed vast domains of reflection and practice where numbers have taken little hold. Prior to Descartes, geometry was not established on a thoroughgoing numerical basis, and many branches of mathematics, like symbolic logic or projective geometry, may be pursued without in-

Reprinted by permission of the author from *Erkenntnis*, II Band, 1931, pp. 313–33. This essay is the second chapter of a larger work on the logic of measurement.
    1. Spaier, *La Pensée et la Qualité*, p. 34.

troducing numbers. In arts like cookery, measurement is not primarily numerical, and the operations used are very often controlled by disciplined judgments on the qualitative alterations of the subject matter.

But the difficulty and uncertainty often experienced in obtaining desired consequences when vaguely defined ideas and crude methods of applying them are used, soon lead wherever possible to the introduction of more or less refined mathematical processes. The immediate, direct evaluation of subject matter secures too little uniformity to be of much value; and direct judgments, e.g., of lengths by the eye, are consequently replaced by more complicated and indirect operations, such as the transporting of unit lengths along the lines to be compared. It is, indeed, because less error, that is, greater uniformity, is obtained in judgments of spatial coincidences than, e. g. in judgments of differences of length or color, that spatial congruence plays so large a role in laboratory practice. The *raison d'être* of numbers in measurement, is the elimination of ambiguity in classification, and the achievement of uniformity in practice.

It is generally only after numerical measurements have been established and standardized, that references to the "real" properties of things begin to appear: those properties, that is, which appear in circumstances allowing for most facility in their measurement. The "real" shape of the penny is round, because from the point of view from which the penny is round, measurements and correlations of other shapes can be carried on most easily. Nevertheless, it must not be overlooked that a numerical evaluation of things, is only one way of making evaluations of certain selected characters, although it is so far the best. It is preeminently the best, because in addition to the obvious advantage they have as a universally recognized language, numbers make possible a refinement of analysis without loss of clarity; and their emotionally neutral character permits a symbolic rendering of invariant relations in a manifold of changing qualities. Mathematics expresses the recognition of a necessity which is not human.

From this last point of view, therefore, the search for a unified body of principles in terms of which the abrupt, the transitory, the unexpected, are to exhibited and in a measure controlled, is the most conscious guiding principle in the application of numerical science. If the search for mathematical equations is the aim of physics, other activities such as experiment, classification, or measurement are subservient to this aim, and are to be understood only in relation to it. Consequently, if we inquire why we measure in physics, the answer will be that if we do measure, and measure in certain ways, then it will be possible to establish the equations and theories which are the goal of inquiry.

It is relevant, therefore, to demand the logical foundations of the mathematical operations which physics constantly uses. For if mathematics is applicable to the natural world, the formal properties of the symbolic operations of mathematics must also be predicable of many segments of that world. And if we can discover what these formal properties are, since mathematics *is* relevant to the exploration of nature, a physical interpretation *must* be found for them. That physical interpretation will constitute, whenever it can be found, the conditions for the measurement of that subject matter. Consequently, if we ask why in measurement we attend to certain characters of objects to the exclusion of others, the answer will be that the selected characters are precisely those with which applied mathematics can cope. It is only by a reference to the function which the numerical measures or magnitudes of things have in equations, that one can remove the apparent arbitrariness in selecting one rather than another set of conditions as fundamental for measurement.

In recent years the formal conditions of their science have been much discussed by mathematicians. The properties which magnitudes must have in order to be capable of the kind of elaboration which the mathematics of physics requires, have been variously formulated as axioms of quantity.[2] The following set, with some modifications, is taken from Hoelder.[3]

1. Either $a > b$, or $a < b$, or $a = b$.
2. If $a > b$, and $b > c$, then $a > c$.
3. For every a there is an a' such that $a = a'$.
4. If $a > b$, and $b = b'$, then $a > b'$.
5. If $a = b$, then $b = a$.
6. For every a there is a b such that $a > b$ (within limits).
7. For every a and b there is a c such that $c = a + b$.
8. $a + b > a'$.
9. $a + b = a' + b'$.
10. $a + b = b + a$.
11. $(a + b) + c = a + (b + c)$.
12. If $a < b$, there is a number n such that $na > b$ (also within limits).[4]

How much more perspicacious such an axiomatic or functional analysis of magnitudes can be than definitions in terms of private, intrinsic properties (e. g., "quantity is the relation between the existence and the

2. It need be mentioned only in passing that mathematics cannot be regarded as exclusively the science of quantity. Its essence is the study of types of order, of which the quantitative one is a single instance.

3. "Die Axiome der Quantität," *Ber. d. Sächs. Gesellsch. d. Wiss., math.-phys. Klasse,* 1901.

4. On the importance of the Archimedean axiom, see Hilbert, "Axiomatisches Denken," *Math. Annalen,* Vol. 78, p. 408. The fact that one can estimate stellar distances by adjoining earthly ones, is not a logical consequence of theorems on congruence, but is an independent experimental conclusion.

nonexistence of a certain kind of being");[5] how much more successfully it satisfies the demand for a formulation which, true for *every* instance of quantity, should not be the exhaustive statement of *any one* instance, will be evident from the sequel. Even Meinong's definition of magnitude as "whatever is capable of being limited toward zero" or as "that which is capable of having interpolations between itself and its contradictory," has little clarity to recommend it.[6] Later on a distinction will be made between magnitudes which satisfy all twelve axioms and those which satisfy only the first six. A magnitude in the most complete sense, however, is whatever is capable of verifying the whole set.

II. Order and Equality

The illustration of these relations by indicating some of the empirical procedure involved in measurement, is the next portion of our task. At the very outset it must be pointed out, however, that an adequate interpretation of the axioms must inevitably lead out of the laboratory where measurements are usually made, and lead on to a consideration of the manufacture of the laboratory *instruments*. When a laboratory experiment is studied behavioristically, the measurements performed consist in the observation of the movement of a pointer on a scale, or of the superposition of lengths.[7] How natural, therefore, to suppose that measurement consists only in the observation of space-time coincidences! It is important to remember, however, that the experimenter, working with marked or calibrated instruments, assumes that the calibrations indicate various qualitative continuities not *explicitly* present. The process of measurement has not been fully exhibited until all those operations of calibration have been noted. When a weight is attached to a spring balance, and the position of a marker on the scale read, only a very small fraction of the process actually necessary to estimate the weight as five pounds has been observed; the operations entering into the construction and *correlation* of scale and spring must be included. It is of the essence of an experiment that it be repeatable. Therefore it is not the particular instrument used any more than it is the unique experiment which has such an overwhelming importance in science; it is rather the repeatable process capable of producing the markings on the instrument which is. Every marked instrument implies the construction and existence of some standard series of magnitudes, correlation with which constitutes the calibration. A wholehearted recognition of this reference of instruments to something beyond

---

5. So Warrain, a follower of Wronski, in *Quantite, Infini, Continu*, p. 9.
6. "Uber die Bedeutung d. Weberschen Ges." in *Gesam. Abhandl.*, Vol. 2, p. 219.
7. Cf. Duhem, *La Theorie Physique*, p. 219.

themselves, is a recognition that other characters of existence besides the spatial are capable of, and are involved in, the process of measurement.[8]

The relation $>$, or its converse $<$, is a transitive assymetrical relation. It finds its exemplification in some discovered qualitative domain which is sufficiently homogeneous to allow identification as a well defined range of a single quality.[9] Within this domain the character studied must be capable of such a serial gradation that a transitive assymetrical relation can be discovered to hold between the discriminated elements. So we find the character of density which liquids manifest in relation to one another to be such a relation. This character may be defined, with somewhat more care than is shown here, as the capacity of a liquid to float upon other liquids. Liquid a will be said to be more dense than b if b can float on a but a cannot float on b. And it can and *must* be shown *experimentally*, that, for a set of liquids distinguishable from each other by all sorts of physical and chemical properties, the relation $>$ (more dense) is a transitive assymetrical relation: if liquid a is more dense than b (i. e., b floats on a) and if b is more dense than c, then a is more dense than c (i.e., c floats on a).

The relation of equality $(=)$ can be defined in terms of $>$. We say that $a = b$, if a is not $> b$, and a is not $< b$, and if $a \gtrless c$, then also $b \gtrless c$. Equality of density may therefore, be defined thus: c has a density equal to that of d if c does not float on d and d does not float on c; and if c floats on e so does d.[10]

The experimental establishment of series of this kind involving the first six axioms is the first step in the introduction of number. The function of the experiments is the careful exhibition of physical relations symbolized by these axioms; without such an experimental exploration of the subject matter, it may turn out that the relation between the objects under survey will not generate the transitive assymetric sequence. If, to

8. Dingler, in *Das Experiment*, p. 51 ff., has an important discussion in this connection, whatever one may think of his characterization of theoretical physics as without a physical meaning.

9. Runge, "Maß u. Messen" in *Enzyk. d. math. Wiss.*, Bd. V, p. 4.

10. This definition of equality in terms of the relation generating the series is due to Campbell, *Measurement and Calculation*, p. 5. The many debts of this paper to Campbell are evident to all his readers, and cannot be made explicit even by repeated citations. The discussion in Duhem, *Theorie Physique*, p. 159 ff., takes similar lines, while Helmholtz's important essay on *Zaehlen u. Maessen*, contains the germ of almost all subsequent works on the subject. The essay is reprinted in his *Schriften zur Erkenntnistheorie*, edited by Hertz and Schlick. "Equality between the comparable properties of two objects is an exceptional occurrence, and can be recognized in empirical observations only by this, that the two equal objects make possible the noting under proper conditions of a special effect, not usually present in the interaction of other pairs of similar objects" (p. 85) . See also Cournot, *Essai . . .*, p. 286 ff.

take an absurd example, we compare the lengths of elastic rubber bands by superposition, without specifying further the manner in which this is to be done, and if we define x as longer than y when x extends beyond y, it may well be that axiom two will not be satisfied.

Numbers may of course be introduced merely for purposes of identification, and many numerical designations have no more arithmetical significance than do the names of individuals. Thus, the policemen of a large city are often known by their number; but the policeman with number 500 is not thereby known to be stronger, or more efficient, or more handsome, or wealthier, or older than the one numbered 475.

This kind of out-and-out arbitrariness can be considerably reduced if a qualitative series is first established. The numbers assigned in the operations of science are more than a conventional tag; it is desirable that an object numbered 50 occupy a position higher or lower in a qualitative series as determined above, than one marked 40.[11] So, for example, the liquids gasoline, alcohol, water, glycerine, hydrochloric acid, carbon bisulphide, and mercury, are arranged in the order of increasing density as defined above; we can establish a one-to-one correspondence between these liquids and a series of real numbers such that the order of numbers will be symbolic of the order of densities. We may assign any one of the following sets of numbers to the series of liquids: (a) 1, 2, 3, 4, 5, 6, 7, or (b) 100, 90, 88, 85, 80, 60, 10, or (c) 22.5, 20, 19.6, 19.3, 11.2, 10.5, 8.9, or (d) .75, .79, 1, 1.26, 1.27, 1.29, 13.6. If, however, we specify further that the order of increasing numerical magnitude must correspond with the order of increasing density, the sets of numbers (b) and (c) are no longer available.

Nevertheless, there is still very much that is arbitrary and often misleading in assigning the numbers, say, 1 to 7 to the above sets of liquids. In choosing the set (a) instead of (d) or other sets having the same order, we exhibit the arbitrariness. The choice becomes positively misleading, if, without further definition and experimental confirmation, we suppose that because alcohol in this scheme has a density of two and gasoline that of one, there is involved a physical meaning (a physical operation defined in terms of density) in talking of alcohol as *twice* as dense as gasoline, or of *adding* densities of gasoline to obtain the density of alcohol. The operational interpretation of the axioms assigns, so far, physical meaning only to the *order* in which numbers are employed; there is no such meaning as yet for numerical *differences*.

It is just such misunderstandings, however, that are at the basis of many of the confusions in psychological and social measurements.

11. Cf. Runge, *op. cit.*, p. 5 ff.

Bogoslovsky's attempt to measure the proportion of "mental activity elements" in various tasks, is one instance of such confusions.[12] He presented to several people twenty descriptions of different situations involving mental and physical activity, and obtained their estimate of the proportion of mental activity "elements" to physical activity "elements" in each. The judges were asked first to arrange the descriptions in order of increasing mental activity (an attempt to generate a qualitative series) and then to express the numerical percentage of mental activity "elements" in each. But "percentage of mental activity elements" has a well defined meaning only on the assumption that the situation can be regarded as the *summed* total of elements; and that assumption must be justified both by an adequate criterion for an "element" and by the exhibition of a process of *addition* for them. Without that assumption and its justification, the results obtained indicate nothing. How little the nature of measurement is understood by the author is clear from the defense he makes for his procedure: "All our scales are arbitrary. There are no special intrinsic reasons for dividing an hour into sixty minutes." An hour may, of course, be divided into any number of intervals. But it is the possibility of definite operations defining more or less time and the addition of intervals, that makes the division of an hour into sixty parts valuable, and the nonexistence of such operations which makes valueless the computation of mental activity percentages.

When, therefore, numbers are correlated with some qualitative spectrum, it is not obvious and not always true, that the differences between the numbers represent differences between qualities definable otherwise than ordinally. Since none of the operations exhibited so far in connection with density have defined anything besides "greater than" and "equality," it is meaningless to talk, at this stage, of addition.

III. Addition

"At this stage" must be amended in the case of some properties to read "ever." For it is a well established character of existence that, although many qualities can be serially ordered and so numbered in accordance with an arbitrary plan, a few of these qualities (and a limited few they are) possess also a capacity for "addition" which the rest do not. This cleavage between additive or extensive qualities, and nonadditive or intensive qualities, is of fundamental importance in the philosophy of physical measurement.

Consequently, new operations and experiments must be introduced to define addition. And for density, defined as above, such operations are

12. *Technique of Controversy*, pp. 162–74.

not obtainable, since there is no clear sense in which two liquids equally dense could be added to produce a liquid twice as dense, and so obtain a "sum" which would possess the formal properties listed in the last six axioms. Other physical properties must be used to exhibit these new operations.[13]

"Illumination" is a very important photometric property which is capable of addition when defined as follows. It will be assumed, in the first place, that the immediate but disciplined judgments, with respect to perceived inequality or equality of the brightness of certain surfaces, can be obtained with sufficient uniformity. Brightness thus forms a qualitative domain within which more or less bright can be distinguished; equality of brightness may therefore be defined in conformity to the formal characters already specified. The surfaces used will have the same shape, same color, and the same reflection and diffusion coefficients. Surfaces will be said to form a "pair" if, when their positions are interchanged and everything else remains the same, no disturbance can be noted in the equality of the brightness of the surfaces.

A pair of surfaces will be said to have equal *illumination*, when they are judged to be equally bright under specified circumstances: e.g., the lines which join the two surfaces to the eye must make equal angles with the perpendiculars to the surfaces. Many more specified conditions must be introduced in practice. When some body can be found such that, by altering its physical state or of the medium between it and the surfaces, the brightness of the surfaces is changed, the body is defined as the *source* of the illumination. It can now be shown experimentally that if a pair of surfaces form an equally illuminated pair under a certain source, a second pair will also be equally illuminated if substituted for the first pair, even though a member of the first set does not form a "pair" with a member of the second set. This is the reason why illumination comes to be regarded as a quality not of the pair of surfaces but of the conditions under which they are placed; these conditions include, as a minimum, the nature of the source and the position of the surfaces with respect to it.

Addition of illumination can now be defined. The illumination of a

13. For the following illustration, see Campbell and Dudding, "Measurement of Light," *Philos. Magazine*, 6 Ser. Vol. 44. The discussion in Lambert's *Photometrie* (translated in Ostwald's Klassiker) is the foundation for the whole science. He shows, incidentally, the self-corrective nature of scientific procedure. "If one wanted to investigate the validity of eye judgments, one must remember these psychological illusions in order to take cognizance of the remaining principles of photometry. But it is these principles that are presupposed in any investigation of the errors of eye judgments. Hence I do not see how a logical circle can be avoided if a rigorously proved photometry is desired. But if this rigor is abated a little, one can obtain the propositions of photometry with some degree of certainty" (Vol. 1. p. 7).

surface S from sources A *and* B is defined to be equal to the *sum* of the illumination of S from A alone and the illumination from B alone, if A and B remain in the same physical state and relative position to S. If we use sources of the same color (strictly, only monochromatic light can be used) it can be shown that axioms 8, 9, and 11, receive an experimental confirmation; but for heterochromatic light, addition in this complete sense no longer exists.[14]

With physical addition defined, numbers can now be introduced, so that all arbitrariness in assigning them, except in the choice of a unit, is removed. Some constant source B_1 is chosen and placed at a fixed, convenient position with respect to the surface S_2, one of the pair S_1 and S_2. The illumination thus obtained is taken as unity. Next, two sources A_1 and A_1' are found and placed on a cone whose axis is perpendicular to S_1, so that either alone makes the illumination on S_1 equal to that on S_2 under B_1, and gives therefore unit illumination on S_1. Now letting A_1 and A_1' illuminate S_1 together, and extinguishing B_1, a source B_2 is found such that, placed with respect to S_2, it makes the illumination on S_2 equal to that on S_1 under both A_1 and A_1'. A source A_2 is found and placed on the cone so that the illumination of S_1 under A_2 is equal to that of S_2 under B_2. This illumination on S_1 is two, and it now has a clear physical meaning to speak of this illumination as *twice* that under A_1. This method of assigning numbers can be extended indefinitely for integral as well as fractional values, so illustrating the last axiom. To define fractional values we must find n sources placed in such positions on the cone, that any one of them makes the illumination on S_1 equal to that on S_2 under a constant source, and such that all n together make the illumination on S_1 equal to a unit illumination. Each of these n sources gives an illumination equal to $1/n$ of the unit illumination.

The construction of such a series of standards, both integral and fractional, in terms of explicit physical operations which take note of qualitative homogeneities and differences, is the logical prius to any other mode of physical measurement. Only now can the theorem, that illumination is inversely proportional to the square of the distance from the source, be given experimental confirmation. Only after the physical meaning of numerical operations has been thus fixed, may mathematical variables be introduced to denote unambiguously portions of subject matter. Only then may the movements of pointers be taken as signs for qualitative differences.

---

14. Walsh, *Photometry*, p. 7. Campbell points out that heat is capable of addition in a very restricted sense only, because no definition of addition will make the commutative and distributive axioms verifiable. *Physics.*, p. 287.

Clearly, the most important of the operations used in the definition of magnitude, are those fixing the meaning of addition. Operations for defining physical addition can be found for mass, length, period, electrical resistance, area, volume, force, and about a dozen others; for these, processes of fundamental measurement can be found, while other properties studied in physics are measurable only in terms of them, that is, derivatively. The example chosen makes clear, as Helmholtz pointed out long ago,[15] how experimental a concept addition is, valid only in so far as axioms 9, 10, and 11 are verified. It should, moreover, help expose the dogma that measurement consists in the observation of pointer coincidences, as well as the belief that addition is exclusively spatial juxtaposition and division. Undoubtedly, spatial juxtaposition is the primitive meaning in the addition of lengths; but in the example used addition involves *conjoint activity* of sources; in measurement of weights, it means the establishment of *rigid connections* between solids; in the estimation of time periods, it requires the *temporal repetition* of certain rhythmus; in the evaluation of volumes it signifies the discovery of liquids which *fill containers* without implying spatial contiguity. The presence of spatial relations is not a sufficient condition for addition; there is necessary a distinctive qualitative context, inclusion in which identifies different instances of addition, as addition of the *same* characters.

When once the standard sets of different magnitudes are constructed, the measurement of even those properties whose standards these sets are, is indirect, and consists in the comparison of the properties with their standards. The comparisons are very often circuitous, because advantage is taken of the relations between variations in the fundamentally measured magnitudes and the variations in the position of a pointer with which they are in some physical connection. It is easier to make the correlation between pointers and weights once for all, than to engage in fundamental measurement whenever one is estimating the tonnage of an elephant.

IV. Some Objections Examined

The preceding analysis and exposition (1) assumed without much question that measurement was the evaluation of the empirical relations between physical objects diversely qualified; and (2) stressed the importance of the distinctions between extensive and intensive qualities. Both of these doctrines have met some opposition.

(1) Impressed, no doubt, by the profound difference in cognitive status between the unmeasured and measured qualitative world, Russell con-

15. Helmholtz, *op. cit.*, p. 89 ff.

verts that difference, achieved in terms of the processes already examined, into a difference between a concrete actuality and a realm of essences: this latter realm is understood not as the ordered relations of and between existences, but as a domain of immaterial entities having no necessary reference to existence. For Russell, therefore, actual footrules are quantity, their lengths are magnitudes. It is only by an ellipsis that two quantities can be said to be equal: they are equal because they possess the same magnitude; and it is improper to say that one of two quantities is greater than the other: what is meant is that the magnitude which the first quantity possesses is greater than the second magnitude. Only quantities can be said to be equal, by having the same magnitude; two magnitudes cannot be equal, since there is only one of each kind.[16]

Russell's first objection to the relative view of quantity (the view that $>$, $<$, and $=$, are relations holding directly between physical things), and his espousal of the absolute theory (for which things must first be referred to an otherwise undefined realm of magnitudes in order to be measured), is based upon the observation that in any proposition asserting $>$, $<$, or $=$, an *equal* quantity may always be substituted anywhere, without altering the truth value of the proposition. It is not the *actual* quantity, but some character which it has with other quantities that is of importance.

It will be granted that this point is well taken, if possession of relations like equality is interpreted without reference to the specific process which defines equality. When two weights are equal, that relation holds by virtue of the way the two weights enter into a complicated evaluating process. Two weights are not equal "in themselves," they are equal as a consequence of the manipulation to which they are subjected. Of course, once having specified the defining operation, whether it is actually performed or not, the things measured have a nature prior to the actual performance which conditions their behavior in it. This observation may be verbal only: if equality is defined in terms of the process, quantities can be called equal prior to the process only proleptically; unless at *some* time the process eventuates, we cannot know that there is such a property as equality.

But it is one thing to say that relations like equality hold between two objects only in specified contexts, and another thing to convert those relations into possession of some third entity or common essence, incapable of every empirical verification. That third entity is the reification of a relation. The absolute space which haunted physics is just such a hypostatization of relations; space and time may be construed more

---

16. *Principles of Mathematics*, p. 164 ff.

simply as pervasive relations between events, rather than as containers, extrinsic and outside the changing qualities. So when magnitudes, which are always found to be relations exhibited in the physical operations of things, are invoked as the locus of those operations, it seems legitimate to ask what empirical difference their existence or nonexistence as "common essences" would make.

Russell's second point is that every transitive symmetrical relation is analyzable into a complex of two assymetric relations and a third term, and that since equality is such a relation, it should be analyzable into the possession of a common magnitude by two quantities. "The decision between the absolute and relative theories can be made at once by appealing to a certain general principle, the principle of abstraction. Whenever a relation of which there are instances is symmetric and transitive, then the relation is not primitive, but analyzable into sameness of relation to some other term, and this common relation is such that there is only one term at most to which a given term can be so related, though many terms may be so related to a given term."[17]

If the analysis given above is sound, all measurement which is not fundamental in the sense defined, or which is not surrogative in the manner to be explained below, consists in the direct or indirect correlation of quantities to be measured with the standard series. It is correct to say, therefore, that two equal footlengths are equal because they possess a common magnitude, *if* this means that they are compared, ultimately, with the same member of a standard series. But in establishing equal magnitudes *within* the standard series itself, no reference to a "common term" was necessary. Equality, it is true, was not taken as a fundamental character, since it was defined in terms of two transitive assymetrical relations. But those relations held or did not hold between *qualities,* and did not relate to some third term *outside* the qualitative series. The weights *a* and *b* are equal, because with respect to the relation "greater than" which defines the series of weights (it may be defined in terms of the sinking or rising of the arms of a lever), *a* is neither greater than nor less than *b,* and if *c* is any other *weight* (a member of the *series,* not *outside* of it) then if *a* is greater or less than *c* so also is *b.* No unique third term outside the physical domain is involved in defining equality; if Occam's razor still can cut, the magnitudes demanded by the absolute theory may be eliminated.[18]

---

17. *Op. cit.,* p. 166.

18. Couturat declares that sound common sense sides for the absolute theory. "It is reasonable to suppose that equal magnitudes have something more in common than those magnitudes of the same kind which are not equal; and it is contradictory to suppose that two equal magnitudes are differentiated from two unequal ones only by

(2) The doctrine that magnitudes are essences, and therefore not divisible or additive even though it is only between them that the relation "greater than" can hold, leads very naturally to the view that the distinction between extensive and intensive magnitudes is purely conventional. Additiveness belongs, on the absolute theory, only indirectly to magnitudes, as a manner of speaking of the addition of quantities whose magnitudes they are. For "addition" of two magnitudes yields two magnitudes, not a new magnitude, while the addition of two quantities does give a new single whole, "provided the addition is of the kind which results from logical addition by regarding classes as the wholes formed by their terms."[19] Addition, for Russell, thus always refers to the conjunction of collections, and to the enumeration of the number of parts in the new whole. And even in the logical addition of classes there is no clear warrant, he believes, for affirming that the "divisibility" of a sum of n units is n-fold that of one unit. "We can only mean that the sum of two units contains twice as many parts, which is an arithmetical, not a quantitative judgment, and is adequate only in the case when the number of parts is finite, since in other cases the double of a number is in general equal to it. Thus even the measurement of divisibility by numbers contains an element of convention."

Now, indeed, if magnitudes express the results of physical measurement, it is not the magnitudes which are added, just as the measure of anything measured is not itself measured. If magnitudes have a logical status, then surely it is only logical addition of which they are capable. For *magnitudes*, therefore, the distinction between extensive and intensive has no *meaning*. Nevertheless, on Russell's theory, it is magnitudes which are measured, and it is because he takes "logical sum" to be the primary sense of addition that he can find only a convention in the addition of spatial distances or time intervals. How the transition from the conceptual to the existential order is effected, how "logical addition" may receive an interpretation in terms of physical operations, is a consideration omitted from his analysis. Is it not, however, more perspicacious to

---

the fact of the unessential relation which connects them: equality in one case, inequality in the others." *Rev. de Met. et de Morale*, 1904, p. 677. The reified concept is very evident in the "something in common." Moreover, on Russell's theory there is a magnitude for every specific kind: a specific magnitude of pleasure for each grade of pleasure, a specific magnitude of density for each grade of density. Is the magnitude of density of two substances, whose densities are the same but are otherwise very dissimilar, *two* magnitudes or *one*? On the absolute theory, since two magnitudes are of the same kind if one can be greater or less than the other, the answer is: one magnitude. But what determines whether two magnitudes can be compared? Is not the decision made, not by appealing to the magnitudes, but to the physical operations on quantities?

19. *Op. cit.*, p. 178; cf. also *Analysis of Matter*, p. 116.

think of mathematical "addition" as a *universal,* whose variable empirical content will be *cases* of addition, but which will require further specific definition and experimental proof of the presence of those formal characters which make those empirical contents instances of that universal?

Nevertheless, the unusual sense in which addition is sometimes used should not be overlooked. The order generated by the relation "male ancester" may be measured in the following fashion: Suppose a is the father of b, b the father of c, c the father of d; then the "relation-distance" father-of is "equal" in all the pairs ab, bc, and cd, and a has the relation of "great-grand-father-of" to d. By an obvious convention, we could express the ancestral relation of a to d as three times the relation of a to b; and we could say that the relation of a to c is the sum of the relations of a to b and b to c. It is clear, however, that the "addition" here defined does not possess *all* the formal properties demanded by the axioms; it does not obey, for example, the commutative rule. The addition here defined is not very much more than the ordinal arrangement of relations. It is the failure to recognize the necessity of obtaining all the formal characters in fundamental measurement, which makes so unsatisfactory the attempt of Spaier to defend the measurement of nonspatial properties, and which enables him so easily to minimize the distinction between intensive and extensive qualities.[20] The introduction of numbers has a function more inclusive than the *identification* of quality.

V. Surrogative Measurement

In the light of what has been said, the dichotomy between primary and secondary qualities becomes more illuminating if we view this distinction not as between objective and subjective, efficient and otiose, pervasive and local, permanent and evanescent, but as between those qualities which are capable of fundamental measurement and those which are not.

No science, certainly not physics, can dispense with qualities that are incapable of addition in the fullest sense, and the progress of modern science has consisted very largely in bringing nonadditive qualities like density, temperature, hardness, viscosity, compressibility, under the sway of numerical determination.[21] There is of course one obvious way, already suggested, how this could be done. That way is to place qualities like density into a serial order, and to assign numbers to points of this qualitative spectrum. It is characteristic of modern science, however, that such is *not* the method which has been adopted, just as in zoology it is with bats and not with fish that whales are classified. For it is the particular virtue

20. Spaier, *op. cit.*, pp. 242-55.
21. Duhem, *L'Evolution de la Mécanique,* p. 199 ff.

of modern science not to be concerned with the grouping of the most obvious qualities, thereby treating them as isolated from and unconnected with other groups; that virtue resides in the persistent attempt to obtain well defined connections, expressed mathematically wherever possible, between qualities measured or measurable fundamentally and those incapable of such measurement.

Unfortunately, the correlation of qualities has been often interpreted as the production of secondary qualities by the primary ones: the latter alone have been endowed with causal efficacy, the former degraded as otiose and epiphenomenal. Thus a distinction which in *operation* is a practical and logical one, has been converted into a distinction between grades of reality, on the ground that causes are more real than their supposed effects. Mathematical physics has been understood to make nonsense of the poet's cry—"Natur hat weder Kern noch Schale." Nonetheless, all that the equations of physics and the method of establishing them do imply, is that nonadditive qualities are inextricably interwoven with additive ones. This dependence is existentially mutual; from the point of view of the *logic* of measurement, it is assymetrical. It is because of this dependence, expressible in the form of numerical laws, that numbers may be correlated unambiguously with nonadditive qualities; it is because of such laws for density, that of the four sets of numbers entertained on a previous page, only the last set is adopted.[22]

How nonadditive qualities may be unambiguously denoted will be clear from an example. In the case of density it is discovered that the weight of an object is intimately connected with its volume, a connection exhibited in the uniform association of these characters. Weight and volume are measurable fundamentally so that independently of the numerical relations which affirm their uniform association, numbers can be assigned to them. There is no special theoretical difficulty, although there may be many practical ones, of determining the value of the constant which expresses this uniform association, and which mathematically is the ratio of the numerical value of the mass to the numerical value of the volume. The *form* which the mathematical equation will take is, of course, dictated only partly by the measurements on the properties studied in any *one* instance: more general considerations will come into play arising from the desire to make the many numerical equations themselves interconnected and parts of a unified doctrine.

In most cases, the order of the constants or ratios which are determined

---

22. "The ascertainment of qualitative features and relations is called measuring. We measure e. g. the length of different chords that have been put into a state of vibration, with an eye to the qualitative difference of the tones caused by their vibration, corresponding to this difference of length." Hegel's *Logic* (Wallace tr.) , p. 200.

for several objects can be shown to be the same order as the serial order of some nonadditive quality which the objects possess in addition to those already measured fundamentally. So the order of the ratios of mass to volume is identically the order of the density of liquids as defined by their floating capacities relative to one another. Consequently, the same set of numbers may be used to denote both the uniform association of mass and volume and the relative buoyancies. It goes without saying that the numbers thus obtained for qualities from numerical equations, are not always amenable to a physical interpretation of addition. When one body is said to be thirteen times as heavy as another, a different meaning must be given to such a statement from the meaning of the statement that mercury is thirteen times as dense as water; only in terms of the numerical law connecting mass and volume has the latter proposition significance.

All equations which define a constant, to be identified perhaps with some property capable of definition independently of the equation, require therefore that the other terms of the equation be measurable without reference to the defining equations of the constant. In the sense that the constants are defined, not *all* magnitudes can be defined without leading to a circle; there must be an ultimate reference to magnitudes obtained by fundamental measurement. The equation $pv = RT$ has meaning only if p, v, T have a meaning outside of this equation; only then may R be determined experimentally. If p, v, T are not measurable fundamentally there must be a chain of equations connecting them with magnitudes which are. It is a testimony to the endless *complexity* of nature, not to her poverty of qualities, that only six independent fundamentally measured magnitudes are required for the investigations of physics. "There are only a few independent magnitudes in physics. But between these and the countless number of independent magnitudes appearing in human life, there is no sharp separation." Strength of wind would be a genuine aspect of some events, even if strength of wind were completely definable, which it is not yet, in terms of the velocity and force of air particles.[23]

Moreover, the power of all symbols, and especially of numerical symbols, to refer simultaneously to several contexts must be recognized if mystification is to be avoided. The unification and identification in statement that follows from the introduction of mathematical methods is due to the pervasiveness of certain formal characters in situations qualitatively different, which are as a consequence capable of a unified treatment. In the interpretation of equations as the literal identification of different qualitative continua, and as the attribution of intrinsic, non-relational

23. *Cf. Handbuch d. Physik*, Bd. II, p. 5.

common characters to diverse subject matter, lies the force of most of the petulant criticism of science. Let us, for example, study the equilibrium conditions of a beam balance. If x, y, z, represent certain bodies A, B, C, then the equation $x + y = z$ will mean that A and B on the same pan will balance C on the other, and it will also mean that the numerical measure of the weight of C is equal to the sum of the measures of A and B. The perhaps less relevant case of a chemical equation like $2H_2 + O_2 = 2H_2O$ represents theoretical conceptions like atom, valence, physical operations like passing a spark through a mixture of two gases, and numerical relations between the measures of weights and volumes.

The confusions which arise from the failure to note how complex the functions of symbols may be, is illustrated in a recent criticism of the achievements of science. The principle of the lever, as expressed in the form of a proportion, is the climax of an extended critique, but this special discussion is reproduced in full.

It will be best to point out that it is the result of two apparently unjustified leaps of imagination. . . . A double leap it is, and the reader can supply whatever theory of revelation, reincarnation, or conventional fiction he prefers, to account for it.

To say that W_1: W_2: : D_2: D_1 is by itself ambiguous. Perhaps it only means that certain numbers stand to each other this way: 2: 4 : : 3 : 6. In that case it is merely a happy discovery in arithmetic. But W stands for weight and D stands for distance. It may therefore mean that the relation between two weights is the same as the relation between two distances. But this is not true for many relations; for instance "heavier than" is a relation between two weights, but not between two distances. The only relation that works is a hybrid combination of these two.

The combination is evidently derived from two previous proportions, namely: $W_1 : W_2 : : 2 : 4$ and $D_2 : D_1 : : 3 : 6$. Then because we already know in arithmetic that 2 : 4 : : 3 : 6 we can finally see how it is that $W_1 : W_2 : : D_2 : D_1$. But why weights and distances are like numbers, to use the simile, has still to be explained. The only answer that I know is that some poet of the commonplace was playing with words, and somebody took him literally.[24]

It cannot be pretended that the actual history of the principle of the lever is being reported. As a deduction or validation of an important numerical law, however, it merits a conspicuous place in some future Budget of Paradoxes. The difficulties which are raised seem to arise partly from a dogma that numbers cannot be the numbers of anything without losing caste, and partly from the failure to realize that in the statement of the law at least two relations are symbolized. In the first place, the relation $W_1 : W_2 : : 2 : 4$ can be intelligibly interpreted as meaning that four weights each equal to W_1 are equal to two weights each equal to W_2,

<hr>

24. Buchanan, *Poetry and Mathematics,* p. 90.

where "equal" is defined in some unambiguous way. Secondly, it must be observed that the *symbol* : : stands for both a *numerical* relation between numbers assigned in the way suggested, and a *physical* relation into which the lever enters in a very specific way. The proportion as it stands expresses *two* sets of relations, and we can use the same *symbol* to represent both because the relations have certain formal properties which are *identical*. Numbers are not *like* weights, indeed, but numbers are, and numerals express, definite properties of weights. It is a great sin to *compare* the statement of a relation with the relation itself. If discourse cannot be literally compared with what it is about, is it not wisdom to recognize that discourse expresses it? The equation, in every case, is a symbolic statement, pointing to several aspects of the subject matter. When once the plural referents of the symbols are made explicit, and when numbers are not regarded either as common qualities or as chaste platonic beings, but rather as the expression of relations or operations between qualities, belief in the power and validity of mathematical physics need not be superstition.

As there are critics who find the application of numbers to additive properties a puzzle, so there are other critics who challenge the validity of the application of numbers to non-additive ones. It is never the non-additive qualities that are measured, it is said; velocity has a unique existential quale as a velocity, and to express it as the ratio of space and time is to measure space and to measure time, but it is not to measure velocity. A twofold reply may be made.

(1) There are many qualities which, as a matter of practice, are measured as derived magnitudes by means of numerical equations, but which can be measured fundamentally. Bridgeman has shown how this may be done for velocity. Areas, volumes, or electric charges, are in the same position: they are usually measured as derived magnitudes, but are capable of fundamental evaluation. Nevertheless, even when they enter numerical equations derivatively and measured therefore in their relation to other characters, *they* are measured.

(2) It is true that most qualities, being incapable of addition, must be measured in terms of their "surrogates" in numerical equations. If the term "measurement" is restricted only to such qualities which are fundamentally measurable, it must be acknowledged that density and acceleration are incapable of measurement. But by calling the process of assigning numerical values to density some other name, the *significance* of what is done is not destroyed. What is beyond much doubt, is that the measure numbers of those characters incapable of addition, or even incapable of sensuous intuition like magnetism, are not mere numerals or formulae for nominal combinations of additive characters; they represent rather

certain coordinated qualities or certain relational properties of the systems studied.[25] If it is the expansion of mercury that we actually measure in the more restricted sense, the measurement is performed because there is a uniform association between this expansion and qualitative temperature changes.[26]

Several modes of surrogative measurement may be obtainable for the same property. So we may define the temperatures of black bodies by using the well known Stefan-Boltzmann law of energy radiation. Two temperatures will then be equal, if the energy of radiation which the surfaces of black bodies send out in a given time into a given space, is the same for the two temperatures; two temperature-differences will be equal, if the differences of energies radiated are the same. Except for the zero point and unity, the temperature scale is defined. Comparison of this scale with the gas thermometer scale would show that the scales cannot be made directly congruent. However, by a proper choice of the zero point, the numbers of the gas thermometer are found proportional to the fourth roots of the numbers of the energy scale; if the fourth roots of the energy scale are used to define the temperature of the black body, the two scales can be used interchangeably.[27] Similarly, velocity may be measured not in terms of space and time, but in terms of the resistance which a given body meets when moving through a specified medium.

It is by discovering the recurrence of certain constants in different numerical laws, that the ideal of a unified science is progressively realized.

---

25. Meinong, *op. cit.*, p. 228, 275. The expression "surrogative measurement" is, of course, due to Meinong.

26. But one need not therefore confuse, as does Hegel, the distinction between temperature as intensive and the correlated expansion, which is extensive. Hegel, *op. cit.*, p. 194.

27. Runge, *op. cit.*, p. 8.

Chapter 2

MEASUREMENT

S. Smith Stevens
Harvard University

The business of pinning numbers on things—which is what we mean by measurement—has become a pandemic activity in modern science and human affairs. The attitude seems to be: if it exists, measure it. Impelled by this spirit, we have taken the measure of many things formerly considered to lie beyond the bounds of quantification. In the process we have scandalized the conservatives, created occasional chaos, and stirred a ferment that holds rich promise for the better ordering of knowledge. Restrictive definitions of measurement have toppled as the practice of measurement, outrunning legislation, has forced us to broaden and generalize our conceptions. The ultimate, perhaps, is reached by those who claim that whenever we can single out a relation between two things we have a scale of measurement of some sort.

It is no new thing, of course, to find practice outrunning legislation,

Reprinted with permission from the author, editor, and publisher from C. West Churchman (ed.), *Measurement: Definitions and Theories*, pp. 18–36. New York: John Wiley & Sons, Inc., 1959.

Prepared under Grant *G–2668* from the National Science Foundation and Contract *Nonr–1866 (15)* between Harvard University and the Office of Naval Research (Project *NR142–201*, Report *PNR–201*). Several friends and critics have helped the writing at some points and have contested it at others. For their helpful discussions, I am especially indebted to L. J. Savage, John W. Tukey, G. A. Miller, Ardie Lubin, W. H. Kruskal, F. M. Lord, and Garrett Birkhoff.

for that is the nub of the story of mathematics. The irrationals, the surds, the imaginaries, and the negatives are numbers that still bear names reminiscent of protest—protest against outlandish practice and against the writing of unauthorized absurdities. But orthodoxy bent to accommodate practice. Mathematicians staved off chaos by rationalizing the use of irrationals, and by imagining a broader domain in which imaginaries and negatives could serve as proper elements. An analogous story can be told of measurement. The reach of this concept is becoming enlarged to include as measurement the assignment of numerals[1] to objects or events according to rule—any rule. Of course, the fact that numerals can be assigned under different rules leads to different kinds of scales and different kinds of measurement, not all of equal power and usefulness. Nevertheless, provided a consistent rule is followed, some form of measurement is achieved.

Perhaps it is unfair to pretend that this liberal and open-handed definition of measurement is the accepted norm, for some practitioners of the physical sciences prefer to cling to the narrower view that only certain of the tidier rules are admissible. This attitude is quite understandable, but just as the acceptance of the absurd, fictitious, and imaginary roots of negative numbers redounded to the enrichment of mathematics, so, too, may the theory of measurement find itself enriched by the inclusion of all orderly numerical assignments. Mathematics has at last freed itself of the earthy perspective under which it formerly sought to justify its laws by manipulations performed upon solid objects, and has taken off into the blue of pure abstraction—where it properly belongs. Measurement, however, must remain anchored here below, for it deals with empirical matters (although it borrows its models from mathematics). In the other hand, there is no requirement that measurement remain confined to the simpler problems of counting which first gave rise to it.

In the beginning, mathematics and measurement were so closely bound together that no one seemed to suspect that two quite different disciplines were involved. Whole-number arithmetic and scales of numerosity grew

---

1. Should we say "numeral" or "number"? N. R. Campbell says numeral. Most other writers say number. Elsewhere I have tried to distinguish among the meanings of these terms and the related terms numerousness and numerosity (see 31, p. 22). It would be nice to adhere to a consistent usage, but, as the reader will discover, I have not done so. The term "numeral" in this paragraph refers to an element in a formal model, not to a particular mark on a particular piece of paper. The term "number" sometimes has this meaning, but, in common usage, it also has many others. I have tried to reserve the term "numerousness" for the "subjective" aspect or attribute which we observe when we look at, but do not count, a collection of objects. Numerosity is the "physical" attribute of a collection which we measure by counting, i.e., by pairing off the items in a collection against the successive positive integers.

up tightly intertwined, and the ancients seem not to have discerned the difference between the formal model on the one hand and the empirical matter of scaling on the other. The numerosity of collections of objects (number in the layman's sense) constitutes the oldest and one of the most basic scales of measurement. It belongs to the class that I have called *ratio scales*. Undoubtedly it was man's early efforts to construct a scale of numerosity that led to his invention of a formal arithmetic, and it is not surprising that many millennia were to come and go before men saw that the test of a formal system need not reside in its ability to reflect what can be done with piles of pebbles.

Modern mathematics, far from concerning itself merely with numerosity, has become so nonquantitative in its abstract reaches that Gödel could suggest that it was purely an historical accident that mathematics developed along quantitative lines (21). Perhaps this is so. But from another point of view the "accident" has a certain inevitability about it. Striving to deal with collections, be they fish, cattle, or warriors, ancient man seemed destined in the nature of things to have hit upon the concept of number and to have made therein his first triumphant abstraction. It is hard to conceive how mathematics could have begun elsewhere than in measurement.

In recent times, however, it has become clear that the formal system of mathematics, the "empty play upon symbols," constitutes a game of signs and rules having no necessary reference to empirical objects or events. As Hardy says, "It is impossible to prove, by mathematical reasoning, any proposition whatsoever concerning the physical world, and only a mathematical crank would now imagine it his function to do so" (18). As human history goes, this hard-won understanding is an affair of very recent times. The final divorcement between the formal, abstract, analytic system and the empirical questions that originally sparked its development has clarified the relation between mathematics and measurement. Under the modern view, the process of measurement is the process of mapping empirical properties or relations into a formal model.[2] Measurement is possible only because there is a kind of isomorphism between (1) the empirical relations among properties of objects and events and (2) the properties of the formal game in which numerals are the pawns and operators the moves. As Russell put it: "Measurement demands some one-one relation between the numbers and magnitudes in question—a relation which may

---

2. It is now common parlance to refer to formal, mathematical systems as "models." The recency of the distinction between the formal model and the empirical operations of science impresses me particularly, for as little as two decades ago some of my colleagues expressed surprise that I should refer to a formal system as a "model" that could be used to represent empirical relations (29).

be direct or indirect, important or trivial, according to circumstances" (23, p. 176).

Before we consider this matter further, let us review some recent history.

The Classical View

What I should like to mean by the classical view of measurement is the conception that grew up in the physical sciences and that received its fullest exposition in the works of N. R. Campbell (4, 6, 7, 14). This is the view that dominated the scene until the 1930s. More recently I have been startled to see some of my own notions described as the "classical measurement theory" (10, 43), but this seems somewhat premature. It is not so much that the term "classical" has the connotation of "out of date and slightly wrong;" it is rather that the point of view I am urging represents a considerable departure from Campbell's tradition. Moreover, its development is still in a state of flux. But more about this later.

The classical "classical theory" grew quite naturally out of the evolution of the number system, which, as we have already noted, apparently began as a primitive project in the scaling of numerosity. The exact history of how this enterprise came about is lost to the record, of course, but our attempts to reconstruct the main features of the story are probably not far wrong. The important point for our present concern is the fact that some long-forgotten genius somehow contrived to build a number system—a formal model—to represent what he did with pebbles, fingers, or cattle. With the aid of this model, he measured the numerosity of his possessions.

Since the formal model was invented for this purpose, it is scarcely surprising that the model turned out to be isomorphic with the empirical operations performable with such things as pebbles (from whose Latin name we derive our word calculus). Thus three pebbles put with four pebbles to make a larger pile is mirrored by the symbolic expression $4 + 3 = 7$. Ultimately, when the axioms of addition were finally exhumed from the long-standing "rules of practice" of whole-number arithmetic, it was readily seen that the property of numerosity "satisfied" these postulates. Thus the commutative law, to take a particular example, which says that $4 + 3 = 3 + 4$ has its analogue in pebble counting, for it makes no difference in what order we take the piles of pebbles.

It was a straightforward matter to extend the scaling procedure developed for numerosity to the scaling of length. Of course, a richer formal model was required for this extension, and the number domain had to be expanded to include fractions and the rest of the rational domain. However, the isomorphism between the model and the physical operations performable with lengths is still close. For example, both the physical addition of lengths and the formal addition of rational numbers are commutative.

The classical view of measurement, as Campbell presents it, is essentially the view that direct or "fundamental" measurement is possible only when the "axioms of additivity" can be shown to be isomorphic with the manipulations we perform upon objects. Only a few properties, such as length, weight, and electric resistance, are measurable in this fundamental way. Most other magnitudes dealt with in physics are measured by indirect or "derived" measurement—a process in which derived magnitudes are defined by means of numerical laws relating fundamental magnitudes. Thus density, the classical example, is measured by the ratio of mass to volume.

Partly as a result of Campbell's masterly exposition, the view has been widely held that the assignment of numerals to objects other than by the procedures involved in fundamental or derived measurement is not measurement at all. The clash of opinion on this issue is well documented by the deliberations of a distinguished British committee appointed in 1932 to consider and report upon the possibility of "quantitative estimates of sensory events" (14, 30). The particular subject of discussion was the sone scale of loudness (39). The committee split wide on the meaning of measurement, and some harsh words were written condemning the alleged measurement of sensation. In the Final Report (14), one member said, "I submit that any law purporting to express a quantitative relation between sensation intensity and stimulus intensity is not merely false but is in fact meaningless unless and until a meaning can be given to the concept of addition as applied to sensation."

Campbell himself was troubled. At times he seemed to suggest that loudness measurement might be possible, if based on monaural-binaural loudness matches, or, alternatively, if estimations of half loudness should turn out to be consistent with other estimations, such as one-third or one-tenth loudness. In the end, however, Campbell went along with the conservative wing of the committee and asked: "Why do not psychologists accept the natural and obvious conclusion that subjective measurements of loudness in numerical terms (like those of length . . .) are mutually inconsistent and cannot be the basis of measurement?" Why, he might have asked, does the psychologist not give up and go quietly off to limbo?

This is how matters stood in 1940. In the meantime, unaware that the British committee was trying to settle the issue, some of us at Harvard were wrestling with similar problems. I remember especially some lively discussions with G. D. Birkhoff, R. Carnap, H. Feigl, C. G. Hempel, and G. Bergmann. What I gained from these discussions was a conviction that a more general theory of measurement was needed, and that the definition of measurement should not be limited to one restricted class of empirical operations. There was also an obvious need for an improved terminology. The terms "fundamental" and "derived," as used by Campbell, were clear enough, but much cloudy discourse revolved around certain other distinctions, such as that between "intensive" and "extensive" magnitudes. Like some other authors, I had tried to redefine these terms (28) but had succeeded only in compounding confusion.

The best way out seemed to be to approach the problem from another point of view, namely, that of invariance, and to classify scales of measurement in terms of the group of transformations that leave the scale form invariant. This approach would seem to get to the heart of the matter, for the range of invariance under mathematical transformations is a powerful criterion of the nature of a scale. A fourfold classification of scales based on this notion was worked out sometime around 1939 and was presented to the International Congress for the Unity of Science in 1941. World War II then came along, and publication was delayed until 1946 (30). A fuller development appeared in 1951 (31).

One consequence of this approach is to make it clear that the classical terms "fundamental" and "derived" describe two classes of operations, not two classes of scales. Both fundamental measurement and derived measurement usually result in ratio scales, which are invariant under multiplication by a constant, or, as is sometimes said, are unique up to a similarity transformation. This is not to deny, of course, that the distinction between fundamental and derived is a useful one in physics, for in many contexts it is important to distinguish between kinds of operations. But there is a sense in which even more powerful distinctions can be made.

Scales of Measurement

Although the definition of measurement could, if we wished, be broadened to include the determination of any kind of relation between properties of objects or events, it seems reasonable, for the present, to restrict its meaning to those relations for which one or another property of the real number system might serve as a useful model. This restriction is implied when we say that measurement is the assignment of numerals to aspects of objects or events according to rule. But even this restriction leaves the concept broader than would be countenanced by the classical view, for all that is required is a consistent rule of assignment. The only procedure excluded is "random" assignment: if there is no criterion for determining whether a given numeral should or should not be assigned, it is not measurement.

Within this framework, we can distinguish four kinds of scales, which I have called nominal, ordinal, interval, and ratio. A fifth kind of scale, called a logarithmic interval scale (37), is also possible, but apparently it has never been put to use. We will return to it in a later section.

The four scales are listed in Table 2.1. An advantage of this tabulation is that it reveals certain interesting relations among the scales. Thus, the second column is cumulative in the sense that to an empirical operation listed opposite a given scale must be added all those operations listed above it. To erect an interval scale, for example, we need a procedure for equating intervals or differences, plus a procedure for determining greater or less, and a procedure for determining equality or equivalence. To these procedures must be added a method for ascertaining equality of ratios if a ratio scale is to be achieved.

The next column lists the kinds of transformations that leave the "structure" of the scale undistorted. In this column, each group of transformations is contained in the one above it.

The last column in Table 2.1 lists some examples of each scale.

It is an interesting fact that the measurement of some quantities may have progressed from scale to scale. We can imagine, for example, that certain Eskimos might speak of temperature only as freezing or not freezing and, thereby, place it on a nominal scale. Others might try to express degrees of warmer and colder, perhaps in terms of some series of natural events, and thereby achieve an ordinal scale. As we all know, temperature became an interval scale with the development of thermometry, and, after thermodynamics had used the expansion ratio of gases to extrapolate to zero, it became a ratio scale.

Since a more complete description of these scales is available elsewhere (31), we will not discuss them more fully. I would merely point out that the oft-debated question whether the process of classification underlying the nominal scale constitutes measurement is one of those semantic issues that depends upon taste. Whitehead (45, p. 30) calls classification a "halfway house" on the road to measurement. I prefer to call it a form of measurement, because the use of numerals to designate classes of objects, such as items in a catalogue, is an example of the assignment of numerals according to rule. The rule is: Do not

Table 2.1. A Classification of Scales of Measurement*

Scale	Basic Empirical Operations	Mathematical Group Structure	Typical Examples
Nominal	Determination of equality	Permutation group $$x' = f(x)$$ where $f(x)$ means any one-to-one substitution	"Numbering" of football players Assignment of type or model numbers to classes
Ordinal	Determination of greater or less	Isotonic group $$x' = f(x)$$ where $f(x)$ means any increasing monotonic function	Hardness of minerals Street numbers Grades of leather, lumber, wool, etc. Intelligence test raw scores
Interval	Determination of the equality of intervals or of differences	Linear or affine group $$x' = ax + b$$ $$a > 0$$	Temperature (Fahrenheit or Celsius) Position Time (calendar) Energy (potential) Intelligence test "standard scores" (?)
Ratio	Determination of the equality of ratios	Similarity group $$x' = cx$$ $$c > 0$$	Numerosity Length, density, work, time intervals, etc. Temperature (Rankine or Kelvin) Loudness (sones) Brightness (brils)

* Measurement is the assignment of numerals to events or objects according to rule. The rules for four kinds of scales are tabulated above. The basic operations needed to create a given scale are all those listed in the second column, down to and including the operation listed opposite the scale. The third column gives the mathematical transformations that leave the scale form invariant. Any numeral x on a scale can be replaced by another numeral x' where x' is the function of x listed in column 3.

assign the same numeral to different classes or different numerals to the same class. This is quite different from a "random" assignment under which no rule would be in force. With no rule in force, the same numeral might be assigned to different classes, and different numerals might be assigned to the same class. A class in this sense may, of course, contain only one member, as when the coach "numbers" his football players.

The forming of classes of equivalent objects or events is no trivial matter. An operation for determining equality is obviously the first step in measurement, but it is more than that. It is the basis of all our categorizing and conceptualizing (2)—of all our coding and recording of information. It underlies the ubiquitous process by which we sort the environment into significant constellations of events and thereby, take the first step toward systematizing the booming con fusion of the universe. Thus, it provides the basis of our identifying, recognizing, and labeling ordinary objects. Without this step, no further measurement would be possible.

In some contexts, what is largely a matter of nominal scaling goes by the heading "detection theory" and a considerable discipline is flowering under this name. A specific issue concerns the determination of a threshold (38), the separation of stimuli into two classes: those that produce a reaction and those that do not. It is true, of course, that the location of the boundary between these two classes is often stated in terms of its position on a ratio scale of some stimulus continuum, but the prior existence of such a ratio scale is not a prerequisite to the nominal process of sorting stimuli into two classes, detectable and not detectable. These are nominal categories, and threshold determination is an instance of nominal scaling.

Another development that has bolstered the scientific importance of nominal scaling is the development of information theory. This theory provides a tool for the treatment of data at the nominal level of measurement (15). It allows us to deal effectively with categories (alternatives) without regard for any ordinal relations that might obtain among the categories. Thus the measure of transmitted information, T, provides a measure of association at the nominal level of scales. Other measures of association useful with nominal scales have also been the object of recent research (16).

Statistics and Measurement

The fourfold classification of scales of measurement provides a convenient framework on which to display some of the common statistical measures. Depending on what type of scale we have constructed, some statistics are appropriate, others not. The group of mathematical transformations permitted on each scale (see Table 2.1) determines which statistical measures are applicable. In general, the more unrestricted the permissible transformations, the more restricted the statistics. Thus, nearly all statistics are applicable to measurements made on ratio scales, but only a very limited group of statistics may be applied to measurements made on nominal scales. A few examples of typical statistics appropriate to the various scales are listed in Table 2.2.

The fact that a given statistic is appropriate, in the sense used here, does not always tell us whether the statistic is the one we should compute. For example, the median, the arithmetic mean, the geometric mean, and the harmonic mean are all appropriate to ratio scales, but which of them should be used in a given circumstance must be decided by other criteria than the fact that measurements were made on a ratio scale (34). The type of scale involved provides a necessary but not a sufficient condition for the choice of a statistic.

The criterion for the appropriateness of a statistic is *invariance* under the transformations permitted by the scale, as listed in Table 2.1.

Table 2.2. Examples of Statistical Measures Appropriate to Measurements Made on the Various Classes of Scales

Scale	Measures of Location	Dispersion	Association or Correlation	Significance Tests
Nominal	Mode	Information, H	Information transmitted, T Contingency correlation	Chi square
Ordinal	Median	Percentiles	Rank-order correlation (31)	Sign test Run test
Interval	Arithmetic mean	Standard deviation Average deviation	Product-moment correlation Correlation ratio	t test F test
Ratio	Geometric mean Harmonic mean	Per cent variation		

This invariance may be of two principal kinds: the numerical value of the statistic may remain constant when the scale is transformed, or else the numerical value may change although the item designated by the statistic remains the same (31). The first type, invariance of numerical value, holds, for example, for such dimensionless statistics as the product-moment correlation coefficient, r, and the coefficient of variations, V, which expresses percent variability. The second type, invariance of reference, holds for such dimensional statistics as the median, mean, standard deviation, etc. For example, the case standing at the median of a distribution remains at the mid-point under all scale transformations that preserve order (isotonic group), and the case standing at the mean remains there under all linear transformations.

Each of the columns in Table 2.2 is cumulative in the sense that a statistic listed opposite a given scale is appropriate not only to that scale but also to all scales listed below it. Thus, the mean is appropriate to an interval scale and also to a ratio scale (but not, of course, to an ordinal or a nominal scale).

This cumulative property of Table 2.2 needs qualification, however, when we go from nominal scales to the other varieties. The nominal scale involves only discrete categories or classes. When the categories are naturally discrete, e.g., *male* or *female,* the problem is straightforward. We can then count men and women and determine which, for example, is the modal class. But when categories are formed by partitioning a continuous variable, like stature, a certain arbitrariness enters in. If we group statures into class intervals 1 inch wide, we may, again, find the modal class for a given sample of people. However, if we change the boundaries of the class intervals, or if we make the intervals different in size, the modal value may change. If we make finer and finer measurements and reduce the size of the class intervals more and more, we may even find that no two statures are the same, i.e., there is no mode.

We see, then, that there is an essential difference between the concept of a mode as applied to naturally discrete classes and as applied to a continuous scale. In the discrete case, the mode remains invariant under all the scale transformations listed in Table 2.1, but, in the continuous case, the mode is not invariant under increasing monotonic transformations.[3]

Returning now to Table 2.2, we note that it contains two empty cells. This merely reflects the fact that I am not aware of any measures of correlation or tests of significance that apply only to ratio scales. Perhaps such statistics exist and will soon be brought to my attention. If they do not exist, the possibility to their invention presents an interesting problem.

As already indicated, the appropriateness of a given statistic is conditioned by the nature of the scale against which measurements are made. Table 2.2 suggests, for example, that, when operations are available to determine only a rank order, it is of questionable propriety to compute means and standard deviations. This conclusion, obvious as it may seem to some (25, 1), met with disapproval by others (3). One author makes his point in a humorous skit on "The statistical treatment of football numbers" (19). One of the objections seems to be that the implications of Table 2.2 place too great a restriction on the usefulness of statistics.

As I see this issue, there can surely be no objection to anyone computing any statistic that suits his fancy, regardless of where the numbers came from in the first place. Our freedom to calculate must re-

---

3. I am indebted to J. W. Tukey, W. H. Kruskal, and L. J. Savage for calling my attention to this point. My earlier discussions neglected the problem and led thereby to some wrong implications (see especially 34).

main as firm as our freedom to speak. The only question of substantial interest concerns the use to which the calculated statistic is intended. What purposes are we trying to serve? When we compute the mean of the numerals assigned to a team of football players, are we trying to say something about the players, or only about the numerals? Obviously, we cannot say much about the players if the original numerals were passed out under a nominal assignment—the sole rule being "one to a customer." The only "meaningful" statistic here would be N, the number of players assigned a numeral.

Or suppose the assignment were made on some ordinal basis. The coach might line the players up in order of stature, and assign successive integers beginning with the shortest man whom he calls 1. We then compute the mean, and it turns out to be 20. Would the player bearing the numeral 20 stand at the mean of the distribution of statures as measured by a yardstick? Only by accident would this be the case. Plainly, then, if we want to interpret the result of averaging a set of data as an arithmetic mean in the usual sense, we need to begin with more than an ordinal assignment of numerals.

Discouraging as it may appear, the outcome of "statisticizing" is no better than the empirical measurements that go into it. The assertion that "statistical technique begins and ends with numbers and with statements about them" (3) is both true and false. At the formal, syntactical level of discourse, where we are concerned only with the mathematical model itself, we can accept this statement as essentially correct. But when, with the aid of semantical rules, we relate certain aspects of the model to certain aspects of the empirical universe, it becomes another story. For when we use a statistical model to reach conclusions about matters of empirical fact, we are no longer concerned merely with numbers. The question then arises, to what do the numbers refer? This question takes us back to the empirical operations that underlie our measurements and that give the numbers meaning, sharp or vague, as the case may be.

The basic principle is this. Having measured a set of items by making numerical assignments in accordance with a set of rules, we are free to change the assignments by whatever group of transformations will preserve the empirical information contained in the scale. These transformations, depending on which group they belong to, will upset some statistical measures and leave others unaffected. In other words, for guidance in setting bounds on the statistical treatment of empirical measurements, we must look to the principle of invariance. The empirical operations that underlie the scale determine what transformations can be made without the sacrifice of information, and the permissible transformations determine, in turn, the appropriate statistical measures, i.e., those that preserve the requisite invariance.

Some of the statistics appropriate to nominal and ordinal scales are sometimes called nonparametric, or distribution-free. These names express the idea that, in using them, we make no assumptions regarding the distribution of the underlying population from which a sample was drawn. To what extent certain of the so-called nonparametric statistics are, in fact, "distribution-free" is perhaps an open question at this time, and not one to be settled here. The only point I would like to make is that, when our empirical procedures limit us to ordinal measurement, it is obviously proper that we forego assuming knowledge of the parameters of the population because the form of a distribution has meaning only in terms of an interval or a ratio scale. Fortunately, many useful statistical procedures have recently been invented to deal with nominal and ordinal measures, and these various procedures could be added to Table 2.2 (27).

In making these additions, we must be careful to ask what type of measurement scale is assumed by the statistical procedure. Usually this question has an obvious answer, but, sometimes, it may not be so clear. Thus, although Wilcoxon's "matched-pairs signed-ranks" test is sometimes classed with statistics applicable to ordinal scales, one of the steps involves taking the difference between the two scores of a pair. If the scores in question are measurements made only on an ordinal scale, it seems evident that the numerical difference between them can have no firm meaning. The numerical differences would be determined only by the accidental manner in which numerals were assigned to the original rank-ordered scores, and this assignment could be changed by any increasing monotonic transformation.

The Logarithmic Interval Scale

Thus far we have discussed the four classes of scales most widely used in the scientific enterprise. It is an interesting question whether the list in Table 2.1 is exhaustive of the possibilities, or whether other classes might be devised. Under the broad definition of measurement—the assignment of numerals according to rule—it would seem probable that other rules than those described might be invented.

One interesting possibility would be to base a scale of measurement on the three empirical operations: determination of equality, determination of greater or less, and determination of equal ratios. An empirical operation for determining equal intervals would be assumed to be lacking.

What kind of scale would this provide? Clearly we would be able to identify items, order them, and set them in the relation $a/b = b/c = c/d$

. . . . The term "logarithmic interval scale," suggests itself as a name for this scale, because log a — log b = log b — log c, etc. The mathematical transformation under which the structure of this scale remains invariant I have called the *power group*. Thus, for any numerical value x, we can substitute x', where $x' = kx^n$, and the constants k and n are positive.

With this class of scales added to the list, the hierarchy of scales, together with their transformation groups, would take the form shown in Figure 2.1. The arrangement of this figure is intended to suggest that a ratio scale is possible only when empirical operations are available to create both types of interval scales—linear *and* logarithmic. If we can determine both equal differences and equal ratios, we can eliminate the additive constant b and the exponent n, and we are left with only the multiplying constant of the similarity transformation, i.e., we are left with a ratio scale.

If we cannot determine equal intervals, we are left with equal, but unknown, ratios, which limits us to a logarithmic interval scale. The formal properties of such a scale may be interesting, but, like many mathematical models, it has thus far proved useless to the empirical

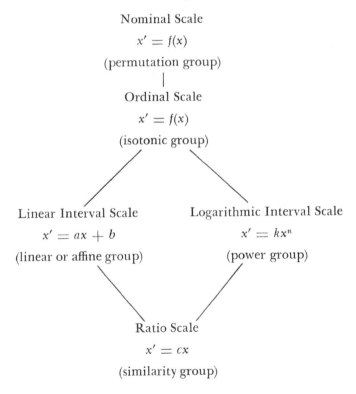

Nominal Scale

$x' = f(x)$

(permutation group)

|

Ordinal Scale

$x' = f(x)$

(isotonic group)

Linear Interval Scale

$x' = ax + b$

(linear or affine group)

Logarithmic Interval Scale

$x' = kx^n$

(power group)

Ratio Scale

$x' = cx$

(similarity group)

Figure 2.1. Hierarchy of Scales

business of science. Scales of equal intervals, however, have had great utility, as witness the scales of temperature (Fahrenheit and Celsius) and the scale of calendar time. There is probably no *a priori* reason why a use could not be found for a logarithmic interval scale built on equated ratios, where the numerical value of the ratio can be chosen only arbitrarily, just as we arbitrarily set the numerical value of a difference on a linear interval scale. Note that both types of interval scales are determined up to two arbitrary constants. In other words, the statement that a scale is determined except for two constants is not always definitive.

As a concrete example of how a set of measurement standards for a logarithmic interval scale might be constructed, let us imagine a situation in which we have an ordinary balance (for determining weight) plus a balance so constructed that the empty pans hang level, but the fulcrum is located at some unknown position, not in the center of the beam. We will also assume that the operation of placing two weights in the same scale pan is for some reason impossible. How then might we proceed? With the balance having the uncentered fulcrum, we could construct a series of weights, A, B, C, . . . , such that $A/B = B/C = C/D$, etc. On the un-centered balance, weight A balances weight B, and then, when weight B is moved to the other pan, it balances weight C placed in the pan formerly occupied by weight B. Whether weight A is heavier or lighter than weight B could be determined by the balance with the centered fulcrum. With these two balances, we could construct an ordered series of standard weights, and to these weights we could assign an appropriate series of numerals. Any two numerals could be chosen arbitrarily, and this choice would fix the values of all the others. Thus, if we set $A = 2$ and $B = 6$, then $C = 18$ and $D = 54$, etc.

It should be mentioned, perhaps, that elsewhere I have attempted to show how, by the addition of a third type of balance with which to create a series of weights spaced by equal intervals, we could proceed to create a standard series of weights answering to the requirements of a ratio scale (31). The point of that demonstration is to try to show that the *physical* addition of weights—the placing of two weights in the same scale pan—is not required to produce a scale isomorphic with Campbell's "fundamental" scale of weight (or of mass, as he prefers to call it).

The statistics applicable to measurements made on a logarithmic interval scale would include those appropriate to a linear interval scale, except that we would need to work with the logarithms of the scale values rather than with the scale values themselves. For example, under the power group of transformations, the item corresponding to the mean would remain invariant, provided we averaged the logarithms of the scale values rather than the scale values themselves.

There is little likelihood, of course, that anyone is going to express the weight of meat and potatoes on a logarithmic interval scales. Our lack of interest in these scales stems from our greater interest in ratio scales, to which we nearly always resort whenever we can.

It is possible, on the other hand, that in certain areas of psychophysics a use may be found for logarithmic interval scales. There is evidence that, on a certain class of "prothetic" continua (subjective brightness, loudness, heaviness, etc.), observers' judgments of equal intervals or differences are subject to a systematic bias that is not present in judgments of equal ratios (40). If this is indeed the case, then scales built on equalized ratios are a promising possibility. Again, we would prefer, if possible, to go a step further and construct ratio scales of these psychological magnitudes. As we shall see, ratio scales are possible, provided we assume that the observer can tell not only when two perceived ratios appear equal but also what the numerical value of the ratio is.

There is a further possibility that the outcome of certain well-known scaling techniques (e.g., pair comparisons) may, when applied to prothetic continua, lead most naturally to logarithmic interval scales. This comes about because discriminal dispersion on prothetic continua, instead of being constant (as generally assumed), tends to be proportional to the psychological magnitude in question. The psychological magnitude, in turn, is a power function of the stimulus magnitude (on at least 14 prothetic continua) (37). It follows, therefore, that psychological values separated by equal units of dispersion on the stimulus scale stand in a constant ratio to each other. In other words, there is reason to believe that some of the "equal-interval scales" constructed by the method of pair comparisons, or by related methods, are, in fact, "equal-ratio scales," i.e., logarithmic interval scales.

Other Types of Scales

The five scales listed in Figure 2.1 would seem to exhaust the possibilities, at least those of scientific interest, except perhaps for the class of scales on which no transformations would be possible. These would be ratio scales having in some sense or other a natural unit, so that not even the similarity transformation (multiplication by a constant) would be admitted.

On the other hand, it has been suggested that certain other scales of measurement, intermediate in power between some of those in Figure 2.1, might be added to the list. The most interesting suggestions along this line are those of Coombs (8, 9, 10). It is not possible to discuss all his suggestions here, but I would like to try to examine one of his proposed scales, the so-called ordered metric scale. This scale has been the object of considerable interest, and ingenious procedures for generating it have been attempted.

The ordered metric scale is said to lie between the ordinal and the interval scales. It assumes that we have an operation for ordering objects on a continuum and also for ordering the intervals between the objects. In other words, we are able to say that the distance between one pair of objects is greater or less than the distance between another pair. This rank-ordering of intervals gives us more information than does the mere rank-ordering of the objects, but it gives us less information than is contained in an interval scale.

The question that naturally suggests itself is this: if our operations can tell us when one interval is greater or less than another, why can they not tell us when the one interval is not greater or less than the other? Why, in other words, can we not determine equal intervals? The operations sufficient for an ordered metric scale ought to be sufficient to determine an interval scale, for otherwise we would be in the odd position of having to argue that an ordinal scale (on intervals) is possible whereas a nominal scale is not.

In what sense, then, is an ordered metric not an interval scale? As I understand it, the argument seems to be that the outcome is an ordered metric scale whenever a procedure sufficient to determine an interval scale is applied to a finite set of objects that do not happen to be equally spaced on the continuum in question. The failure to achieve interval measurement is not caused by an inadequacy of the available procedures but by a paucity of things to measure.

To illustrate this point, let me try once again to construct an imaginary example in an area with which we are all familiar. Suppose we were to try to scale, in terms of their weights, a set of objects A, B, C, D, ... All we have to work with, let us say, is a well-constructed balance and plenty of fine sand. We will assume that we can put one object and some sand in the same scale pan, but that we are not permitted to use the sand to construct a graded series of weights having interval or ratio properties. In other words, we cannot weigh the objects in the ordinary sense, because we have no series of standard weights. How then might we proceed?

First, we use the balance to determine the rank-order of the objects, which we will assume turns out to be $A < B < C < D$ . . . . Next we place A in one pan and B in the other, and proceed to add to the pan holding object A an amount of sand a required to achieve a balance. We then set aside sand a and proceed to do the same with B and C, adding to B enough sand b to balance; similarly, with C and D, etc. We can then proceed to rank-order the differences between the objects by using the balance to determine the rank-order of the weights of the various amounts of sand, *a*, *b*, *c*, ... In this manner we achieve an ordered metric scale.

It is clear that the operations we have envisaged would really be sufficient to determine an interval scale, provided objects of appropriate degrees of heaviness were available or could be manufactured. We could then set up a series of weights (objects) for which the required incremental amounts of sand *a*, *b*, *c*, . . . would all be equal. Such a series might serve as a set of standards in terms of which unknown weights could be measured on a scale unique up to a linear transformation, i.e., on a linear interval scale.

Why, if we can achieve an interval scale, should we settle for less? It seems probable that, in a really serious enterprise in which the establishment of an interval scale was of sufficient moment, we would not be content to stop short of it. For the particular experimental undertakings (8, 26) that have been presented as examples of ordered metric scaling, it seems possible to suggest ways in which the procedures could be modified or extended to produce interval or ratio scales. This is especially true when the scale in question concerns subjective magnitudes, for then, as we shall see, the position of the stimuli on the subjective scale can be ascertained, even though the stimuli do not happen to mark off equal distances. Under such circumstances, perhaps it is fair to ask for evidence that interval or ratio scaling is impossible before we settle for a weaker form. The ordered metric scale appears in practice to be a kind of unfinished interval scale. It is probably not necessary to regard it as a new class of scale in the sense of the listing in Figure 2.1.

References

1. BEHAN, F. L., and R. A. BEHAN. Football numbers (continued). *Amer. Psychologist*, 9, 1954, pp. 262–263.
2. BRUNER, J. S., J. S. GOODMAN, and G. A. AUSTIN. *A study of thinking*. New York: Wiley, 1956.
3. BURKE, C. J. Additive scales and statistics. *Psychol. Rev.*, 60, 1953, pp. 73–75.
4. CAMPBELL, N. R. *Physics. The elements*. Cambridge: The University Press, 1920.
5. ————. *What is science?* 1921; reprinted New York: Dover Publications, 1952.
6. ————. *An account of the principles of measurement and calculation*. London: Longmans, Green, 1928.
7. ————. *Symposium: measurement and its importance for philosophy*. Aristotelian Society, Suppl. Vol. 17. London: Harrison, 1938.
8. COOMBS, C. H. Psychological scaling without a unit of measurement. *Psychol. Rev.*, 57, 1950, pp. 145–158.
9. ————. *A theory of scaling*. Ann Arbor: University of Michigan Press, 1952.
10. ————, H. RAIFFA, and R. M. THRALL. Some views on mathematical models and measurement theory. *Psychol. Rev.*, 61, 1954, pp. 132–144.

11. DAVIDSON, D., P. SUPPES, and S. SIEGEL. *Decision making.* Stanford: Stanford University Press, 1957.
12. EDWARDS, W. The theory of decision making. *Psychol. Bull.,* 51, 1954, pp. 380–417.
13. FECHNER, G. T. *Elemente der Psychophysik.* Leipzig: Breitkopf & Hartel, 1907.
14. Final report. *Advanc. Sci.,* 1940, No. 2, pp. 331–349.
15. GARNER, W. R., and W. J. McGILL. The relation between information and variance analyses. *Psychometrika,* 21, 1956, pp. 219–228.
16. GOODMAN, L. A., and W. H. KRUSKAL. Measures of association for cross classifications. *J. Amer. stat. Assn.,* 49, 1954, pp. 732–764.
17. GUILLIKSEN, H. Measurement of subjective values. *Psychometrika,* 21, 1956, pp. 229–244.
18. HARDY, G. N. Theory of numbers. *Science,* 56, 1922, pp. 402–405.
19. LORD, F. M. On the statistical treatment of football numbers. *Amer. Psychologist,* 8, 1953, pp. 750–751.
20. NEUMANN, J. VON, and O. MORGENSTERN. *Theory of games and economic behavior* (2nd ed.). Princeton: Princeton University Press, 1947.
21. OPPENHEIMER, R. Analogy in science. *Amer. Psychologist,* 11, 1956, pp. 127–135.
22. PLATEAU, J. A. F. Sur la mesure des sensations physiques, et sur la loi qui lie l'intensité de ces sensations à l'intensité de la cause excitante. *Bull. Acad. roy. Belg.,* 33, 1872, pp. 376–388.
23. RUSSELL, B. *The principles of mathematics* (2nd ed.). New York: Norton, 1937.
24. SAVAGE, L. J. *The foundations of statistics.* New York: Wiley, 1954.
25. SENDERS, VIRGINIA L. A comment on Burke's additive scales and statistics. *Psychol. Rev.,* 60, 1953, pp. 423–424.
26. SIEGEL, S. A method for obtaining an ordered metric scale. *Psychometrika,* 21, 1956, pp. 207–216.
27. ————. *Nonparametric statistics.* New York: McGraw-Hill, 1956, p. 76.
28. STEVENS, S. S. On the problem of scales for the measurement of psychological magnitudes. *J. unif. Sci.,* 9, 1939, pp. 94–99.
29. ————. Psychology and the science of science. *Psychol. Bull.,* 36, 1939, pp. 221–263. Reprinted in M. H. Marx (ed.), *Psychological theory, contemporary readings.* New York: Macmillan, 1951, pp. 21–54; P. P. Wiener (ed.), *Readings in philosophy of science.* New York: Scribner's, 1953, pp. 158–184.
30. ————. On the theory of scales and measurement. *Science,* 103, 1946, pp. 667–680.
31. ————. Mathematics, measurement and psychophysics. In S. S. Stevens (ed.), *Handbook of experimental psychology.* New York: Wiley, 1951.
32. ————. Pitch discrimination, mels, and Kock's contention. *J. acoust. Soc. Amer.,* 26, 1954, pp. 1075–1077.
33. ————. The measurement of loudness. *J. acoust. Soc. Amer.,* 27, 1955, pp. 815–829.
34. ————. On the averaging of data. *Science,* 121, 1955, pp. 113–116.
35. ————. Calculation of the loudness of complex noise. *J. acoust. Soc. Amer.,* 28, 1956, pp. 807–832; *see also* Calculating loudness. *Noise Control.,* 3 (5), 1957, pp. 11–22.

36. ————. The direct estimation of sensory magnitudes—loudness. *Amer. J. Psychol.,* 69, 1956, pp. 1–25.
37. ————. On the psychophysical law. *Psychol. Rev.,* 64, 1957, pp. 153–181.
38. ————. Problems and methods of psychophysics. *Psychol. Bull.,* 54, 1958, pp. 177–196.
39. ————, and H. DAVIS. *Hearing: its psychology and physiology.* New York: Wiley, 1938.
40. STEVENS, S. S., and E. H. GALANTER. Ratio scales and category scales for a dozen perceptual continua. *J. exp. Psychol.,* 54, 1957, pp. 377–411.
41. STEVENS, S. S., and J. VOLKMANN. The relation of pitch to frequency: a revised scale. *Amer. J. Psychol.,* 53, 1940, pp. 329–353.
42. STIGLER, G. J. The development of utility theory. *J. polit. Econ.,* 58, 1950, pp. 307–327, 373–396.
43. THRALL, R. M., C. H. COOMBS, and R. L. DAVIS (eds.). *Decision processes.* New York: Wiley, 1954.
44. THURSTONE, L. L. The indifference function. *J. soc. Psychol.,* 2, 1931, pp. 139–167.
45. WHITEHEAD, A. N. *Science and the modern world.* New York: Macmillan, 1925.
46. WIENER, N. A new theory of measurement: a study in the logic of mathematics. *Proc. London math. Soc.,* Ser. 2, 19, 1920, pp. 181–205.

Chapter 3

MEASUREMENT SCALES AND STATISTICAL MODELS

Cletus J. Burke

California State College, Hayward

During the past fifteen or twenty years, there have been several flurries of papers about issues related to statistics and measurement. A controversy has arisen over the interrelations of statistics and measurement; the two opposed viewpoints can be named the measurement-directed position and the measurement-independent position. Proponents of the measurement-directed position, who will be referred to in the sequel as "measurement-directeds," hold that statistical techniques are directed by measurement considerations. Proponents of the measurement-independent position, referred to in the sequel as "measurement-independents," hold that measurement and statistics are separate, independent domains and that therefore measurement considerations do not influence statistical techniques. We shall define the issue between these two positions and summarize their views on important questions, with special reference to practical applications of statistics. Finally, we shall try to re-examine the whole issue and propose a resolution.

The Measurement-Directed Position

Briefly, adherents of this position hold that measurement scales are frequently subject to certain laws of regularity which we shall call measurement models. Thus the measurement model for physical length or for physical weight is a system of ten or so axioms and the theorems which can be derived from them. The power and utility of a measurement scale derives, at least for certain problems, from the properties of the measurement model and its applicability to data. Statistical usage involves operations with numbers, such as the addition of numbers, which may be valid with some kinds of measurement scales and invalid with others. The validity of any statistical operation is to be decided on the basis of an underlying measurement model; for example, with any given measurement scale, the statistical operation of taking the mean may or may not be valid. Since the operation of taking the mean involves the addition of numbers, it will be valid only when the measurement scale has an additive property.

The measurement-directed view has its roots in writings in physics and in the philosophy of physics but takes form in psychology essentially in the writings of Stevens. Stevens (1951) classifies scales of measurement into four classes, namely, nominal, ordinal, interval, and ratio scales. The nominal scale is simply a classification based on what Stevens calls a determination of equality (in other words, of common class membership), and the only permissible statistical measure of central tendency is the mode. The ordinal scale has order based on the empirical operation of the determination of greater or less, and the median is a permissible measure. The interval scale is based on the determination of the equality of intervals or of differences, and the arithmetic mean is a permissible measure. The ratio scale is based on the empirical operation of the determination of the equality of ratios, and the geometric mean or the harmonic mean are permissible measures of central tendency. We can readily see that this classification of scales is based on properties of a measurement model but that recommendations based on the classification are carried over into the domain of statistics.

The philosophical origins of the measurement-directed view seem to be Platonic. Until a few short years ago, much of psychology rested in Plato's special world. Psychological concepts and dimensions were conceived as existing independently of the body of psychological knowledge. Among the denizens of the Platonic world were entities such as intelligence, will, and feeling. The psychologist's task was to discover in the complex world of reality the laws governing the operation and interaction of such entities. We might remark parenthetically that physics, too, has spent much of its career in this world. In the past half century, a number of psychological workers from Watson through Kantor, Stevens, and

Tolman to Skinner, Graham, and Spence have accomplished a revolution in psychological thought. As a result, the concepts of the psychologist are frankly recognized as scientific constructions made to organize the data of the behavioral domain. In contemporary literature such words as *construct, operation,* and *criterion* replace words such as *true* or *real.* Consequently, the Platonic world is now primarily of archaeological interest, but I suspect that living fossils lurk and defend themselves in the area of psychological measurement.

Much of the writing of the measurement-directed school can be understood on the basis of the following hypothesis, namely there exists a Platonic world of real lengths and real temperatures, and in that world our present measurements of length and absolute temperature correspond to the truth, but our measurements of centigrade temperature do not. This Platonic world is composed of real intelligences which, as scientific measures, are every bit as good as the real lengths. In scientific practice, however, the intelligences are inferior to lengths, because we have not yet found the way to measure them which corresponds to the underlying truth. However, a Platonic world of real loudnesses also exists, and here in recent years we have found the proper way to measure. That this sort of hypothesis is found in the writings of Stevens is curious, for he is one of the psychologists who has promoted the revolution of the past thirty years away from such concepts in psychology.

The Measurement-Independent Position

Proponents of this position hold that the results of measurement operations are sets of numbers and, further, that statistical techniques are methods for drawing inferences about sets of numbers or making comparisons between sets of numbers. Hence, once the numbers are available, the statistician is free to proceed with his methods without bothering about any outside considerations. In particular, no properties of a measurement model can have any relevance for statistical operations.

On the philosophical side, the measurement-independent position rejects Platonism. So far as measurements are concerned, the scales may be more or less adequate for the scientific jobs they are expected to do. A completely adequate scale is a scale which exists within a comprehensive and successful scientific theory and which permits accurate and useful computations and predictions. According to this hypothesis, length has been a successful scale in physics because Euclidian geometry has been successfully applied, and temperature has been a successful scale because thermodynamics is a valid and enduring science.

Every empirical domain now covered by polished and viable theory whose history I have studied exhibits in its early development very crude measuring techniques. The workers in every field have reserved complete freedom in performing numerical and algebraic operations on the numbers resulting from the crude measurements. Such freedom of operation has often led to the discovery of important empirical laws which point the way to the final polished theory. Incidentally, empirical laws also point the way to the final adequate measurement scale, but one cannot reach even the first signpost on this way unless he has taken some crude measurement as a starting point and unless he has reserved for himself freedom for operating with the numbers resulting from the crude measurement. Such freedom, however, is categorically denied the scientific worker by the measurement-directed position.

The origins of the measurement-independent position are also partly in statistics..The view of statistics embodied in that position is, as has been previously asserted, that statistics is a set of methods which begins and ends with numbers, a set of methods which is concerned with inferences about sets of numbers, comparisons between sets of numbers, and ultimately statements about sets of numbers.

The Issues

There are a number of issues on which the two positions should be compared. They are in agreement with respect to 1) the definition of measurement and 2) the philosophical status of measurement properties. They are in disagreement with respect to 1) their view of statistics, 2) their view of the role of numbers in scientific work, 3) the resolution of certain difficulties in the application of statistical methods to monotonic transformations of simple measures, and 4) the criteria according to which selection is to be made from alternative non-parametric or distribution-free tests. These differences in the positions are, of course, all related to the fundamental difference on the question of the relevance or irrelevance for statistical procedures of measurement considerations. We shall proceed to compare the two positions with respect to the points of similiarity and difference.

The points of agreement are easily disposed of.

1. DEFINITION OF MEASUREMENT
Proponents of the two positions have agreed, so far as one can tell, on a definition of measurement as the assignment, according to fixed rules, of numbers to objects.

2. PHILOSOPHICAL STATUS OF MEASUREMENT PROPERTIES
Proponents of the two positions agree that the measurement properties of a scale refer to semantic relations between the numbers of the scale and certain phenomena outside the scale.

The points of difference require somewhat more discussion.

1. THE VIEW OF STATISTICS

The picture of statistics which characterizes the measurement-directed position makes statistics directly dependent upon measurement. Statistics is conceived as a group of techniques not for comparing sets of numbers but for comparing sets of objects. The comparison of the sets of objects must somehow involve more than the comparison of the sets of numbers which represent the objects. The "more" which is involved is a measurement model which sets out certain correspondences between properties of the objects and properties of the numbers. The position may be summarized in the following quotation from Stevens (1955[b]):

> The kind of scale we work with depends, of course, upon the concrete empirical operations we are able to perform, and, as we might expect, the character of the operations determines the kind of statistics that are permissible.
> This comes about because a scale erected by a given set of operations can be transformed in certain permissible ways without doing violence to the essential nature of the scale. As a matter of fact, the best way to specify the nature of the scale is in terms of its "group structure"—the group of mathematical transformations that leave the scale form invariant. And it follows quite naturally that the statistics applicable to a given scale are those that remain appropriately invariant under the transformation permitted by the scale.

The measurement-independents maintain that statistical techniques are techniques for the comparison of sets of numbers as numbers and, therefore, that measurement properties of a scale are irrelevant for statistical tests.

2. THE ROLE OF NUMBERS IN SCIENCE

Adherents of both positions recognize the power of numbers in scientific work. The measurement-directeds ascribe this power entirely to the properties of numbers as they occur in measurement models. Whenever gains result from the use of numbers, attempts are made to derive the gains from properties of a measurement model. The measurement-independents hold that numbers have had great use and success in science simply because, after measurements have been made, difficult, clumsy, and sometimes impossible comparisons between objects can be replaced by easy, and often elegant, comparisons between numbers. Such comparisons will be statistical when statistical questions are involved. They may be comparisons of order or other properties when measurement models are involved. They may be, in addition, almost any sort of comparison which the ingenuity of a scientific theorist operating with a theory finds relevant.

3. STATISTIC METHODS AND MONOTONIC TRANSFORMATIONS

Proponents of the measurement-independent position hold that, for statistical purposes, the important property of the measurement scale is order and, further, that in making statistical comparisons one deals with the numbers as numbers alone. The consequences of these two assumptions when applied to monotonic transformations undoubtedly influence the measurement-directed position.

One can obtain a set of measurements and run a test without finding statistical significance. One may then make a monotonic transformation of the numbers by looking up their logarithms, say, or taking their reciprocals. The monotonic transformation changes nothing about the order of the numbers, yet the same test on the transformed set may yield statistical significance. Stevens' writings, as well as those of Senders (1958) and Siegel (1956), make clear that their common viewpoint toward measurement and statistics is brought about at least in part by a desire for strictures to prohibit statistical turpitude. Without some sort of stricture, whether a person uses a set of numbers, or their logarithms, or some other monotonic transformation, is arbitrary. Thus the measurement-directeds seek to legislate statistical honesty by imposing measurement restrictions.

The measurement-independent position admits these difficulties. Of course, the investigator can reach one conclusion by applying a statistical test to a set of measures but another by applying the same test to a set of monotonic transforms. He can also reach different conclusions by applying different tests of the same hypothesis to the same set of measures. Either of these dilemmas is a statistical dilemma. For statistical purposes selection of the proper measurement with a given test or the proper test with a given measurement must be based on statistical grounds. A vast literature defines optimal statistical procedures and gives the conditions under which given procedures are optimal. The conditions are usually concerned with the form of the underlying population. Empirical information about the populations dealt with is necessary for making good choices of either a test or a transformation. Introducing irrelevancy to ignorance does not help solve the problems.

4. SELECTION OF NONPARAMETRIC OR DISTRIBUTION-FREE TESTS

In distribution-free statistics a number of alternative tests for the same hypothesis frequently exist. To choose among them is difficult. The measurement-directed position invokes measurement properties for making the selection. The entire theoretical burden of Siegel's book on nonparametric statistics is concerned with the use of measurement properties for making such a choice.

The measurement-independents hold that the choice among alternative distribution-free tests involves difficult statistical problems. The tests are distribution-free only when the hypotheses being tested are true. Thus the probability of a type I error can be calculated without recourse to the distribution. But the power of a distribution-free test is scarcely non-parametric. In other words, when the hypothesis being tested is false, selection of a distribution-free test should be motivated by a consideration of the most likely alternative hypotheses. This observation points up a number of unsolved statistical problems. Since the problems are unsolved, the suggestion given here is of little practical value for a person wishing to make a selection at the present time. However, the proper problem is the determination of statistical criteria for the selection and is only hidden by the introduction of irrelevant measurement considerations. What must be carried forward is work on the type II error in testing various sorts of hypotheses by distribution-free methods.

There are several papers which deal with various aspects of the issues in general or philosophical terms. Bergmann and Spence (1944) have dealt excellently with the concomitants of the two viewpoints, giving special reference to the history of physics. The emphasis of the remainder of the present paper is on the statistical aspects of the issues.

Statistical Practice

As the positions of the measurement-directeds and the measurement-independents have been described, some differences certainly appear to exist. Whether the differences are of more than philosophical import depends upon whether they have any effects on statistical practice. We shall show concretely that the proponents of both positions approach their statistical problems in the same way but that the adherents of the measurement-independent position will work with greater statistical efficiency.

The operations of the measurement-independents in statistical inference are easily described. 1) They assess the sample data. Obviously, no statements of scientific interest can be made about populations without looking at the data at least in samples. 2) The relation of the sample to the population from which it arises is considered. Specifically, the sampling method is taken into account; usually a random sampling will have been made. 3) Statistical theorems are employed to make statements about the population on the basis of the sample data and the relation of the sample to the population. 4) A scientific interpretation of the results is made. For the interpretation the only important requirement is that the order of the numbers preserve some order in the underlying objects. A fundamental question is whether the measurement-directeds proceed in any different way. To see that they do not, the simplest procedure is to consider a few examples.

EXAMPLE 1: THE USE OF THE RESPONSES OF CHILDREN
AS MEASURING INSTRUMENTS

In Table 3.1 is given the simplest batch of data which could be collected to illustrate the point. There are two courses, A and B, and we are concerned with the lengths of the courses. We ask four children to walk the courses naturally, counting their steps. No child is told how many steps any other child takes. This is a physical measurement, to be sure, but a bad one. I suspect that the scale is not an interval scale, for parenthood has taught me that even a fairly large child will take anywhere from four to ten times as many steps in walking two miles as in walking one. Our problem is to find out whether one course is longer than the other. We proceed by testing the hypothesis that they are of the same length. For purposes of argument we shall agree that the scale is not interval; however, as we proceed, we shall see that this is of no great importance.

Table 3.1. Number of Paces Needed by Each Child to Measure Each Course

Course	child			
	M	T	S	K
A	22	24	30	30
B	66	75	86	111
(B-A)	44	51	56	81

The measurement-independent proceeds according to the following steps: 1) He sets up the physical or geometrical hypothesis that the courses are of the same length. 2) He assumes that the number of steps taken by any child preserves the order of length but does not necessarily have any stronger measurement properties. 3) From 1 and 2 he deduces that any number assigned by any child is as likely to be assigned to course *A* as to course *B*. 4) From 3 he deduces his statistical hypothesis, namely, that the population mean of the difference between the numbers assigned by the children to the two courses, *A* and *B*, is zero.

Note that the statistical hypothesis of step 4 has reference only to the population mean of a set of numbers. Furthermore, this so-called null hypothesis has been derived from considerations of order alone.

The measurement-independent will probably make an assumption of normality and test the hypothesis by using the *t*-test, since the absence of an interval scale does not, for him, prohibit the calculation of means and variances. He will obtain a value of 3.6 for t on the given data, which is significant with 3 df at beyond the five-percent level.

A measurement-directed will insist on a nonparametric test of the hypothesis if he agrees that the scale is not interval. He goes though the following steps: 1) He hypothesizes that the two courses have the same length. 2) He assumes that order in the numbers reflects order in the lengths. 3) From 1 and 2 he deduces that any number assigned by any child is as likely to be assigned to course A as to course B. 4) From 3 he concludes that for any child the number assigned to course A is as likely to be larger as to be smaller than the number assigned to course B.

As with the proponents of the other position, the statistical hypothesis of step 4 is a statement about, and only about, a population of numbers. Again it has been derived solely from a physical hypothesis and considerations of order.

Proponents of the measurement-directed position would work up the data by means of the sign test, obtaining a level of significance of $\frac{1}{8}$, which is as strong a level as is possible for a double-ended sign test on these data.

What should be clear is that: 1) The statement of the physical or geometrical hypothesis is the same in the two cases. 2) The assumption of the preservation of order is common to the two cases. 3) The deduction of the indifference of the assignment of numbers with respect to courses is the same in the two cases. 4) The relationship between the indifference of assignment of numbers and the statistical hypothesis is equivalent in the two cases, differing only as the hypotheses differ in detail of statement. Therefore, the only difference between the two positions in this example is the question of the statistical efficiency of the test selected.

The conclusion based on the statistical test can, of course, be wrong. The population may not be normal, in which case the method of the measurement-independent would be biased, but surely this merely statistical question is not among the points at issue. Or the sampling might not be random, in which case both methods would be biased, not necessarily equally. Finally, the experiment might be bad in that the order of step-numbers might not reflect order of length (for example, course A might be level but course B sharply inclined, or each child might be fatigued after walking course A, so that he would take more steps on course B), but such lack of preservation of order is equally damaging to both procedures.

Before passing to the next example, we remark that A and B can be any objects—physical, psychological, or whatnot. Our four children can provide the numbers of the slide as measures of any property of the objects *A* and *B*. The measures may be test scores, ratings of aggression, or anything. Whatever they represent, if we agree that their order is correct, aside from possible errors of measurement, then by the argument just concluded, the procedures characteristic of both positions are still defensible. The decision on procedure should be based on statistical criteria, entirely within the realm of numbers.

EXAMPLE 2: THE MEAN AND MEDIAN FROM A NORMAL POPULATION

This example is more abstract than the first, but it makes the same point. Suppose we have two random samples from populations known to be normal and of unit variance. We wish to test the hypothesis that the two populations have the same mean—or median, since the distribution is symmetrical—under the assumption that the scale is ordinal at best.

Measurement-independents will emphasize the normal form of the population and, seeking statistical stability, will use the *t*-test to compare the sample means. Measurement-directeds will emphasize the weakness of the measurement scale and will compare the sample medians.

The interpretive question on whether the measures reflect some desired extra-numerical property is not at issue, since adherents of the two positions test equivalent hypotheses by assessing information from identical samples of numbers. Hence, the measurement properties of the scale are simply irrelevant. But a great gain in statistical efficiency obtains for the measurement-independents. When the population is normal, the sample mean is more stable than the sample median. For every 100 subjects run by a measurement-independent, the measurement-directed must run 250 subjects to obtain the same experimental precision. His statistical luxury is again experimentally wanton.

EXAMPLE 3: LATENCIES IN HULLIAN THEORY

When a variable exists within the framework of a systematic theory, the structure of the theory tells us how to manipulate numbers representing the variable. To show the importance of theory for the present discussion, let us make the friendly assumption that the learning theory of Hull and Spence is comprehensive and successful and that all parameters have been evaluated. Then let us suppose that we have the problem of camparing two sets of response latencies.

Among the basic relations of the theory is one between response latency and momentary effective reaction potential. Response latency as a time measure has several well-understood and presumably desirable measurement properties but has a badly skewed and poorly understood distribution; in fact, if the theory of Hull is accurate, the distribution of latencies is very bad—none of the moments exist. Momentary effective reaction potential, on the other hand, may be an abominable variable from the point of view of measurement models, but it has a simple, well-understood normal distribution. If a proponent of the measurement-directed position were faced with this problem in an experimental interpretation of the learning theory, I am convinced that he would use the t-test on the momentary effective reaction potential rather than mess with the difficult and unstable distribution of response latencies.

That the random variables on which statistical tests and estimates are based are themselves mathematical entities is simply a matter of fact. It follows therefore that statistical decisions cannot depend on measurement properties.

Re-examination and Re-evaluation

If the statistical practices of proponents of the two viewpoints are as we have described, the question arises as to why any disagreement exists at all. Why has not the position of the measurement-independents carried the day? A possible answer is that the issues have not yet been explored at sufficient depth and that adequate exploration will reveal weaknesses in the position. A second possibility is that misunderstandings have arisen through unclarity of language.

In connection with the second possibility, we might note that in the literature on measurement and statistics the term "measurement" is used with several meanings. We shall here distinguish four concepts, each of which is often called *measurement,* namely, *measurement model, measurement operation, measurement result,* and *measurement.* A *measurement model* is a theory based on a measurement scale. A *measurement operation* is a collection of empirical and numerical manipulations which leads to the assignment of a single number to a unique object. A *measurement result* is the single number assigned through a measurement operation. The fourth term, *measurement,* will be discussed at length below. We shall not review in detail the literature separating the four meanings of the term but will state a conviction that much misunderstanding can be avoided if the four meanings are kept clearly separated.

More serious misunderstandings may have arisen because of the first possibility, the possibility that the fundamental definition of measurement, although agreed upon by the two positions, has been given in insufficient depth. The definition given for measurement corresponds really to the definition we have just given for measurement operation. Yet, in interesting cases, whether in physics, psychology, or whatever science, a measurement operation clearly falls short of what we mean by measurement. If we examine practice instead of homily, we discover that scientific measurement is the attempt to assign a statistical population to a unique object.

Examples abound. The velocity of light is given as two numbers, an index of central tendency accompanied by an index of variability. Our first attempts to weigh a chemical object with great accuracy were guided by instructions to perform several weighings (*three* is the magic number of my own recollection) and to base the final weight on all. Nor is there any indication in finding or precept that repetition of precisely the same operations on presumably the same object will always lead to the same result. On the contrary, variability is expected as well as encountered whenever we try to make very accurate determinations of quantity.

This has been recognized before. But it has been described Platonically. The stick has been described as having a fixed but unknown number for its length and the scientist as making errors in trying to evaluate the number. Surely the length of the stick is more accurately described as a population of numbers whose properties depend upon the stick and the technique of measurement. In point of fact, one never assigns the population. At best, one performs finitely many measurement operations and thus draws a sample of finite size. Indeed, experience has shown that with certain scales and for certain purposes a sample of unit size suffices.

The stochastic view of measurement as the assignment of a population to an object is consistent with much of the language of measurement. When one speaks of making enough observations to get a stable result or describes the mean of several observations as better than any single observation, one is clearly in the realm of statistical estimation.

For many purposes—testing an additivity property of a measurement model, for example—it is necessary to represent each object by a single number. Such necessity does not contradict the stochastic view of measurement. The number to be used is simply an estimate of some parameter of the population, most often an index of central tendency. Historically, the population mean has almost invariably been estimated in measurement problems. There is no a priori reason, however, why one should not find an additive property characterizing a measurement scale which is based on population modes or medians.

In recognition of the needs set out in the previous paragraph, we distinguish between two classes of stochastic measurement. *Extended measurement* is the attempt to assign a statistical population to an object. *Restricted measurement* is the assignment of a single number to the object via extended measurement and statistical estimation. Restricted measurement can occur with respect to any parameter which can be estimated. Historically the concepts named here as *measurement operation* and *restricted measurement* have been inextricably confused.

The important point to be made is that, when scientific measurement is considered in proper depth, statistical models intervene between the measurement operations and the measurement models. The statistical models deal with the measurement results, which are already samples of numbers. Hence, measurement scales and models are based on statistical models.

Returning to the two basic positions on measurement and statistics, we see that they are both inadequate when viewed from the standpoint of stochastic measurement. The statistical practices advocated by the measurement-independents are the correct ones, but the independence of measurement and statistics postulated by adherents of this position is incorrect. The measurement-directed position is correct in asserting an interdependence of measurement and statistics, but the dependence which they postulate is in the wrong direction. Measurement scales are dependent upon statistical models which are themselves dependent upon measurement results. Each position has a partial, but only a partial, truth. However, the arguments of earlier sections of this paper showing the correctness of the measurement-independent view of statistical practice remain valid. Statistical problems should be solved with statistical theorems. Measurement models are irrelevant.

References

BERGMANN, G., AND SPENCE, K. W. Logic of psychophysical measurement. *Psychol. Rev.*, 1944, 51, 1–24.

BORING, E. G. The logic of the normal law of error in mental measurement. *Amer. J. Psychol.*, 1920, 31, 1–33.

BURKE, C. J. Additive scales and statistics. *Psychol. Rev.*, 1953, 60, 73–75.

———. The gazelle and the hippopotamus. *Worm Runner's Digest* (in press).

CAMPBELL, N. R. *Physics, the elements*. London: Cambridge Univ. Press, 1920.

COMREY, A. L. An operational approach to some problems in psychological measurement. *Psychol. Rev.*, 1950, 57, 217–228.

DAVIDSON, D., SIEGEL, S., and SUPPES P., *Some experiments and related theory on the measurement of utility and subjective probability*. Rep. 4, Stanford Value Theory Project. 1955.

GULLIKSEN, H. Paired comparisons and the logic of measurement. *Psychol. Rev.*, 1946, 53, 199–213.

HULL, CLARK L. *Principles of behavior*. New York: D. Appleton-Century Co., 1943.

SENDERS, VIRGINIA L. A comment on Burke's additive scales and statistics. *Psychol. Rev.*, 1953, 60, 423–424.

———. *Measurement and statistics*. New York: Oxford, 1958. Chapter 2 provides a discussion of "Numbers, things, and measurement."

SIEGEL, S. *Nonparametric statistics for the behavioral sciences*. New York: McGraw-Hill, 1956.

STEVENS, S. S. A scale for the measurement of a psychological magnitude: loudness. *Psychol. Rev.*, 1936, 43, 405–416.

————. On the problem of scales for the measurement of pschological magnitudes. *J. Unif. Sci.*, 1939, 9, 94–99.

————. On the theory of scales of measurement. *Science*, 1946, 103, 677–680.

————. Mathematics, measurement, and psychophysics. In S. S. Stevens (ed.), *Handbook of experimental psychology*. New York: Wiley, 1951, pp. 1–49.

————. The measurement of loudness. *J. Acoust. Soc. Amer.*, 1955ª, 27, 815–829.

————. On the averaging of data. *Science*, 1955ᵇ, 121, 113–116.

————. On the psychophysical law. *Psychol. Rev.*, 1957, 64, 153–181.

STEVENS, S. S., and GALANTER, E. H. Ratio scales and category scales for a dozen perceptual continua. *J. Exp. Psychol.*, 1957, 54, 377–411.

PART II

Thurstonian Methods

L. L. Thurstone is directly responsible for the development of three widely used methods of scaling: paired-comparisons, successive intervals, and equal-appearing intervals. These three methods are described and discussed in Part II. The articles describe the scaling procedures required by each method, the theoretical background, and some of the various tests of fit and signficance used to appraise each technique.

Part II begins with two articles by Thurstone that provide the underpinnings of the method of paired comparisons. The first of these, "Psychophysical Analysis," was Thurstone's first paper in the field of psychophysics and is, in his own judgment, his most important contribution to psychology. He felt that it had more implications for psychological science than any other paper he had written.[1] In this paper he provides a foundation for psychological measurement based on dispersion in the discriminal processes by which individuals are able to differentiate between stimuli. The second paper in Part II, also by Thurstone, describes the law of comparative judgment and presents the assumptions leading to the five cases of the law.

The particular procedures for constructing a paired comparison scale under the assumptions of Case V are presented in Chapter 6 by Bert Green. In addition to treating both the complete and incomplete data formats, this section describes a test that permits the evaluation of the simplifying assumptions. The fourth selection, by Allen L. Edwards, provides procedures for calculating and appraising the consistency of judges and the extent of their agreement. In both cases significance tests are included and described.

1. Thurstone, L. L., *The Measurement of Value*, University of Chicago Press, 1959, p. 15.

An interesting literature has developed that considers the importance of the order in the presentation of stimuli. It appears that certain orders of presentation produce biases and thus influence scale results. Discrimination is more difficult for some pairs than it is for others. For example, the more similar the stimuli paired, the more difficult it is for the individual to make the required judgment. Furthermore, when a stimulus is presented predominately in a particular position, such as when it is always the first of the set of paired stimuli or when it is presented repeatedly in close proximity to itself, such as pairing it with two different stimuli sequentially, judgment bias can result. More explicitly the various biases are: space errors—cases in which the stimulus is in the same position twice in a row; balance errors—cases in which a stimulus appears predominately in the first or second position in the pairs; time errors—cases in which the stimulus is involved in pairs too close together. In addition to removing such errors, the more difficult judgments should be spaced evenly throughout the judgment tasks. The general effect of fatigue in making many judgments of pairs can be reduced by reversing the order of presentation for half the judges. Rational procedures for avoiding these errors and obtaining an optimum arrangement of pairs are described in the selection by Robert T. Ross.

The sixth selection in Part II, by Allen L. Edwards, describes two procedures for obtaining attitude scores of individuals for statements that have been scaled by the method of paired comparison. This is a step necessary in scaling attitudes with paired comparisons that is not involved when one is only interested in obtaining scale scores for objects or stimuli.

The second Thurstone scale procedure, the method of equal-appearing intervals is described in Chapter 10. This method greatly reduces the data production task for the subjects. This selection also describes the application of the method to attitude scaling. The last selection in this section presents the method of successive intervals. This method overcomes some of the difficulties of both the method of paired-comparison and the method of equal-appearing intervals. The particular advantages, such as reducing the data production task for the subject and using less demanding assumptions are discussed by Green in this final selection.

Chapter 4

PSYCHOPHYSICAL ANALYSIS

Louis L. Thurstone

The purpose of this paper is to present a new point of view in psycho-physics and to trace some of its implications. In the determination of a difference limen, the psychophysical judgment, no matter which of the classical methods is followed, is traditionally considered to be a function of *two* factors, namely, (1) the separation or difference between the two physical stimulus magnitudes, and (2) a discriminatory power measured in terms of sense-distances or just noticeable differences. The psychologi-cal continuum, no matter what it may be called, is supposedly determined by these just noticeable differences or equal appearing intervals, which are by definition assumed to be equal. The stimulus magnitudes are laid out on this continuum as landmarks, and the psychological separa-tion between them is stated in terms of just noticeable differences or equal appearing intervals.

It will lead to a rather more flexible and illuminating analysis if we start out a little differently. I shall suppose that every psychophysical judgment is mainly conditioned by four factors, namely the two stimulus magnitudes or the separation between them, the dispersion or variability of the process which identifies the standard stimulus, and the dispersion

Reprinted by permission from the *American Journal of Psychology*, Vol. XXXVIII, 1927, pp. 368–89.

or variability of the process which identifies the variable stimulus. The present analysis will concern these variables and finally the experimental procedures by which they may be isolated.

At the outset it may be well to make clear some things that will *not* be assumed. I shall not assume that the process by which an organism differentiates between two stimuli is either psychic or physiological. I suppose it must really be either, or perhaps both, but it is indifferent for the present argument whether the processes by which we identify or discriminate grays and loudnesses and handwriting specimens are mental or physiological. Hence this analysis has nothing really to do with any psychological system. I shall try not to disturb the main argument with systematic irrelevances or with my personal notions regarding the psychic or physiological nature of the psychophysical judgment.

Further, I shall not assume that sensations, or whatever the identifying and discriminating functions may be called, are magnitudes. It is not even necessary for the present argument to assume that sensations have intensity. They may be as qualitative as you like, without intensity or magnitude, but I shall assume that sensations differ. In other words, the identifying process for red is assumed to be different from that by which we identify or discriminate blue.

A term is needed for that process by which the organism identifies, distinguishes, discriminates, or reacts to stimuli, a term which is innocuous and as noncommittal as possible, because we are not now interested in the nature of the process. Sensations, or more generally, subjective conditions would be good terms but physiological states or intraorganic conditions would also be satisfactory. In order to avoid any implications I shall call the psychological values of psychophysics *discriminal processes. The psychophysical problem concerns, then, the association between a stimulus series and the discriminal processes with which the organism differentiates the stimuli.*

In figure 4.1 let the circles R_1, R_2, R_3, R_n represent a series of stimuli which constitutes a continuum with regard to any prescribed stimulus attribute. It is not necessary to limit psychophysical analysis to stimuli which have intensity or magnitude as their principal attribute. For example, a series of handwriting specimens may be arranged in a continuum on the basis of general excellence. They would of course arrange themselves in a different continuum if some other attribute were specified such as size of letters, legibility, coarseness of pen, or what not. Similarily a series of spectral colors may be arranged in a continuum for discrimination of brightness, chroma, saturation, apparent remoteness from red, or what not. Psychologically some of these attributes can be measured, while physically the measurement may even be impossible. We are

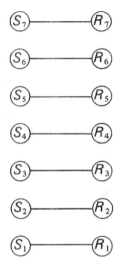

Figure 4.1

assuming, then, that a series of stimuli have been arranged in a continuum according to any attribute about which one can say "more" or "less" and that psychophysics need not be limited to stimuli which have magnitude or size, such as lifted weights and the brightnesses of grays.

Referring again to Figure 4.1, suppose that each stimulus in the series has a discriminal process which is a psychic or physiological function of the organism. Thus the stimulus R_5 has a discriminal process S_5 with which the discrimination of the stimulus takes place. These discriminal processes, whatever be their nature, can be labeled only in terms of their corresponding stimuli so that the discriminal process S_5 is labeled by the stimulus R_5 with which it is associated. In the same manner the other discriminal processes in the series may be labeled by the stimuli which produce them. Naturally the discriminal processes would arrange themselves in a totally different order by changing the attribute of the stimuli by which they are arranged in a continuum. We have then two continua, one for the stimuli and one for the discriminal processes of these stimuli. The stimulus continuum must of course be defined in terms of some definite stimulus attribute. The discriminal continuum is a qualitative one which does not necessarily have either magnitude or intensity.

There is of course no possibility of recording experimentally in any direct way these discriminal processes that correspond to a series of stimuli. It is possible, however, to make some interesting inferences about the psychological continuum indirectly. The stimuli may be used to designate locations in the psychological scale just as though the stimuli, or their names, were used as tags or landmarks in a continuum which has otherwise no identifying marks or mile posts. It is the relative separations between these landmarks on the qualitative psychological continuum which it is the central problem of psychophysics to survey. In the figure there is no attempt to indicate quantitatively the relative separations between the stimuli, or between their psychological correlates. The diagram indicates only that for each of the stimuli in the stimulus continuum one may postulate a discriminal correlate and that these psychological correlates also form a continuum of some kind. Nothing more is known, for the purposes of measurement, about the psychological continuum except that a discrete series of discriminal processes of unknown nature can be used as landmarks along its course and that these processes or landmarks are experimentally controlable or identifiable only in terms of the physical stimuli that produce them.

So far the argument has proceeded as though there were a fixed one-to-one relation between the stimuli and their respective psychological correlates. It may be assumed that this relation is not so fixed as might be indicated by Fig. 4.1. It undoubtedly happens that stimulus R_5, for example, does not always produce the same discriminal process S_5. The present method of psychophysical analysis rests on the assumption that constant and repeated stimuli are not always associated with exactly the same discriminal process but that there is some qualitative fluctuation from one occasion to the next in this process for a given stimulus. This raises an interesting possibility. It might happen for example, that stimulus R_5 has ordinarily S_5 as its discriminal process but that sometimes the qualitative fluctuations would spread to S_4 or to S_6. It might even happen, although rather seldom, that the stimulus R_5 would have as its process S_3 or S_7. It should be recalled that each of these processes or qualities is identified by that stimulus which most frequently produces it so that S_4, for example, is habitually associated with R_4 and so on. This is the fundamental idea of the psychophysical analysis of the present paper.

The variability of this connection between the stimulus and its discriminal process works both ways. A given process S_5 would be associated most frequently with R_5 but occasionally also with adjacent and closely similar stimuli in the stimulus continuum such as R_3, R_4, R_6, R_7. Similarly, the stimulus R_5 can be thought of as most frequently associated with the process or quality S_5 but occasionally with the adjacent qualities such as S_3, S_4, S_6, S_7. Since the discrimination between stimuli is made in the processes of the psychological continuum we shall be concerned with the latter of these two regressions, namely the qualitative fluctuations in the discriminal processes that are associated with a constant and repeated stimulus.

The psychophysical relations may be summarized, so far, in the following propositions.

1. A series of stimuli R_1, R_2, R_3 ... R_n can be arranged in a continuum, with reference to any prescribed quantitative or qualitative stimulus attribute.

2. These stimuli are differentiated by processes of the organism of unknown nature and they are designated S_1, S_2, S_3 ... S_n respectively. Every stimulus R_k is identified by the organism with the process S_k. These processes may be either psychic or physiological or both. In this discussion they are referred to as the discriminal processes or qualities.

3. When the discriminal processes S_1 ... S_n are considered in the same serial order as the corresponding stimulus series they constitute what may be called the discriminal continuum or the psychological continuum. This continuum is the correlate of the already postulated stimulus continuum.

4. It is assumed that the correspondence R_n—S_n is subject to noticeable fluctuation so that R_n does not always produce the exact process S_n but sometimes nearly similar processes S_{n+1} or S_{n-1} and sometimes even S_{n+2} or S_{n-2}. It goes without saying that the numerical subscripts are here used to denote qualitative similarity and that no quantitative attributes are thereby necessarily injected into the discriminal processes. This fluctuation among the discriminal processes for a uniform repeated stimulus will be designated the *discriminal dispersion*.

In Figure 4.2 are represented the two continua, one for the stimulus series and one for the corresponding discriminal processes. Let R_5 be one of the stimuli in the stimulus series. It is assumed that some discriminal process S_5 occurs more frequently with this stimulus than any of the other processes. Hence it is designated the *modal discriminal process* for that stimulus. In this sense S_5 is the modal discriminal process for the stimulus R_5, and so on.

The relative frequencies of the different processes are represented for stimulus R_5 in a rough diagrammatic way. Thus there are three lines connecting R_5 with S_5 to indicate the relation between the stimulus and its modal discriminal process. There are only two lines connecting the adjacent processes with the same stimulus R_5 and this represents the rela-

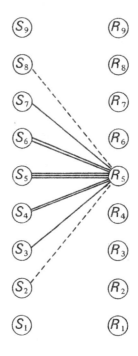

Figure 4.2

tively lower frequency of this association. The processes S_3 and S_7 are connected with the same stimulus with only one line to represent relatively infrequent association. Finally the dotted lines represent in the same manner very infrequent association between the processes so marked and the stimulus R_5. The extreme processes without connection with R_5 represent, then, those processes which are so different from the modal process for R_5 that they never occur in association with the given stimulus or that such association would take place only under unusual conditions as affected by practice, fatigue, sensory adaptation, successive or simultaneous contrast, and so on.

The simplest and perhaps the most obvious plan for scaling would be to assign linear values to the discriminal processes, with reference to a given stimulus, inversely proportional to the frequencies with which these processes occur with the given stimulus. With R_5 as the given stimulus in Figure 4.2 the reckoning would start with the corresponding modal process S_5 as an origin or datum. For this stimulus the other processes could be assigned distance-values from S_5 inversely proportional to their frequencies of occurrence with the given stimulus. Any plan that might be adopted is subject to experimental test in that the separations between the processes can be scaled with reference to each of the various stimuli. Naturally these scale distances between the processes should remain practically constant, no matter what the stimulus may be, in order to have a valid measuring method. Experimental test shows that the plan just suggested of assigning distance values on the psychological continuum breaks down. It is found that the separations between the processes do not retain stable values when they are determined for different stimuli. Therefore some other plan must be adopted.

The normal probability curve has been so generally abused in psychological and educational measurement that one has reason to be fearful of criticism from the very start in even mentioning it. The only valid justification for bringing in the probability curve in this connection is that its presence can be experimentally tested. The writer has found experimentally that the normal probability curve was not applicable for certain stimuli. In most of the experiments the distributions are reasonably close to normal.

Since the assumption of a normal distribution for the discriminal dispersion can be experimentally verified and limited to those stimulus series where its reality can be tested, it will be reasonable to make this assumption subject to verification in every case. The hypothesis can be stated as follows. *The discriminal dispersion which any given repeated stimulus produces on the psychological continuum is usually normal. The frequencies with which the discriminal processes occur for a given stimulus ordinarily describe a normal distribution when plotted on the psychological continuum as a base.* In experimental practice the procedure is the reverse of this hypothesis because the frequencies are known first experimentally and from these frequencies we *construct* the psychological continuum. The writer has found in several studies that the separation between any pair of processes remains practically constant no matter which of the neighboring stimuli is used as a base for the calculation. Such is not the case, however, when the separation of any pair of processes is assigned values directly or inversely proportional to their frequency of occurrence.

In Figure 4.2 where R_5 is chosen as the stimulus we should therefore, according to this hypothesis, assign scale values to the various processes as distances from S_5 as an origin. These distances would be assigned in terms of the standard deviation of the distribution of process-frequencies. There is of course no further unit in terms of which this standard deviation can be expressed. It is itself a unit of measurement because all that we can do with the psychological continuum is to lay off linear separations between the processes proportional to their true value since, so far as we know, there is in the nature of the case no further absolute unit of measurement

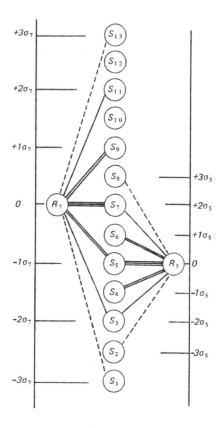

Figure 4.3

for the psychological continuum. But we shall see that it is possible to compare the discriminal dispersions for two stimuli and to determine experimentally the ratio of any two of these dispersions. *Psychological measurement depends, then, on the adoption of one of these dispersions as a base, and the use of its standard deviation as a unit of measurement for the psychological continuum under investigation.*

In Figure 4.3 let the column of thirteen circles represent so many discriminal processes, each of them being a modal discriminal process for a stimulus with the same numerical designation. Two of these thirteen stimuli are indicated in the figure, namely R_5 and R_7. Suppose that these stimuli are arranged in a continuum according to any prescribed stimulus attribute and let R_7 be more ambiguous, or less sharply defined, than R_5. An example would be two specimens of handwriting, one of which would be a beautiful but unusual handwriting, or perhaps it might be written in a foreign language, or it might be in German script which would possibly call forth judgments influenced by prejudice from factors other than those of the handwriting characteristics. If the experiment involves the comparison of loudnesses, a variation of the certainty or ambiguity of judgment for a particular stimulus might be caused by variations in timbre or pitch. Ordinarily psychophysical experiments are so set up as to avoid, as completely as possible, the introduction of extraneous factors to influence the ambiguity of judgment and the stimuli are made into as homogeneous a series as may be experimentally possible.

In Figure 4.3, the two stimuli are represented as differing in the certainty with which they can be judged as to the prescribed attribute for the stimulus continuum and R_7 is indicated as the more variable or uncertain of the two. The modal discriminal process for R_7 is S_7 as before, and the discriminal processes S_5, S_3, S_1 might be assigned deviation values of 1σ, 2σ, 3σ respectively from S_7 as a datum. These deviation values would be assigned on the basis of the frequency with which each of these processes occur with R_7 as a stimulus. With the same diagrammatic representation let the other processes be assigned their deviation values from S_7 as a base and let the same processes be assigned frequency-deviation values from S_5 as a base for stimulus R_5. In Figure 4.3 these hypothetical deviations are given numerical values. Note from the figure that the discriminal process S_5 which is modal for stimulus R_5 has a deviation value of $-1\sigma_7$ for stimulus R_7. Similarly the discriminal process S_1 has a deviation value of $-1.5\sigma_7$ for stimulus R_7 while it has a deviation value of $-1\sigma_5$ for stimulus R_5. If this analysis is correct it should happen not infrequently that the stimuli which constitute a continuum according to any prescribed stimulus attribute are subject to varying degrees of dispersion when they are perceived or judged. Some stimuli are probably placed with reference to the prescribed attribute more accurately and consequently with a smaller discriminal or subjective dispersion than other stimuli. It is probably true that this variability of the discriminal dispersion on the psychological continuum is of relatively less serious importance in dealing with strictly homogeneous stimulus series but it becomes a serious factor in dealing with less conspicuous attributes or with less homogeneous stimulus series such as handwriting specimens, English compositions, sewing samples, Oriental rugs. In measurements of the type known as judgment scales the discriminal dispersion on the psychological continuum becomes one of the unknowns to be determined as well as the scale value of the specimen. Every specimen in such a series presents two unknown values to be determined: namely, the scale value of its modal discriminal process on the psychological continuum, and its discriminal dispersion.

Instead of the diagrammatic representation of Figure 4.3 two normal

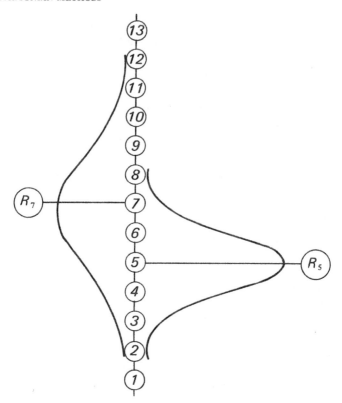

Figure 4.4

probability curves may be substituted, subject of course to subsequent experimental verification. This has been done in Figure 4.4. Here the psychological continuum has been constructed on the hypothesis that the discriminal processes describe a normal distribution when plotted on that continuum. When R_7 and R_5 are presented for a comparative judgment, each of the stimuli produces a discriminal process of some kind and the certainty of the discrimination may be assumed to be mainly a function of the difference between these two processes. If R_7 happens to be associated with one of the processes at the upper range of its discriminal dispersion and if R_5 happens to be associated with one of the processes at the lower end of its discriminal dispersion, then the discrimination is made with ease and the judgment is correct. If these conditions are reversed so that R_7 has a process slightly below its modal process while R_5 happens to have a process slightly above its modal process, then the two stimuli may even have the same discriminal process and there would be no possibility of a confident discrimination. Finally, if on some occasion R_7 happens to have a process unusually low in its scale, while R_5 has a process higher in the psychological scale, then the judgment would be made, perhaps even with confidence, that R_5 is greater than R_7 and the judgment would be recorded as incorrect. The discrimination is considered, then, as a function of the *discriminal difference* between the two processes that happen to be associated in the same judgment. By the discriminal difference is meant the linear separation on the psychological continuum between the two processes involved in any particular judgment. It may be designated $S_{7.5}$ or more generally S_{ka}. The discriminal difference is the same as the *sense distance* if we allow that the sense distance for two stimuli fluctuates from one occasion to the next.

If in a long series of experimental judgments it were possible to isolate the two discriminal processes for every judgment and if the separation between these two processes for every judgment were recorded, one could tabulate them in the form of a frequency table of discriminal differences. These differences would of course be expressed in terms of some unit of measurement on the psychological continuum. Let the standard stimulus be A and the variable stimulus K. The mean of the distribution of discriminal differences would be the mean or true difference $(S_k - S_a)$ and its standard deviation would be

$$\sigma_{ka} = \sqrt{\sigma^2{}_k + \sigma^2{}_a}$$

on the assumption that deviations from the modal processes for the two stimuli are not correlated. This distribution is represented in Figure 4.5. The base line of this distribution represents discriminal differences in

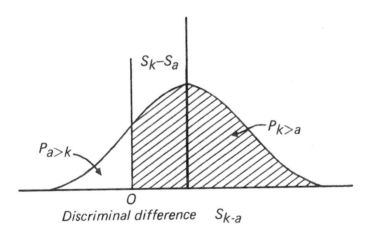

Figure 4.5

terms of any desired unit of measurement on the psychological continuum. The mean is $(S_k - S_a)$ because that is the difference between the two modal processes. The origin represents a difference of zero. This would occur when the two stimuli happen to be associated with the same discriminal process in which case there is no discrimination possible. The points to the right of the origin on the base line represent positive values for the differences S_{k-a} in which R_k has a process higher in the scale than R_a. Similarly the points to the left of the origin represent negative values for the discriminal difference S_{k-a} in which R_k happens to have a process lower in the psychological scale than R_a. It should be recalled that S_{k-a} or S_{ka} represents the sense distance between two stimuli on any particular occasion whereas $S_k - S_a$ represents the mean sense distance for several hundred judgments and it is in scale construction called the "true" sense distance or scale distance between the two stimuli.

For the present it will simplify analysis to assume that any discriminal difference, no matter how small, is directly reflected in the judgment. A correction may be inserted for this approximation by which a *discriminal difference limen* can be calculated but this correction will not seriously alter the results. It may be assumed for the present that all positive discriminal differences, S_{k-a}, result in the judgment "R_k greater than R_a" and that all negative discriminal differences result in the judgment "R_a greater than R_k." If the two paired stimuli are presented N times there will of course be observed N discriminal differences and their expected distribution is represented in Figure 4.5. The shaded portion of that figure represents the expected proportion of judgments "R_k greater than R_a" and these judgments would be correct if K is higher than A. The unshaded portion of the surface represents the expected proportion of judgments "R_a greater than R_k." The proportion of correct judgments will of course increase if the two stimuli are chosen farther apart. Also, the proportion of correct judgments will increase if stimuli are chosen with smaller discriminal dispersions. If the shaded area is greater than $N/2$ it represents correct judgments. If it is less than $N/2$ it represents the proportion of incorrect judgments.

A discriminal difference, S_{k-a}, is not necessarily a magnitude. It is a pair of processes, a pair of qualities. The only way in which numerical value is assigned to it is by placing each of these processes on a measured continuum by means of the frequency with which each of them is associated with the same stimulus. The difference between these two assigned linear values is the discriminal difference, S_{k-a}. The scale distance, $S_k - S_a$, can be defined as the most common discriminal difference, S_{k-a}.

At this point we have arrived at a measure which can be experimentally verified. By the method of constant stimuli it is readily possible to ascertain the actual proportion of judgments "R_7 greater than R_5" for the two stimuli. This proportion is a function of four variables namely S_7, S_5, σ_7, σ_5. If there are n stimuli in the stimulus series there will be $2n$ unknowns to be evaluated, namely n scale values for the modal discriminal processes, and n scale values for the discriminal dispersions. If every stimulus is used in turn as a standard the number of possible pairs of stimuli will be

$$T = \frac{n(n-1)}{2}. \tag{1}$$

Since there is an experimental proportion "a greater than b" for every possible pair of stimuli, it follows that there will be $n(n-1)/2$ observation equations and $2n$ unknowns. One of the modal discriminal processes can be chosen as a datum or origin for the psychological scale, and one of the discriminal dispersions can be chosen as a unit of measurement for the construction of a psychological scale. This reduces the number of unknowns to $(2n-2)$ or $2(n-1)$.

Table 4.1

Number of stimuli in the series, n	Total number of unknowns, $2(n-1)$	Number of observation equations, $T = n(n-1)/2$
1	0	0
2	2	1
3	4	3
4	6	6
5	8	10
6	10	15
7	12	21
8	14	28
9	16	36
10	18	45

Table 4.1 shows for stimulus series of varying length from 1 to 10, the number of available observation equations and the total number of unknowns. When the stimulus series has less than four stimuli, the number of unknowns is greater than the number of observation equations and the problem therefore cannot be solved. When there are four stimuli in the series the number of observation equations exactly equals the total number of unknowns and the problem can then be solved by simultaneous equations. When the stimulus series has more than four stimuli there are more observation equations than there are unknowns and the problem must then be solved by the method of least squares or by some other method of balancing errors of observation.

The fundamental psychophysical equation can then be stated in the following form.

$$S_k - S_a = X_{ka}\sqrt{\sigma_k^2 + \sigma_a^2}$$ [2]

in which S_k and S_a are the two modal scale values on the psychological continuum for the two stimuli R_k and R_a.

X_{ka} is the sigma value for the experimentally observed proportion of judgments "R_k greater than R_a." When these proportions are greater than .50 the stimulus R_k is higher in the psychological continuum than R_a.

$\sigma_k =$ the discriminal dispersion of R_k on the psychological continuum.

$\sigma_a =$ the discriminal dispersion of R_a on the psychological continuum.

The assumptions underlying this psychological equation are as follows:

1. That every stimulus in the stimulus-series is associated with a modal discriminal process with which the organism identifies the stimulus for a prescribed attribute.

2. That the modal discriminal process for any given stimulus retains at least some of its identity even when the stimulus is combined with other stimuli into a single perceptual judgment.

3. That the modal processes may be arranged in a linear psychological continuum in the same serial or rank order as the corresponding stimulus series.

4. In addition to arranging the discriminal processes in rank or serial order, *linear separations* between them are assigned on the assumption that the discriminal dispersion for any stimulus is normal on the psychological continuum. This assumption is subject to experimental verification.

5. That the discriminal deviations for the different stimuli are uncorrelated. This is a fairly safe assumption but if they are correlated, the psychophysical equation [2] becomes

$$S_k - S_a = X_{ka}\sqrt{\sigma^2_k + \sigma^2_a - 2 \cdot r_{ka} \cdot \sigma_k \cdot \sigma_a}$$

in which case the numerical solution becomes unwieldy.

6. That all positive discriminal differences S_{k-a} give the judgment "$k>a$," that all negative discriminal differences S_{k-a} give the judgment "$k<a$," and that discriminal differences of zero $S_{k-a} = O$, are equally distributed between "higher" and "lower" if only two judgments are allowed. This is a close approximation to truth but a correction can be introduced in terms of a discriminal difference limen for judgments "equal" and "doubtful." This correction is left for a separate paper.

Experimental Procedure for Verifying Assumed Normality of Discriminal Dispersion

Assumption 4, that the discriminal dispersion for any stimulus is normal on the psychological continuum, may be experimentally tested by ascertaining whether the separation between any two modal processes (sense distance) remains constant no matter which of the stimuli is used as a base. Consider R_k as the base or standard for equation [2]. Then the proportion of judgments $k > a$ will be controlled by the relation

$$S_k - S_a = X_{ka}\sqrt{\sigma^2_k + \sigma^2_a} \qquad [2]$$

Similarly for the proportion of judgments $k > b$,

$$S_k - S_b = X_{kb}\sqrt{\sigma^2_k + \sigma^2_b}$$

Subtracting,

$$S_b - S_a = X_{ka}\sqrt{\sigma^2_k + \sigma^2_a} - X_{kb}\sqrt{\sigma^2_k + \sigma^2_b} \qquad [3]$$

If the same equation is written with R_l, R_m, R_n as standards, we have

$$S_b - S_a = X_{la}\sqrt{\sigma_l^2 + \sigma_a^2} - X_{lb}\sqrt{\sigma_l^2 + \sigma_b^2}$$
$$= X_{ma}\sqrt{\sigma_m^2 + \sigma_a^2} - X_{mb}\sqrt{\sigma_m^2 + \sigma_b^2}$$
$$= X_{na}\sqrt{\sigma_n^2 + \sigma_a^2} - X_{nb}\sqrt{\sigma_n^2 + \sigma_b^2}$$

If every separation such as $S_b - S_a$ remains constant when determined by different stimuli such as R_k, R_l, R_m, R_n as standards, then internal consistency for the measurements has been demonstrated and the validity of Assumption 4 is thereby established. Such internal consistency depends on the nature of the assumed distribution of discriminal processes by which the psychological continuum is constructed.

The point of view that I am describing has many implications bearing on well known psychophysical principles. One of the conclusions to which the present analysis leads is that Fechner's law and Weber's law are really independent and that it is consequently incorrect to speak of these two laws jointly as "the Weber-Fechner law." Another important conclusion relates to the well known hypothesis that equally often noticed differences are equal. The present analysis shows that hypothesis is incorrect because it is possible for two differences to *seem unequal* on the average and yet be equally often discriminated. Other implications concern the limitations of the phi-gamma hypothesis in psychophysical experimentation and the distribution of judgments of equality. A few applications of the concept of discriminal dispersion are described below.

Fechner's Law

Fechner's law is usually phrased as follows:

$$S = K \log R$$

in which S represents sensation intensity which we have here called scale value. The notation R refers to stimulus intensity or magnitude. It will be noticed that in writing our psychophysical equation nothing has been said about stimulus magnitudes or intensities because of the fact that many stimulus series that are subjected to psychological measurement are not capable of quantitative measurement on their objective side. For example, the relative excellence of a series of handwriting specimens may be measured on a psychological continuum but the corresponding physical "magnitudes" probably do not exist as a single variable. The physical handwriting specimens cannot be readily measured as to the stimulus variable "excellence."

Fechner's law can be applicable only to those stimulus series in which the attribute which is being judged can also be physically isolated. Then, if the discriminal separations of the psychological continuum are plotted against the physical stimulus attribute and if this plot is logarithmic, Fechner's law is verified.

In many cases there is no possibility of making sure that the physical variable really corresponds to the psychological one. For example, a series of circles can be arranged in a stimulus series in accordance with their diameters. The discriminal experiments may then be carried out with instructions to indicate which of two exposed circles is the larger without specifying further what is meant by larger. The circles would no doubt arrange themselves in the same serial order in the psychological continuum as in the stimulus continuum so that the two series would have exactly the same rank orders. Now, if we want to verify Fechner's law, we should plot the separations between the modal processes for the circles along the psychological continuum against the corresponding physical stimulus variable. Shall we plot diameters on the base line or shall we plot areas? These two plans would arrange the stimulus series in the same rank order, but the relation between diameter and area is not linear. Both diameter and area would be physical variables covariant with the apparent bigness of the circles. Now, if Fechner's law is verified for one of these physical variables, it could not possibly be verified for the other because of the non-linear relation of diameter and area. If we should find experimentally that Fechner's law is satisfied by plotting the psychological continuum against the diameters, for example, that would not justify the conclusion that Fechner's law applies. We could artificially force Fechner's law everywhere by merely selecting that particular stimulus variable which does give a logarithmic relation with the psychological continuum. Fortunately the law has been shown to hold true for many stimulus series in which there is hardly any possibility of an ambiguous stimulus variable and its universality therefore commands our respect.

Weber's Law as Independent of Fechner's Law

In the present discussion Weber's law is interpreted broadly for the frequently observed relation between the stimulus magnitudes and the scale distances on the psychological continuum. I am not here limiting myself to those particular applications of the law by which it is restricted to sensory intensities. The law is not always verified for sensation intensities, but, on the other hand, I have found it applicable to some other stimulus series that are not sensory intensity magnitudes. The present discussion of Weber's law concerns the functional relation between stimulus magnitudes and psychological scale distances without implying that the law is limited to sensory stimulus intensities.

Weber's law and Fechner's law are often described together and they are frequently called jointly "the Weber-Fechner law." The two laws are independent so that either one of them may be applicable without the other being verified for a particular set of data. The two laws must be separately verified for any given set of data.

Weber's law is usually stated as follows: The just noticeable increase of a stimulus is a constant fraction of the stimulus. The term "just noticeable" is ambiguous so that it is necessary to specify how often a stimulus increase must be correctly noted in order for the stimulus increase to be called "noticeable." This frequency is often placed arbitrarily at 75 percent of the judgments when two judgments are allowed. Restating Weber's law with this provision so as to remove the ambiguity of the term "just noticeable" we have the following statement of the law: The stimulus increase which is correctly discriminated in 75 percent of the attempts, when only two judgments "higher" and "lower," or their equivalents, are allowed, is a constant fraction of the stimulus magnitude. With reference to Fechner's law there are two cases under which Weber's law may be verified. In *Case 1* Fechner's law is postulated, and in *Case 2* it is not postulated.

Case 1. Let the stimulus magnitude be designated R_a and let it be increased to the magnitude R_b at which separation the two stimuli are correctly discriminated in 75 percent of the attempts by the constant method and with two judgments allowed. At this separation between the two stimuli our psychophysical equation [2] takes the following form:

$$S_b - S_a = X_{ab}\sqrt{\sigma_a^2 + \sigma_b^2}$$

which, when stated explicitly for the required proportion of 75 percent judgments "b greater than a," becomes

$$\frac{S_b - S_a}{\sqrt{\sigma_b^2 + \sigma_a^2}} = X_{ab} = 0.674. \qquad [4]$$

Weber's law states that any pair of stimuli, R_a and its increased magnitude R_b, corresponding to the two modal processes S_a and S_b in the above equation, are such that the fraction R_b/R_a remains a constant no matter what

the absolute magnitudes of the stimuli may be. It is clear from the above
equation that the separation between the two stimuli which gives a result
of 75 percent correct judgments is a function not only of the two stimulus
magnitudes and their corresponding modal processes but also of the discriminal dispersions for the two stimuli. Weber's law may be verified
under *Case 1* if an additional condition is satisfied, namely, that the discriminal dispersions are the same for all the stimuli. If the discriminal
dispersions are not constant, then it is possible for Fechner's law to be
applicable when Weber's is not. If the discriminal dispersions are equal
for all the stimuli, then equation [4] may be written as follows:

$$\frac{S_b - S_a}{\sigma\sqrt{2}} = X_{ab} = 0.674 \qquad [5]$$

and since the discriminal dispersions which is here assumed to be constant
may be taken as a unit of measurement on the psychological continuum,
we have

$$\frac{S_b - S_a}{\sqrt{2}} = 0.674$$

$$\text{or } S_b - S_a = \sqrt{2 \cdot 0.674}. \qquad [6]$$

This relation is obtained by Condition 2 above. But Weber's law states
a constant relation in terms of the stimuli. This transformation can be
made by Fechner's law as follows:

$$S_a = K \log R_a$$

$$S_b = K \log R_b$$

$$S_b - S_a = K \, [\log R_b - \log R_a]$$

$$S_b - S_a = K \log \frac{R_b}{R_a} \qquad [7]$$

From equations [6] and [7] we have

$$S_b - S_a = \sqrt{2} \cdot 0.674 = K \log \frac{R_b}{R_a}$$

or simply

$$\log \frac{R_b}{R_a} = \text{constant}$$

and hence

$$\frac{R_b}{R_a} = constant$$

thus verifying Weber's law. But in order to verify Weber's law under *Case 1* it was necessary to make two assumptions, namely that Fechner's law applies and also that the discriminal dispersions are constant. If stimuli were used of varying degrees of homogeneity or ambiguity the discriminal dispersions would not be constant and it would then be possible to discover that Fechner's law is applicable when Weber's law is not.

Case 2. It is possible for Weber's law to be applicable when Fechner's law is not verified and when the discriminal dispersions are not all equal. This is best illustrated by a short list of stimuli with hypothetical discriminal dispersions. For the purpose of this illustration we can assume any relation between S and R except the logarithmic relations of Fechner's law. Let us tabulate some paired values for S and R such that $S = R^2$. This is clearly, then, a case in which Fechner's law does not apply. In Table 4.2 the first column identifies the six stimuli in the hypothetical series. Column R designates the stimulus magnitudes. Column S shows the scale values of the corresponding modal processes (sensation intensities). Column σ shows a hypothetical series of discriminal dispersions. By

Table 4.2

Stim. Series	R	$S = R^2$	σ	$P\,(_{n+1}) > _n$
1	10.00	100.	20.00	.75
2	11.00	121.	23.94	.75
3	12.10	146.	29.16	.75
4	13.31	177.	35.40	.75
5	14.64	215.	43.40	.75
6	16.11	259.	51.90	.75

means of the fundamental psychophysical equation [2] it can then be shown that the stimuli 1 and 2 are correctly discriminated in 75 percent of the judgments, that stimuli 2 and 3 are correctly discriminated in 75 percent of the judgments, and so on. Since the ratio of each stimulus magnitude to the next lower stimulus magnitude is always 1.10 in this table and since these successive pairs of stimuli are correctly differentiated in 75 percent of the observations, we conclude that Weber's law has been verified by these hypothetical data. The only new factor that we have introduced is the plausible assumption that the discriminal dispersion may not be constant throughout the whole stimulus range. With an assumed variation in the discriminal dispersion we find that it is logically possible to have a set of data in which Weber's law is verified but in which Fechner's law is not verified. All that is necessary for the discriminal dispersion to vary from one stimulus to another is that the stimuli be unequal in the ambiguity or difficulty with which they are judged and this surely must happen much more often than we suspect when the stimuli consist in such qualitative values as handwriting specimens or specimens of English composition. It is quite probable that the variation in discriminal dispersion is rather slight and perhaps negligible when the stimulus series is rather homogeneous. A good example of a homogeneous stimulus series is a set of cylinders for the lifted weight experiment in which size, color, texture, shape, and even temperature are ruled out of the experiment by keeping them constant. In such experiments it is probable that the discriminal dispersion stays constant.

Finally, if the discriminal dispersions can be assumed to be equal throughout the whole stimulus range, then Fechner's law and Weber's law become identical. The frequent association of these two laws as though they were always identical depends on the constancy of the discriminal dispersion. It may be expected in psychophysical experiments with stimuli that are not experimentally kept constant in all but one stimulus variable, that one or two stimuli in the series are more difficult to judge than the rest. In such a case these one or two stimuli will have larger discriminal dispersions than the other stimuli and the consistency of the psychological continuum is thereby disturbed if these variations are not accounted for in the derivation of the scale values.

Equally Often Noticed Differences are not Necessarily Equal

It is usually assumed that equally often noticed differences are equal on the psychological continuum. They are rarely assumed to be equal on the stimulus continuum. It is however incorrect to assume that pairs of stimuli are equally distant on the psychological scale even though all the pairs are equally often discriminated. It is not even correct to say that stimulus differences *seem* equal, or that they are subjectively equal, just because the differences are equally often noticed. Two pairs of stimuli may be equally often discriminated while one of the separations may on the average actually *seem* greater than the other.

Referring again to the psychophysical equation [2] the psychological or *apparent* separation between two stimuli R_a and R_b is expressed by the difference $(S_b - S_a)$, measured on the psychological scale which is a scale of appearances or impressions. The frequency with which the two stimuli can be discriminated is, however, a function their respective discriminal dispersions as well as their modal discriminal processes. The separation between the modal processes can also be called the mean sense distance. Here again, if we can assume that the discriminal dispersions are constant, then it is correct to say that equally often noticed differences are psychologically equal but that assumption should be tested before constructing a psychological continuum or scale by means of this assumption.

A Possible Effect of Practice

It is probable that practice has the effect of reducing the discriminal dispersions and that this may account for the shifts in the proportions of correct judgments in psychophysical experiments. If two stimuli are presented to an unpracticed subject for whom these stimuli have relatively large discriminal dispersions, the denominator of equation [4] will be relatively large while the numerator remains constant. Graphically the situation can be represented in Fig. 4.5 by increasing the standard deviation of that probability curve while the separation $(S_k - S_a)$ remains constant. This produces a low proportion of correct judgments. With practice, the subject reduces the discriminal dispersions and this might be represented in Fig. 4.5 by reducing the standard deviation of that curve while the separation between the two modal processes remains constant. The effect is to increase the proportion of correct judgments. Naturally, stable results for the construction of a psychological scale depend on reaching such a practice level that the discriminal dispersions will remain practically constant throughout the experiments. The interpretation of the psychophysical equation in connection with the effect of practice would be that two lights, for example, seem just about as bright to the practiced laboratory subject as to an unpracticed subject. Practice in psychophysical experimentation does not make one of the lights seem brighter or the other one weaker. The two lights retain their same general level of brightness except for sensory adaptation and contrast which are momentary effects. But there is a practice effect in the capacity to discriminate between the two lights. This is determined by the discriminal dispersion or subjective observational error. Here again, equally often noticed differences are not necessarily equal subjectively or psychologically.

Experimental Test

The simplest experimental procedure for verifying the assumption that the discriminal dispersions are constant for any particular stimulus series is probably to arrange a table showing the proportion of judgments, $P_{a\ b}$, for all the possible pairs of stimuli. If there are N stimuli, such a table will contain $N(N - 1)$ entries if identical stimuli are not experimentally compared. From such a table the stimuli can readily be arranged in rank order. From the table of proportions of judgments, a corresponding table of sigma values can be prepared. One can then plot a graph for X_{ka} against X_{kb} in which a and b are standards. If the discriminal dispersions are equal through the stimulus series, the graph should give a linear plot with a slope of unity. This may be demonstrated as follows:

If in the psychophysical equation

$$S_k - S_a = X_{ka} \sqrt{\sigma_k^2 + \sigma_a^2} \qquad [2]$$

we assume that the discriminal dispersions are equal, the equation becomes

$$S_k - S_a = X_{ka}\sqrt{2\sigma^2}$$
$$= X_{ka} \cdot \sigma \cdot \sqrt{2} \qquad [5]$$

and if we use the discriminal dispersion as a unit of measurement on the psychological scale, we have

$$S_k - S_a = X_{ka} \cdot \sqrt{2} \qquad [6]$$

By symmetry it follows that

$$S_k - S_b = X_{kb} \cdot \sqrt{2} \qquad [7]$$

Subtracting and transposing,

$$X_{ka} = X_{kb} + \frac{S_b - S_a}{\sqrt{2}} \qquad [8]$$

This equation is in linear form and if X_{ka} is plotted against X_{kb} we should have a linear plot. The slope should be unity and

$$Y\text{-intercept} = \frac{S_b - S_a}{\sqrt{2}} \qquad [9]$$

If the plot is linear, it proves that the assumed normal distribution of discriminal processes is correct. If the slope is unity, it proves that the discriminal dispersions are equal. It is left for a separate paper to apply this method to educational judgment scale data.

Chapter 5

A LAW OF COMPARATIVE JUDGMENT

Louis L. Thurstone

The object of this paper is to describe a new psychophysical law which may be called the *law of comparative judgment* and to show some of its special applications in the measurement of psychological values. The law of comparative judgment is implied in Weber's law and in Fechner's law. The law of comparative judgment is applicable not only to the comparison of physical stimulus intensities but also to qualitative comparative judgments such as those of excellence of specimens in an educational scale and it has been applied in the measurement of such psychological values as a series of opinions on disputed public issues. The latter application of the law will be illustrated in a forthcoming study. It should be possible also to verify it on comparative judgments which involve simultaneous and successive contrast.

The law has been derived in a previous article and the present study is mainly a description of some of its applications. Since several new concepts are involved in the formulation of the law it has been necessary to invent several terms to describe them, and these will be repeated here.

Reprinted from *Psychological Review*, Vol. XXXIV, 1927, pp. 273–86. Copyright 1927 by the American Psychological Association, and reproduced by permission.

This is one of a series of articles by members of the Behavior Research Staff of the Illinois Institute for Juvenile Research, Chicago, Herman M. Adler, Director, Series B, No. 107.

Let us suppose that we are confronted with a series of stimuli or specimens such as a series of gray values, cylindrical weights, handwriting specimens, children's drawings, or any other series of stimuli that are subject to comparison. The first requirement is of course a specification as to what it is that we are to judge or compare. It may be gray values, or weights, or excellence, or any other quantitative or qualitative attribute about which we can think "more" or "less" for each specimen. This attribute which may be assigned, as it were, in differing amounts to each specimen defines what we shall call the *psychological continuum* for that particular project in measurement.

As we inspect two or more specimens for the task of comparison there must be some kind of process in us by which we react differently to the several specimens, by which we identify the several degrees of excellence or weight or gray value in the specimens. You may suit your own predilections in calling this process psychical, neural, chemical, or electrical but it will be called here in a non-committal way *the discriminal process* because its ultimate nature does not concern the formulation of the law of comparative judgment. If then, one handwriting specimen *seems* to be more excellent than a second specimen, then the two discriminal processes of the observer are different, at least on this occasion.

The so-called "just noticeable difference" is contingent on the fact that an observer is not consistent in his comparative judgments from one occasion to the next. He gives different comparative judgments on successive occasions about the same pair of stimuli. Hence we conclude that the discriminal process corresponding to a given stimulus is not fixed. It fluctuates. For any handwriting specimen, for example, there is one discriminal process that is experienced more often with that specimen than other processes which correspond to higher or lower degrees of excellence. This most common process is called here *the modal discriminal process for the given stimulus.*

The psychological continuum or scale is so constructed or defined that the frequencies of the respective discriminal processes for any given stimulus form a normal distribution on the psychological scale. This involves no assumption of a normal distribution or of anything else. The psychological scale is at best an artificial construct. If it has any physical reality we certainly have not the remotest idea what it may be like. We do not assume, therefore, that the distribution of discriminal processes is normal on the scale because that would imply that the scale is there already. We *define* the scale in terms of the frequencies of the discriminal processes for any stimulus. This artificial construct, the psychological scale, is so spaced off that the frequencies of the discriminal processes for any given stimulus form a normal distribution on the scale. The separation on the scale between the discriminal process for a given stimulus on any particular occasion and the modal discriminal process for that stimulus we shall call *the discriminal deviation* on that occasion. If on a particular occasion, the observer perceives more than the usual degree of excellence or weight in the specimen in question, the discriminal deviation is at that instant positive. In a similar manner the discriminal deviation at another moment will be negative.

The standard deviation of the distribution of discriminal processes on the scale for a particular specimen will be called its *discriminal dispersion*.

This is the central concept in the present analysis. An ambiguous stimulus which is observed at widely different degrees of excellence or weight or gray value on different occasions will have of course a large discriminal dispersion. Some other stimulus or specimen which is provocative of relatively slight fluctuations in discriminal processes will have, similiarly, a small discriminal dispersion.

The scale difference between the discriminal processes of two specimens which are involved in the same judgment will be called *the discriminal difference* on that occasion. If the two stimuli be denoted A and B and if the discriminal processes corresponding to them be denoted a and b on any one occasion, then the discriminal difference will be the scale distance $(a - b)$ which varies of course on different occasions. If, in one of the comparative judgments, A seems to be better than B, then, on that occasion, the discriminal difference $(a - b)$ is positive. If, on another occasion, the stimulus B seems to be the better, then on that occasion the discriminal difference $(a - b)$ is negative.

Finally, the scale distance between the modal discriminal processes for any two specimens is the separation which is assigned to the two specimens on the psychological scale. The two specimens are so allocated on the scale that their separation is equal to the separation between their respective modal discriminal processes.

We can now state the law of comparative judgment as follows:

$$S_1 - S_2 = x_{12} \cdot \sqrt{\sigma_1{}^2 + \sigma_2{}^2 - 2r\sigma_1\sigma_2}, \qquad [1]$$

in which

S_1 and S_2 are the psychological scale values of the two compared stimuli.

$x_{12} =$ the sigma value corresponding to the proposition of judgments $p_1 {}_{>2}$. When $p_1 {}_{>2}$ is greater than .50 the numerical value of x_{12} is positive. When $p_1 {}_{>2}$ is less than .50 the numerical value of x_{12} is negative.

$\sigma_1 =$ discriminal dispersion of stimulus R_1.

$\sigma_2 =$ discriminal dispersion of stimulus R_2.

$r =$ correlation between the discriminal deviations of R_1 and R_2 in the same judgment.

This law of comparative judgment is basic for all experimental work on Weber's law, Fechner's law, and for all educational and psychological scales in which comparative judgments are involved. Its derivation will not be repeated here because it has been described in a previous article.[1] It applies fundamentally to the judgments of a *single observer* who compares a series of stimuli by the method of paired comparison when no "equal" judgments are allowed. It is a rational equation for the method of constant stimuli. It is assumed that the single observer compares each pair of stimuli a sufficient number of times so that a proportion, $p_{a>b}$, may be determined for each pair of stimuli.

For the practical application of the law of comparative judgment we shall consider five cases which differ in assumptions, approximations, and degree of simplification. The more assumptions we care to make, the simpler will be the observation equations. These five cases are as follows:

Case I. The equation can be used in its complete form for paired comparison data obtained from a single subject when only two judgments are allowed for each observation such as "heavier" or "lighter," "better" or "worse," etc. There will be one observation equation for every observed proportion of judgments. It would be written, in its complete form thus:

$$S_1 - S_2 - x_{12} \cdot \sqrt{\sigma_1{}^2 + \sigma_2{}^2 - 2r\sigma_1\sigma_2} = 0. \qquad [1]$$

According to this equation every pair of stimuli presents the possibility of a different correlation between the discriminal deviations. If this degree of freedom is allowed, the problem of psychological scaling would be insoluble because every observation equation would introduce a new unknown and the number of unknowns would then always be greater than the number of observation equations. In order to make the problem soluble, it is necessary to make at least one assumption, namely that the correlation between discriminal deviations is practically constant throughout the stimulus series and for the single observer. Then, if we have n stimuli or specimens in the scale, we shall have $\frac{1}{2} \cdot n(n-1)$ observation equations when each specimen is compared with every other specimen. Each specimen has a scale value, S_1, and a discriminal dispersion, σ_1, to be determined. There are therefore $2n$ unknowns. The scale value of one of the specimens is chosen as an origin and its discriminal dispersion as a unit of measurement, while r is an unknown which is assumed to be constant for the whole series. Hence, for a scale of n specimens there will be $(2n-1)$ unknowns. The smallest number of specimens for which the problem is soluble is five. For such a scale there will be nine unknowns, four scale values, four discriminal dispersions, and r. For a scale of five specimens there will be ten observation equations.

1. Thurstone, L. L., "Psychophysical Analysis," *Amer. J. Psychol.*, July, 1927.

The statement of the law of comparative judgment in the form of equation 1 involves one theoretical assumption which is probably of minor importance. It assumes that all positive discriminal differences $(a - b)$ are judged $A > B$, and that all negative discriminal differences $(a - b)$ are judged $A < B$. This is probably not absolutely correct when the discriminal differences of either sign are very small. The assumption would not affect the experimentally observed proportion $p_{A>B}$ if the small positive discriminal differences occurred as often as the small negative ones. As a matter of fact, when $p_{A>B}$ is greater than .50 the small positive discriminal difference $(a - b)$ are slightly more frequent than the negative perceived differences $(a - b)$. It is probable that rather refined experimental procedures are necessary to isolate this effect. The effect is ignored in our present analysis.

Case II. The law of comparative judgment as described under Case I refers fundamentally to a series of judgments *of a single observer.* It does not constitute an assumption to say that the discriminal processes for a single observer give a normal frequency distribution on the psychological continuum. That is a part of the definition of the psychological scale. But it does constitute an assumption to take for granted that the various degrees of an attribute of a specimen perceived in it by *a group* of subjects is a normal distribution. For example, if a weight-cylinder is lifted by an observer several hundred times in comparison with other cylinders, it is possible to define or construct the psychological scale so that the distribution of the apparent weights of the cylinder for the single observer is normal. It is probably safe to assume that the distribution of apparent weights for *a group* of subjects, each subject perceiving the weight only once, is also normal on the same scale. To transfer the reasoning in the same way from a single observer to a group of observers for specimens such as handwriting or English Composition is not so certain. For practical purposes it may be assumed that when *a group* of observers perceives a specimen of handwriting, the distribution of excellence that they read into the specimen is normal on the psychological continuum of perceived excellence. At least this is a safe assumption if the group is not split in some curious way with prejudices for or against particular elements of the specimen.

With the assumption just described, the law of comparative judgment, derived for the method of constant stimuli with two responses, can be extended to data collected from a group of judges in which each judge compares each stimulus with every other stimulus only once. The other assumptions of Case I apply also to Case II.

Case III. Equation [1] is awkward to handle as an observation equation for a scale with a large number of specimens. In fact the arithmetical labor of constructing an educational or psychological scale with it is almost prohibitive. The equation can be simplified if the correlation r can be assumed to be either zero or unity. It is a safe assumption that when the stimulus series is very homogeneous with no distracting attributes, the correlation between discriminal deviations is low and possibly even zero unless we encounter the effect of stimultaneous or successive contrast. If we accept the correlation as zero, we are really assuming that the degree of excellence which an observer perceives in one of the specimens has no influence on the degree of excellence that he perceives in the comparison specimen. There are two effects that may be operative here and which are antagonistic to each other.

1. If you look at two handwriting specimens in a mood slightly more generous and tolerant than ordinarily, you may perceive a degree of excellence in specimen A a little higher than its mean excellence. But at the same moment specimen B is also judged a little higher than its average or mean excellence for the same reason. To the extent that such a factor is at work the discriminal deviations will tend to vary together and the correlation r will be high and positive.

2. The opposite effect is seen in *simultaneous contrast*. When the correlation between the discriminal deviations is negative the law of comparative judgment gives an exaggerated psychological difference $(S_1 - S_2)$ which we know as simultaneous or successive contrast. In this type of comparative judgment the discriminal deviations are negatively associated. It is probable that this effect tends to be a minimum when the specimens have other perceivable attributes, and that it is a maximum when other distracting stimulus differences are removed. If this statement should be experimentally verified, it would constitute an interesting generalization in perception.

If our last generalization is correct, it should be a safe assumption to write $r = 0$ for those scales in which the specimens are rather complex such as handwriting specimens and children drawings. If we look at two handwriting specimens and perceive one of them as unusually fine, it probably tends to depress somewhat the degree of excellence we would ordinarily perceive in the comparison specimen, but this effect is slight compared with the simultaneous contrast perceived in lifted weights and in gray values. Furthermore, the simultaneous contrast is slight with small stimulus differences and it must be recalled that psychological scales are based on comparisons in the subliminal or barely supraliminal range.

The correlation between discriminal deviations is probably high when the two stimuli give simultaneous contrast and are quite far apart on the scale. When the range for the correlation is reduced to a scale distance comparable with the difference limen, the correlation probably is reduced nearly to zero. At any rate, in order to simplify equation 1 we shall assume that it is zero. This represents the comparative judgment in which the evaluation of one of the specimens has no influence on the evaluation of the other specimen in the paired judgment. The law then takes the following form.

$$S_1 - S_2 = x_{12} \cdot \sqrt{\sigma_1^2 + \sigma_2^2}. \tag{2}$$

Case IV. If we can make the additional assumption that the discriminal dispersions are not subject to gross variation, we can considerably simplify the equation so that it becomes linear and therefore much easier to handle. In equation [2] we let

$$\sigma_2 = \sigma_1 + d,$$

in which d is assumed to be at least smaller than σ_1 and preferably a fraction of σ_1 such as .1 to .5. Then equation [2] becomes

$$
\begin{aligned}
S_1 - S_2 &= x_{12} \cdot \sqrt{\sigma_1^2 + \sigma_2^2} \\
&= x_{12} \cdot \sqrt{\sigma_1^2 + (\sigma_1 + d)^2} \\
&= x_{12} \cdot \sqrt{\sigma_1^2 + \sigma_1^2 + 2\sigma_1 d + d^2}.
\end{aligned}
$$

If d is small, the term d^2 may be dropped. Hence

$$
\begin{aligned}
S_1 - S_2 &= x_{12} \cdot \sqrt{2\sigma_1^2 + 2\sigma_1 d} \\
&= x_{12} \cdot \sqrt{2\sigma}(\sigma_1 + d)^{\frac{1}{2}}.
\end{aligned}
$$

Expanding $(\sigma_1 + d)^{\frac{1}{2}}$ we have

$$
\begin{aligned}
(\sigma_1 + d)^{\frac{1}{2}} &= \sigma_1^{\frac{1}{2}} + \frac{1}{2}\sigma_1^{-(\frac{1}{2})}d - \frac{1}{4}\sigma_1^{-(3/2)}d^2 \\
&= \sqrt{\sigma_1} + \frac{d}{2\sqrt{\sigma_1}} - \frac{d^2}{4\sqrt{\sigma_1^3}}.
\end{aligned}
$$

The third term may be dropped when d^2 is small. Hence

$$(\sigma_1 + d)^{\frac{1}{2}} = \sqrt{\sigma_1} + \frac{d}{2\sqrt{\sigma_1}}.$$

Substituting,

$$
\begin{aligned}
S_1 - S_2 &= x_{12} \cdot \sqrt{2\sigma_1}\left[\sqrt{\sigma_1} + \frac{d}{2\sqrt{\sigma_1}} \right] \\
&= x_{12}\left[\sigma_1\sqrt{2} + \frac{d}{\sqrt{2}} \right]
\end{aligned}
$$

But $d = \sigma_2 - \sigma_1$;

$$\therefore S_1 - S_2 = x_{12}\frac{\sigma_2}{\sqrt{2}} + x_{12}\frac{\sigma_1}{\sqrt{2}}$$

or

$$S_1 - S_2 = .707x_{12}\sigma_2 + .707x_{12}\sigma_1. \tag{3}$$

Equation [3] is linear and very easily handled. If $\sigma_2 - \sigma_1$ is small compared with σ_1, equation [3] gives a close approximation to the true values of S and σ for each specimen.

If there are n stimuli in the scale there will be $(2n - 2)$ unknowns, namely a scale value S and a discriminal dispersion σ for each specimen. The scale value for one of the specimens may be chosen as the origin or zero since the origin of the psychological scale is arbitrary. The discriminal dispersion of the same specimen may be chosen as a unit of measurement for the scale. With n specimens in the series there will be $\frac{1}{2}n(n - 1)$ observation equations. The minimum number of specimens for which the scaling problem can be solved is then four, at which number we have six observation equations and six unknowns.

Case V. The simplest case involves the assumption that all the discriminal dispersions are equal. This may be legitimate for rough measurement such as Thorndike's handwriting scale or the Hillegas scale of English Composition. Equation [2] then becomes

$$S_1 - S_2 = x_{12} \cdot \sqrt{2\sigma^2}$$

$$= x_{12}\sigma \cdot \sqrt{2.}$$

But since the assumed constant discriminal dispersion is the unit of measurement we have

$$S_1 - S_2 = 1.4142x_{12}. \tag{4}$$

This is a simple observation equation which may be used for rather coarse scaling. It measures the scale distance between two specimens as directly proportional to the sigma value of the observed proportion of judgments $p_1 >_2$. This is the equation that is basic for Thorndike's procedure in scaling handwriting and children's drawings although he has not shown the theory underlying his scaling procedure. His unit of measurement was the standard deviation of the discriminal differences which is $.707\sigma$ when the discriminal dispersions are constant. In future scaling problems equation [3] will probably be found to be the most useful.

Weighting the Observation Equations

The observation equations obtained under any of the five cases are not of the same reliability and hence they should not all be equally weighted. Two observed proportions of judgments such as $p_1 \gg_2 = .99$ and $p_1 >_3 = .55$ are not equally reliable. The proportion of judgments $p_1 >_2$ is one of the observations that determine the scale separation between S_1 and S_2. It measures the scale distance $(S_1 - S_2)$ in terms of the standard deviation, σ_{1-2}, of the distribution of discriminal differences for the two stimuli R_1 and R_2. This distribution is necessarily normal by the definition of the psychological scale.

The standard error of a proportion of a normal frequency distribution is

$$\sigma_p = \frac{\sigma}{Z} \cdot \sqrt{\frac{pq}{N}}^{\,2} .$$

in which σ is the standard deviation of the distribution, Z is the ordinate corresponding to p, and $q = 1 - p$ while N is the number of cases on which the proportion is ascertained. The term σ in the present case is the standard deviation σ_{1-2} of the distribution of discriminal differences. Hence the standard error of $p_1 >_2$ is

$$\sigma_{p1>2} = \frac{\sigma_{1-2}}{Z} \cdot \sqrt{\frac{pq}{N}} . \qquad [5]$$

But since, by equation [2]

$$\sigma_{1-2} = \sqrt{\sigma_1^2 + \sigma_2^2} \qquad [6]$$

and since this may be written approximately, by equation [3], as

$$\sigma_{1-2} = .707(\sigma_1 + \sigma_2) \qquad [7]$$

we have

$$\sigma_{p1>2} = \frac{.707(\sigma_1 + \sigma_2)}{Z} \cdot \sqrt{\frac{pq}{N}} . \qquad [8]$$

The weight, w_{1-2}, that should be assigned to observation equation [2] is the reciprocal of the square of its standard error. Hence

$$w_{1-2} = \frac{1}{\sigma_{p1>2}^2} = \frac{Z^2 N}{.5(\sigma_1 + \sigma_2)^2 p \cdot q} . \qquad [9]$$

It will not repay the trouble to attempt to carry the factor $(\sigma_1 + \sigma_2)^2$ in the formula because this factor contains two of the unknowns, and because it destroys the linearity of the observation equation [3], while the only

2. See Kelley, T. L., "Statistical Method," p. 90, equation 43.

advantage gained would be a refinement in the weighting of the observation equations. Since only the weighting is here at stake, it may be approximated by eliminating this factor. The factor .5 is a constant. It has no effect, and the weighting then becomes

$$w_{1-2} = \frac{Z^2 N}{pq}. \qquad [10]$$

By arranging the experiments in such a way that all the observed proportions are based on the same number of judgments the factor N becomes a constant and therefore has no effect on the weighting. Hence

$$w_{1-2} = \frac{Z^2}{pq}. \qquad [11]$$

This weighting factor is entirely determined by the proportion, $p_{1>2}$ of judgments "1 is better than 2" and it can therefore be readily ascertained by the Kelley-Wood tables. The weighted form of observation equation [3] therefore becomes

$$wS_1 - wS_2 - .707wx_{12}\sigma_2 - .707wx_{12}\sigma_1 = 0. \qquad [12]$$

This equation is linear and can therefore be easily handled. The coefficient $.707wx_{12}$ is entirely determined by the observed value of p for each equation and therefore a facilitating table can be prepared to reduce the labor of setting up the normal equations. The same weighting would be used for any of the observation equations in the five cases since the weight is solely a function of p when the factor σ_{1-2} is ignored for the weighting formula.

Summary

A· law of comparative judgment has been formulated which is expressed in its complete form as equation [1]. This law defines the psychological scale or continuum. It allocates the compared stimuli on the continuum. It expresses the experimentally observed proportion, $p_{1>2}$ of judgments "1 is stronger (better, lighter, more excellent) than 2" as a function of the scale values of the stimuli, their respective discriminal dispersions, and the correlation between the paired discriminal deviations.

The formulation of the law of comparative judgment involves the use of a new psychophysical concept, namely, the *discriminal dispersion.* Closely related to this concept are those of the *discriminal process,* the *modal discriminal process,* the *discriminal deviation,* the *discriminal difference.* All of these psychophysical concepts concern the ambiguity or qualitative variation with which one stimulus is perceived by the same observer on different occasions.

The psychological scale has been defined as the particular linear spacing of the confused stimuli which yields a normal distribution of the discriminal processes for any one of the stimuli. The validity of this definition of the psychological continuum can be experimentally and objectively tested. If the stimuli are so spaced out on the scale that the distribution of discriminal processes for one of the stimuli is normal, then these scale allocations should remain the same when they are defined by the distribution of discriminal processes of any other stimulus within the confusing range. It is physically impossible for this condition to obtain for several psychological scales defined by different types of distribution of the discriminal processes. Consistency can be found only for one form of distribution of discriminal processes as a basis for defining the scale. If, for example, the scale is defined on the basis of a rectangular distribution of the discriminal processes, it is easily shown by experimental data that there will be gross discrepancies between experimental and theoretical proportions, $p_1 > _2$. The residuals should be investigated to ascertain whether they are a minimum when the normal or Gaussian distribution of discriminal processes is used as a basis for defining the psychological scale. Triangular and other forms of distribution might be tried. Such an experimental demonstration would constitute perhaps the most fundamental discovery that has been made in the field of psychological measurement. Lacking such proof and since the Gaussian distribution of discriminal processes yields scale values that agree very closely with the experimental data, I have defined the psychological continuum that is implied in Weber's Law, in Fechner's Law, and in educational quality scales as that particular linear spacing of the stimuli which gives a Gaussian distribution of discriminal processes.

The law of comparative judgment has been considered in this paper under five cases which involve different assumptions and degrees of simplification for practical use. These may be summarized as follows.

Case I. The law is stated in complete form by equation [1]. It is a rational equation for the method of paired comparison. It is applicable to all problems involving the method of constant stimuli for the measurement of both quantitative and qualitative stimulus differences. It concerns the repeated judgments of a single observer.

Case II. The same equation [1] is here used for *a group* of observers, each observer making only one judgment for each pair of stimuli, or one serial ranking of all the stimuli. It assumes that the distribution of the perceived relative values of each stimulus is normal for the group of observers.

Case III. The assumptions of Cases I. and II. are involved here also and in addition it is assumed that the correlation between the discriminal deviations of the same judgment are uncorrelated. This leads to the simpler form of the law in equation [2].

Case IV. Besides the preceding assumptions the still simpler form of the law in equation [3] assumes that the discriminal deviations are not grossly different so that in general one may write

$$\sigma_2 - \sigma_1 < \sigma_1$$

and that preferably

$$\sigma_2 - \sigma_1 = d$$

in which d is a small fraction of σ_1.

Case V. This is the simplest formulation of the law and it involves, in addition to previous assumptions, the assumption that all the discriminal dispersions are equal. This assumption should not be made without experimental test. Case V. is identical with Thorndike's method of constructing quality scales for handwriting and for children's drawings. His unit of measurement is the standard deviation of the distribution of discriminal differences when the discriminal dispersions are assumed to be equal.

Since the standard error of the observed proportion of judgments, $p_{1>2}$, is not uniform, it is advisable to weight each of the observation equations by a factor shown in equation [11] which is applicable to the observation equations in any of the five cases considered. Its application to equation [3] leads to the weighted observation equation [12].

Chapter 6

PAIRED COMPARISON SCALING PROCEDURES

BERT GREEN

John Hopkins University

If only a few items are to be scaled, the method of paired comparisons is appropriate (Thurstone, 1927; Guilford, 1936, Chapter VII). Each of several judges (preferably 25 or more) is presented with every possible pair of items, and is asked which of the pair is more favorable to the issue in question. For each pair of items we obtain the proportion of times one statement was judged to be more favorable than the other. In the mathematical model it is assumed that the perceived differences between stimuli have normal distributions. Specifically, the model states that

$$S_i - S_j = X_{ij}\, \sigma_{(i-j)}, \qquad [1]$$

where S_i and S_j are the scale values of items i and j, $\sigma_{(i-j)}$ is the standard deviation of the hypothetical distribution of differences, and X_{ij} is the unit normal deviate corresponding to p_{ij}, the proportion of times item i is judged to be more favorable than item j, i.e.,

$$p_{ij} = \int_{-\infty}^{x_{ij}} \frac{1}{\sqrt{2\pi}} e^{-t2/2}\, dt. \qquad [2]$$

Reprinted with permission of the author and the publisher from the *Handbook of Social Psychology*, Vol. 1, edited by Gardner Lindzey, pp. 344–47. Addison-Wesley, Reading, Massachusetts, 1954.

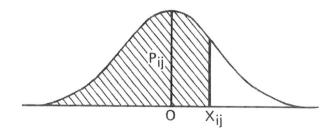

Figure 6.1. The unit normal deviate transformation. The abscissa is measured in units of one standard deviation.

The relation between X_{ij} and p_{ij} is illustrated in Figure 6.1.

In the simplest case of the model, the so-called Case V, all $\sigma_{(i-j)}$ are assumed to be equal and are set equal to unity. In this case the model becomes

$$S_i - S_j = X_{ij}. \qquad [3]$$

This simple model will usually suffice.

The data may or may not be consistent with the model. For a set of scale values, we can use equation [3] to compute an expected set of X_{ij}. These values can then be compared with the observed data. Since there are many more experimental data than there are scale values, some discrepancies between the expected X_{ij} and the observed X_{ij} are inevitable. Our problem is to find a set of scale values for which the discrepancies are minimum. Mosteller (1951a) has shown that Thurstone's method of estimation, described below, is a least-squares solution—the sum of squared discrepancies is minimum.

Mosteller (1951b) has also provided a chi-square test of the fit of the model. Both the expected X_{ij} and the observed X_{ij} are converted into proportions by means of the unit normal deviate transformation—equation [2]. These proportions, labeled P''_{ij} for the expected proportions and P'_{ij} for the actual data, are transformed by the arc sin transformation

$$\theta_{ij} = \text{arc} \quad \sin \sqrt{p_{ij}}. \qquad [4]$$

A table of this transformation is available in the appendix (Table I). This gives us a set of θ'_{ij} and a set of θ''_{ij} which we use with equation [5] to compute chi-square:

$$\chi^2 = \frac{\displaystyle\sum_{i < j} (\theta'_{ij} - \theta''_{ij})^2}{821/n}. \qquad [5]$$

Here n is the number of judgments on which each observed proportion is based, and the summation is over all values of i and j in which i is less than j, a total of $\frac{1}{2}k$ $(k-1)$ terms, where k is the number of items. This chi-square value has $\frac{1}{2}$ $(k-1)$ $(k-2)$ degrees of freedom.

The first step in obtaining the scale values is to convert the observed proportions p_{ij} into X_{ij} by equation [2], using a table of the normal distribution. See Table II in the appendix. These X_{ij} are arranged in a two-way table, or data matrix, with a row for each item i and a column for each item j, as in Table 6.1.

Note that X_{ij} is logically zero, and that $X_{ij} = -X_{ij}$. If any X_{ij} is greater than $+2.0$, or less than -2.0, it should be rejected as unstable. If no values of X_{ij} have been rejected by this rule, the scale value of item i is the average of all entries in the ith column of the data matrix, as shown in Table 6.1 (b).

Table 6.1   Paired Comparisons
Computations—Complete Matrix
(4 Items)

		(a) Matrix of P_{ij} Item j			
		1	2	3	4
	1	—	.58	.83	.95
Item i	2	.42	—	.80	.92
	3	.17	.20	—	.65
	4	.05	.08	.35	—

		(b) Matrix of X_{ij} Item j			
		1	2	3	4
	1	0	.20	.95	1.65
Item i	2	−.20	0	.84	1.41
	3	−.95	−.84	0	.39
	4	−1.65	−1.41	−.39	0
Column sums		−2.80	−2.05	1.40	3.45
Scale values		−.70	−.51	.35	.86

If some X_{ij} have been rejected, as indicated in Table 6.2 (a), then for each pair of successive columns of the data matrix, we subtract the entries in one column from the corresponding entries in the other column, and record these differences in a second matrix, as in Table 6.2 (b). Of course, no difference can be obtained whenever one of the values has been rejected. Since we are using successive columns, it may be possible to arrange the order of the columns so that there are very few impossible differences. The average of these differences for a pair of columns is the difference between the scale value for those two items. Thus, by arbitrarily setting $S_1 = 0$, the remaining scale values can be determined.

Note that both the zero point and the scale factor are arbitrary in the scale. We set the scale factor by letting all standard deviations equal unity. In the case of the complete matrix, the zero point was set by letting the average scale value be zero. In the case of the incomplete matrix, the zero point was set by letting $S_1 = 0$. Any other arbitrary values for the zero point and scale factor may be used.

The application of the chi-square test is illustrated in Table 6.2 (c).

Table 6.2. Paired Comparisons Computations—Incomplete Matrix (6 Items)

		(a) Matrix of P_{ij}						(b) Matrix of X_{ij}				
			Item j						Item j			
		1	2	3	4	5	6	1	2	3	4	5	6
	1		.55	.75	.87	.98	1.00	0	.13	.67	1.13		
	2	.45		.65	.79	.95	1.00	—.13	0	.39	.81	1.65	
Item i	3	.25	.35		.75	.91	.99	—.67	—.39	0	.67	1.34	
	4	.13	.21	.25		.84	.92	—1.13	—.81	—.67	0	.99	1.41
	5	.02	.05	.09	.16		.78	—1.65	—1.34	—.99	0	.77
	6	.00	.00	.01	.08	.22				—1.41	—.77	0

(c) Matrix of successive differences

Items

		2-1	3-2	4-3	5-4	6-5
	1	.13	.54	.46		
	2	.13	.39	.42	.84	
Item i	3	.28	.39	.67	.67	
	4	.32	.14	.67	.99	.42
	5		.31	.35	.99	.77
	6				.64	.77
Column sum		.86	1.77	2.57	4.13	1.96
Column average		.22	.35	.51	.83	.65

$S_1 = 0$, $S_2 = .22$, $S_3 = .57$, $S_4 = 1.08$, $S_5 = 1.91$, $S_6 = 2.56$

(d) Chi-square computation

ij	p'_{ij}	p''_{ij}	θ'_{ij}	θ''_{ij}	$\theta'_{ij} - \theta''_{ij}$
12	.55	.587	47.87	50.01	—2.14
13	.75	.716	60.00	57.80	2.20
14	.87	.860	68.87	68.03	.84
23	.65	.637	53.73	52.95	.78
24	.79	.805	62.72	63.79	—1.07
25	.95	.955	77.08	77.75	— .67
34	.75	.695	60.00	56.48	3.52
35	.91	.910	72.54	72.54	.00
45	.84	.797	66.42	63.22	3.20
46	.92	.931	73.57	74.77	—1.20
56	.78	.742	62.03	59.47	2.56

$$n = 100$$
$$\sum_{i<j} (\theta'_{ij} - \theta''_{ij})^2 = 42.95$$
$$\chi^2 = 5.23$$
$$\text{d.f.} = 6$$
$$.70 > p > .50$$

References

GUILFORD, J. P. *Psychometric methods.* New York: McGraw-Hill, 1936.

MOSTELLER, F. Remarks on the method of paired comparisons: I. The least squares solution assuming equal standard deviations and equal correlations *Psychometrika,* 1951a, 16, 3–10.

MOSTELLER, F. Remarks on the method of paired comparisons: III. A test of significance for paired comparisons when equal correlations and equal standard deviations are assumed. *Psychometrika,* 1951b, 16, 207–218.

THURSTONE, L. L. Psychophysical analysis. *Amer. J. Psychol.,* 1927, 38, 368–389.

Chapter 7

CIRCULAR TRIADS, THE COEFFICIENT OF CONSISTENCE, AND THE COEFFICIENT OF AGREEMENT

ALLEN L. EDWARDS
University of Washington

In making paired comparison judgments, a subject may sometimes be inconsistent. An inconsistency in judgments occurs whenever there is a *circular triad* present in the $n(n-1)/2$ judgments. As an illustration of what is meant by a circular triad, consider three statements, i, j, and k, included in a set of n statements judged on a psychological continuum from least to most favorable. If Statement i is judged more favorable than Statement j, and Statement j is judged more favorable than Statement k, then, to be consistent, the subject should also judge Statement i to be more favorable than Statement k. If Statement k, on the other hand, is judged more favorable than Statement i, these three comparative judgments would constitute a circular triad. The greater the number of circular triads occurring in the set of $n(n-1)/2$ comparative judgments of a given subject, the more inconsistent the subject may be said to be.

Inconsistencies in comparative judgments may occur for a number of reasons. The subject may be disinterested in the task and therefore care-

Reprinted from *Techniques of Attitude Scale Construction* by Allen L. Edwards, pp. 66–72, 76–81. Copyright, © 1957. Reprinted with permission of the author and Appleton-Century-Crofts, Educational Division, Meredith Corporation.

less in his judgments. Some of the statements may fall so close together on the psychological continuum that the judgments are exceedingly difficult to make. Still another possibility is that the statements do not fall along the single dimension on which we are trying to scale them. If statements differ with respect to attributes or dimensions other than the one in which we are interested, these additional attributes may play a part in influencing the comparative judgments. It may also be true that inconsistencies in comparative judgments reflect a general personality or ability trait, that is, that there are some individuals who show a high degree of consistency, regardless of the nature of the comparative judgments they are asked to make, whereas others show a marked degree of inconsistency. Regardless of the conditions producing inconsistencies, it may often be desirable to obtain some measure of the degree of consistency a subject shows in making comparative judgments and this can be done in terms of Kendall's (1948) coefficient of consistence.

Kendall (1948) has shown that when the number of stimuli to be judged is odd, then the maximum number of circular triads that can occur is $(n^3 - n)/24$. When the number of stimuli is even, then the maximum number of circular triads is $(n^3 - 4n)/24$. If we let d be the observed number of circular triads for a given subject, then the coefficient of consistence, *zeta*, may be defined as

$$\zeta = 1 - \frac{24d}{n^3 - n} \quad \text{when } n \text{ is odd} \quad [1]$$

and

$$\zeta = 1 - \frac{24d}{n^3 - 4n} \quad \text{when } n \text{ is even} \quad [2]$$

For example, if we have $n = 10$ stimuli, then the maximum number of circular triads will be $[(10^3 - (4)(10)]/24 = 40$. If a subject makes the maximum number, that is, 40, then the coefficient of consistence would be

$$\zeta = 1 - \frac{(24)(40)}{10^3 - (4)(10)} = 0$$

If the subject does not have a single circular triad in his comparative judgments, then the coefficient of consistence will be 1.00. Zeta can thus range between 0, indicating the maximum number of circular triads, and 1.00 indicating the absence of any circular triads.

The number of circular triads made by a subject can be obtained from a table such as that shown in Table 7.1. When the column stimulus is judged more favorable than the row stimulus (the same arrangement

Table 7.1. Comparative Judgments for a Judge with No Circular Triads

Statements	1	2	3	4	5	6	7
1	—	1	1	1	1	1	1
2	0	—	1	1	1	1	1
3	0	0	—	1	1	1	1
4	0	0	0	—	1	1	1
5	0	0	0	0	—	1	1
6	0	0	0	0	0	—	1
7	0	0	0	0	0	0	—
a	0	1	2	3	4	5	6
a^2	0	1	4	9	16	25	36

we have used previously for showing comparative judgments) we have entered a 1 in the corresponding cell of the table. If the column stimulus is judged less favorable than the row stimulus, we have entered a 0 in the cell. If we let a equal the sum of the entries in a given column of Table 7.1, then the number of circular triads d will be given by

$$d = \left(\frac{1}{12}\right)(n)(n-1)(2n-1) - \frac{1}{2}\sum a^2 \qquad [3]$$

For Table 7.1, it is obvious that the subject has been completely consistent in his comparative judgments and therefore d should be equal to 0. This is true, as substitution in formula [3] shows. Thus

$$d = \left(\frac{1}{12}\right)(7)(7-1)(14-1) - \frac{1}{2}91$$
$$= 45.5 - 45.5$$
$$= 0$$

and ζ therefore equals 1.00.

Table 7.2 illustrates the case of a subject who made some inconsistent judgments. Substitution in formula [3] shows that the number of circular triads is equal to

$$d = 45.5 - \frac{1}{2}73$$
$$d = 9$$

Using the value of d obtained above, we can substitute in formula [1] to find the coefficient of consistence. Thus

$$\zeta = 1 - \frac{(24)(9)}{7^3 - 7} = .357$$

Table 7.2. Comparative Judgments for a Judge with 9 Circular Triads

STATEMENTS	1	2	3	4	5	6	7
1	—	1	0	1	1	1	1
2	0	—	1	1	0	1	1
3	1	0	—	0	1	0	1
4	0	0	1	—	0	1	1
5	0	1	0	1	—	1	0
6	0	0	1	0	0	—	1
7	0	0	0	0	1	0	—
a	1	2	3	3	3	4	5
a^2	1	4	9	9	9	16	25

Significance Test for the Coefficient of Consistence

Often we are interested not only in knowing the degree of consistency or inconsistency in a set of comparative judgments for a given subject, but also in knowing the probability of obtaining a given value of ζ under the hypothesis that the subject's judgments were made at random. We might assume, for example, that a given subject is completely incompetent and that any degree of consistency shown in his comparative judgments is a matter of chance. Kendall (1948) gives tables showing the probability of obtaining a given value of ζ under the hypothesis that the comparative judgments were a matter of chance for stimuli varying in number from 2 to 7. Since it is unlikely that we would, in general, have fewer stimuli than 7, it is fortunate that when $n \geq 7$, then ζ can also be tested for significance in terms of the χ^2 distribution.

We calculate χ^2 in terms of the following formula, given by Kendall (1948):

$$\chi^2 = \left(\frac{8}{n-4} \right) \left(\frac{1}{4} {}_nC_3 - d + \frac{1}{2} \right) + df \qquad [4]$$

where n = the number of stimula

$\quad {}_nC_3$ = the number of combinations of n things taken 3 at a time or $n!/3!(n-3)!$

$\quad d$ = the observed number of circular triads

$\quad df$ = the number of degrees of freedom associated with χ^2

The number of degrees of freedom for the χ^2 of formula [4] will be given by

$$df = \frac{n(n-1)(n-2)}{(n-4)^2} \qquad [5]$$

For the judgments of the subject shown in Table 7.2 we have already found d equal to 9. Since we have 7 stimuli, we have degrees of freedom, as given by formula [5],

$$df = \frac{7(7 - 1) \ (7 - 2)}{(7 - 4)^2} = \frac{210}{9} = 23.33$$

or 23 rounded to the nearest integer. Then substituting in formula [4] with d, the number of circular triads equal to 9, $n = 7$, and $df = 23.33$, we obtain

$$\chi^2 = \left[\frac{8}{(7 - 4)} \right] \left[\frac{1}{4} \ 35 \ - \ 9 \ + \ \frac{1}{2} \right] + 23.33 = 24$$

To evaluate the obtained value of χ^2 equal to 24, we enter Table III in the appendix with df equal to 23. Since the distribution of χ^2 as given by formula [4] is from high to low values of d, the probability that d will be equaled or exceeded is the complement of the tabled probability for χ^2. For the data of Table 7.2 we have $\chi^2 = 24$, with $df = 23$. From the table of χ^2 in the appendix we find that the probability associated with χ^2 is approximately .41.[1] Then the probability of obtaining a value of d equal to or greater than the observed value of 9, is $1 - .41 = .59$.[2] Thus we may conclude that this subject did not make a significantly large number of circular triads, that is, he showed a certain degree of consistency in his judgments, despite a lack of perfection.

The Coefficient of Agreement

It should be clear that several subjects may each have a coefficient of consistence of 1.00 for their comparative judgments of a set of stimuli, and yet not agree in the judgments they have made. A statistic developed by Kendall (1948) which he designates as u, the coefficient of agreement, provides a means of determining the extent to which a group of judges agree in their comparative judgments.

Suppose that we have m judges each making $n(n - 1)/2$ comparative judgments. If there is complete agreement among the judges, then using the recording system of Table 7.1, where we have entered a 1 if the column stimulus is judged more favorable than the row stimulus and a 0 if it is not, we would have $n(n - 1)/2$ cells in which the frequency of judgments "i more favorable than j" was m and all of the other cells would be 0. Table 7.3 repeats the frequencies reported previously for 7 attitude

1. The probability of .41 was obtained by approximate interpolation in the table of χ^2.
2. Kendall's (1948) tables give .580 as the exact probability of $d \geqq 9$, when $n = 7$.

statements judged by 94 judges. We wish to determine the extent of agreement among the 94 judges with respect to their comparative judgments of these 7 statements.

We consider only the entries *below the diagonal* of Table 7.3. Then we may define T as

$$T = (\Sigma f_{ij}^2 - m\, \Sigma f_{ij}) + (_mC_2)(_nC_2) \qquad [6]$$

where Σf_{ij}^2 = the sum of the squared f_{ij} entries *below the diagonal*
$\quad m$ = the number of judges
$\quad \Sigma f_{ij}$ = the sum of the f_{ij} entries *below the diagonal*
$\quad _mC_2$ = the number of combinations of the m judges taken 2 at a time or $m(m-1)/2$
$\quad _nC_2$ = the number of combinations of the n stimuli taken 2 at a time or $n(n-1)/2$

The first row at the bottom of Table 7.3 gives the sum of the f_{ij} values in each column, remembering that the summation extends over the entries below the diagonal only. The last entry at the right of this row is the sum of the row entries below the diagonal and is equal to Σf_{ij} in formula [6]. The second row at the bottom of the table gives the sum of squares of the f_{ij} values in each column, again remembering that the summation extends over the entries below the diagonal only. The last entry in the second row is the sum of all of the squared f_{ij} values below the diagonal and is equal to Σf_{ij}^2 in formula [6].

Substituting with the values of $\Sigma f_{ij}^2 = 20{,}998$ and $\Sigma f_{ij} = 616$ in formula [6] we obtain

$$T = \left[\, 20{,}998 - (94)\,(616) \,\right] + \left[\, \frac{94(94-1)}{2} \,\right]\left[\, \frac{7(7-1)}{2} \,\right]$$

$$= -\,36{,}906 + 91{,}791$$

$$= 54{,}885$$

Kendall's coefficient of agreement is then defined as

$$u = \frac{2T}{(_mC_2)\,(_nC_2)} - 1 \qquad [7]$$

where T = the value obtained from formula [6]
$\quad _mC_2$ = the number of combinations of the m judges taken 2 at a time
$\quad _nC_2$ = the number of combinations of the n stimula taken 2 at a time.

We have T equal to 54,885, and in calculating T we found $_mC_2 = 4{,}371$, and $_nC_2 = 21$. Substituting with these values in formula [7], we obtain

$$u = \frac{2(54{,}885)}{(4{,}371)\,(21)} - 1$$

$$= 1.196 - 1$$

$$= .196$$

The value of u can be 1.00 only if there is perfect agreement among the m judges. The greater the departure from complete agreement (as measured by agreement among pairs of judges), the smaller the value of u. If the number of judges is even, the minimum value of u is $-1\,(m-1)$, and if m is odd, then the minimum value of u is $-1/m$. Thus, only if $m = 2$, can u be equal to -1.00. If u takes any positive value whatsoever, then there is a certain amount of agreement among the judges. To determine whether this agreement is greater than the agreement expected if the judgments of the m judges were made at random, we can make use of tables published by Kendall (1948).

χ^2 Test For the Coefficient of Agreement

For m greater than 6 and for n greater than 4, and this will usually be true of data in which we are interested, then we can use Kendall's test of significance for u based upon the χ^2 distribution. We calculate

$$\chi^2 = \left[\frac{4}{m-2} \right] \left[T - \frac{1}{2}\,(_nC_2)\,(_mC_2)\left(\frac{m-3}{m-2} \right) \right] \quad [8]$$

where $T =$ the value obtained from formula[6]
  $m =$ the number of judges
  $n =$ the number of stimuli

For the data of Table 7.3 we have already found $T = 54{,}885$, $_mC_2 = 4{,}371$, and $_nC_2 = 21$. Then substituting with these values in formula [8] we have

$$\chi^2 = \left[\frac{4}{94-2} \right] \left[54{,}885 - \frac{1}{2}\,(4{,}371){\cdot}(21)\left(\frac{94-3}{94-2} \right) \right]$$

$$= \left(\frac{4}{92} \right)\,(9{,}494.35)$$

$$= 412.8$$

The degrees of freedom available for evaluating the χ^2 of formula [8] will be given by

$$df = \left(_nC_2 \right)\frac{m\,(m-1)}{(m-2)^2} \quad [9]$$

With $n = 7$, we have previously found $_nC_2 = 21$. Then, with $m = 94$ judges, we have

Table 7.3. Frequency with which the Column Statement Was Judged More Favorable than the Row Statement by 94 Judges

STATEMENTS	1	2	3	4	5	6	7
1	—	65	75	80	75	86	88
2	29	—	51	54	62	68	81
3	19	43	—	49	59	60	63
4	14	40	45	—	49	63	67
5	19	32	35	45	—	51	55
6	8	26	34	31	43	—	57
7	6	13	31	27	39	37	—
f_{ij}	95	154	145	103	82	37	$\Sigma f_{ij} = 616$
f_{ij}^2	1,859	5,318	5,367	3,715	3,370	1,369	$\Sigma f_{ij}^2 = 20,998$

$$df = 21 \, \frac{94 \, (94 - 1)}{(94 - 2)^2} = 21.69$$

which rounded to the nearest whole number is 22.

Entering the table of χ^2 in the appendix with degrees of freedom equal to 22, we find a χ^2 of 412.8 is highly significant, and that the probability of a value of u as great as .196 is much less than .01 if the comparative judgments of all of the judges were made at random. We conclude, therefore, that the 94 judges do show significant agreement in their comparative judgments.

Such a finding, of course, does not imply that there are no inconsistencies in the comparative judgments, and this would be true even though u was equal to 1.00, indicating perfect agreement among the judges. If inconsistencies occur and u is equal to 1.00, this merely means that the judges are in agreement in their inconsistencies as well as their consistencies.

The table of χ^2 in the appendix gives values of χ^2 only for degrees of freedom equal to or less than 30. As n, the number of stimuli, increases, and m, the number of judges, decreases, the number of degrees of freedom given by formula will tend to exceed 30. When this is the case, we can evaluate the significance of u by finding

$$z = \sqrt{2\chi^2} - \sqrt{2df - 1} \qquad [10]$$

The value of z obtained from formula [10] is approximately normally distributed with unit variance and can be evaluated in terms of Table I in the appendix.

Reference

KENDALL, M. G. *Rank correlation methods.* London: Griffin, 1948.

Chapter 8

OPTIMAL ORDERS IN THE METHOD OF PAIRED COMPARISONS

ROBERT T. ROSS

California College of Medicine

(Editor's Note: This excerpt presents a matrix and the rules for its use. Its use will provide an ordering of items, objects, or stimuli for presentation in the method of paired comparisons. After one has assigned numbers to the items that one wishes to have scaled by this method, the order for presentation is secured by applying the rules and reading down columns and moving to the right, column by column, using only the number of rows and columns specified by the rules.)

Table 8.1. The Ross Matrix

I	II	III	IV	V	VI	VII	etc.
1–2	2–3	1–3	3–4	1–4	4–5	1–5	etc.
3–n	n–4	4–2	2–5	5–3	3–6	6–4	etc.
4–$(n-1)$	$(n-1)$–5	5–n	n–6	6–2	2–7	7–3	etc.
5–$(n-2)$	$(n-2)$–6	6–$(n-1)$	$(n-1)$–7	7–n	n–8	8–2	etc.
6–$(n-3)$	$(n-3)$–7	7–$(n-2)$	$(n-2)$–8	8–$(n-1)$	$(n-1)$–9	9–n	etc.
7–$(n-4)$	$(n-4)$–8	8–$(n-3)$	$(n-3)$–9	9–$(n-2)$	$(n-2)$–10	10–$(n-1)$	etc.
8–$(n-5)$	$(n-5)$–9	9–$(n-4)$	$(n-4)$–10	10–$(n-3)$	$(n-3)$–11	11–$(n-2)$	etc.
9–$(n-6)$	$(n-6)$–10	10–$(n-5)$	$(n-5)$–11	11–$(n-4)$	$(n-4)$–12	12–$(n-3)$	etc.
10–$(n-7)$	$(n-7)$–11	11–$(n-6)$	$(n-6)$–12	12–$(n-5)$	$(n-5)$–13	13–$(n-4)$	etc.
11–$(n-8)$	$(n-8)$–12	12–$(n-7)$	$(n-7)$–13	13–$(n-6)$	$(n-6)$–14	14–$(n-5)$	etc.
etc.	etc.	etc.	etc.	etc.	etc.	etc.	etc.

Reprinted with permission of the author and the publisher from the *Journal of Experimental Psychology*, Vol. 25, 1939, pp. 417–21. Copyright 1939 by the American Psychological Association.

The rules for its use are as follows:

Use $\frac{n+1}{2}$ rows and $n-1$ columns. In the $\frac{n+1}{2}$ row the fixed number in the first column will be $\frac{n+3}{2}$ and the number on the variable scale will also be $\frac{n+3}{2}$, so that at this point the number on the fixed scale will be identical with that on the variable scale. Similiar identities will appear in all odd-numbered columns. The following rule governs the use of the $\frac{n+1}{2}$ row: In the odd-numbered columns (where the repetition occurs) replace the number in the variable column by one. In the even-numbered columns disregard the entry in this row altogether. These rules are the same as those stated in 1934.

A consideration of the properties of this matrix will indicate the characteristics and limitations of the lists which may be derived from it. In the first place, it may be well to consider whether the spacing is maximal; that is, whether the lists are optimal with regard to "time errors." It is obvious that if we have n items, we will have $n(n-2)$ spaces which separate pairs involving the same item. Since in this discussion n is always odd, it is obvious that if the items are arranged in 1–2, 3–4, 5–6, etc., order, the greatest possible spacing will be insured for the 1 and 2. The number of such complete pairs will be $\frac{n-1}{2}$ and there will be one item left over. With this extra item we must combine either a 1 or a 2. If we put in the 1, then the spacing between 1's is $\frac{n-3}{2}$, and the 2 appearing in the next pair will be separated from the 2 in the first pair by $\frac{n-1}{2}$ pairs. A moment's consideration will show the maximal spacing between pairs involving the same item will be maintained if the pairs are arranged in 1–2, 3–4, 5–6, etc., order until the single term is reached, combining the single term with one and continuing the series $n-1$, 2–3, 4–5, etc., until the final term $(n-1)$–n is reached, and then continuing the series in 1–2, 3–4, 5–6 order again. Of course, this order makes no attempt at presenting all the possible pairs, but it obviously has the characteristic of maximal spacing.

A study of this "maximal" list will show that in its arrangement the spacing $\frac{n-1}{2}$ is maintained half the time and the spacing $\frac{n-3}{2}$ is maintained for the remainder of the time. Since this represents maximal spacing, any other order which maintains the same total of spacings between pairs obviously represents the optimal situation. It may be seen from an inspection of the matrix that the orders obtained from it maintain the spacing $\frac{n-1}{2}$ for half the pairs and the spacing $\frac{n-3}{2}$ for the other half. It therefore must represent the optimal spacing obtainable for the list.

Having demonstrated the "optimal" nature of the lists with respect to the "time" criterion, it is now necessary to investigate their conformity to the "space" criterion. An inspection of the matrix will show that it is impossible to reverse the order of any pair without causing four "space errors" if the pair reversed is in the body of the table, and two "space errors" if the pair is either in the first or last columns or a pairing involving 1 as one of its items. Any attempt at compensating for these errors leads to changes in adjacent columns, these changes cause more errors, and a continued attempt at compensation results either in reversing the order of all the pairs in the table or of coming back to the original form. Since this characteristic of the matrix is determined by the *a priori* balance of the table, it follows that as it stands, the matrix gives the minimum number of "space errors" and is therefore "optimal" in that regard.

The next question concerns the number of "space errors" which we may expect from the use of the matrix. A consideration of the nature of the matrix and the rule for its use will show that since at the point of identity in the first column the number in the variable column which was replaced by 1 was $\frac{n+3}{2}$, it follows that in the first column the variable number which precedes the 1 is $\frac{n+5}{2}$, the one before that $\frac{n+7}{2}$, and so on to n. A further inspection of the matrix will show that all variable entries are balanced *a priori*, with the exception of the first entries in the 3rd, 5th, 7th and remaining odd-numbered columns. Also it will be seen that the last entry on the fixed side of the third column is $\frac{n+5}{2}$, the corresponding number in the fifth column $\frac{n+7}{2}$, etc. But it is likewise obvious that to the left of any one of these terminal items, the item occurs only in the variable part of the matrix, and that the fixed and variable portions are balanced in the opposite directions. It therefore follows that in passing from one arrangement to the other, there is an unavoidable "space error," any attempt to correct which by changing the order within any given pair increases the number of errors in the matrix as has been shown above.

It would appear, then, that the errors are inescapable and that they involve all numbers from $\frac{n+5}{2}$ to n, once each, in their combination with 1. A consideration of the first row of the matrix will show that any number m which is greater than three will find itself in an anamolous position in the $(2m - 3)$ column. Now we know from above that all numbers greater than $\dfrac{n + 5}{2}$ will be involved in "space errors" in the last row of the table. We have only to investigate the numbers less than $\dfrac{n + 5}{2}$ for any possible exception.

Since we use $(n - 1)$ columns, and the first "space error" for any given item occurs in the $(2m - 3)$ column, all items less than $\dfrac{(n + 5)}{2}$ for which $(2m - 3)$ is less than $(n - 1)$ will escape from "space errors." It can be shown that the first error for $\dfrac{n + 3}{2}$ is in the n column, which lies outside the matrix, hence its error does not occur. For all items numbered less than this, the column of first error occurs within the matrix (excepting 1 and 2).

The items 1 and 2 appear uniquely in the matrix. It is obvious that in the arrangement of any series of numbers into lists such as those here considered it is possible to balance items 1 and 2 perfectly, inasmuch as they are dependent on each other only in one common pair and are restrained in their positions by no other conditions.

It follows from the above considerations that in any list of n items there will be $(n - 3)$ "space errors," and that the three exceptions will be for items 1, 2, and $\dfrac{n + 3}{2}$.

One other point is of note. From the above discussion it would appear that the item in error always appears paired with one. This is true if the lists are read in the order in which they appear in the matrix. If they are read in reverse order, the "error" apparently moves. For example, in the list for seven items in Table 8.1 where the pairs 1–3, 1–4, 1–6, 1–7 were considered as those having "errors," if the list is read in reverse order we mark the pairs 2–3, 3–4, 5–6 and 6–7 as apparently involved in the error. It is necessary to keep in mind that in a strict sense no pair or item is involved in an *error* but that the error" is in their arrangement. Which pair is out of order is a purely relative matter.

From the above considerations we may conclude that a "rational method" for finding optimal orders has been found; that this method is merely a restatement of Ross's method; that the matrix proposed gives "optimum" conditions for "time" and "space" errors; that the optimal "time" condition involves a separation of $\dfrac{n - 1}{2}$ half the time and of $\dfrac{n - 3}{2}$ the other half; and that the optimal "space" condition has $(n - 3)$ "errors" involving all items except 1, 2, and $\dfrac{n + 3}{2}$.

Chapter 9

PAIRED COMPARISON ATTITUDE SCALES

ALLEN L. EDWARDS
University of Washington

Having obtained scale values for a set of statements, how can we use these statements and their scale values to obtain estimates of the attitudes of individuals? So far we have not been concerned with measuring the degree of affect that individuals associate with a psychological object, but rather with the judged degree of affect represented by the statements. But we might argue that the manner in which an individual responds to these statements, that is, the particular statements that he accepts or rejects would, in turn, enable us to infer something about his location on the same psychological continuum as the one on which the statements have been scaled.

Assume that we have a set of statements relating to some psychological object and that these statements have been scaled on a psychological continuum from least to most favorable. These statements are now presented in some random order to individuals with instructions to indicate whether they agree or disagree with each one. It is assumed that these agree and disagree responses are a function of the degree of affect associated with the psychological object by the subjects. An individual who has a highly favorable attitude toward the psychological object, in other words, is believed to be more likely to agree with statements that

Reprinted from *Techniques of Attitude Scale Construction* by Allen L. Edwards, pp. 47–50. Copyright, © 1957. Reprinted with permission of the author and Appleton-Century-Crofts, Educational Division, Meredith Corporation.

have highly favorable scale values than he is with statements that do not. And, similarly, individuals who have the least favorable attitudes toward the psychological object are believed to be more likely to endorse or agree with statements that are scaled near their own positions than they are with statements that have highly favorable scale values.

An attitude score for each individual can be obtained by finding the median of the scale values of the statements with which he agrees. This score is assumed to be an indication of the individual's location on the same psychological continuum as that represented by the scaled statements. For example, if an individual has agreed with statements with scale values of 2.4, 2.9, and 3.3, his attitude score would be taken as 2.9, the median or middle-scale value of the three statements. If he has agreed with four statements with scale values of 2.2, 2.4, 2.8, and 3.3, then his attitude score would be taken as the midpoint of the interval between the two middle-scale values, that is, 2.6. This method of obtaining attitude scores, based upon reactions to scaled statements, has been widely used with statements that have been scaled by the method of paired comparisons or one of the other scaling methods to be discussed later.

Certain variations in the procedure described above for obtaining attitude scores have been suggested. For example, after a subject has checked all of the statements with which he agrees, he might then be asked to indicate the one statement that best expresses how he feels about the psychological object. The scale value of this single statement might then be taken as the attitude score of the subject. A major disadvantage of this method would be that the scores would be determined by single scale values and therefore would probably not be as reliable as those obtained by the median method of scoring. A better procedure would be to ask subjects to check the three statements that best express how they feel about the psychological object. Scores could then be taken as the median or middle-scale value of these three statements.

Another procedure has been introduced by Edwards (1956) that departs considerably from the median method of scoring. He obtained scale values for a number of statements relating to attitude toward psychology. He then selected 9 statements whose scale values were fairly equally spaced along the psychological continuum. Each of the 9 statements was then paired with every other statement to give $9(9 - 1)/2 = 36$ pairs. For each pair of statements, one statement had a higher or more favorable scale value than the other. The statement with the higher scale value he designated as A and the one with the lower scale value as B. These pairs of statements comprised the items in the attitude scale.

The attitude scale was given to students in introductory psychology classes at the University of Washington. Each student was asked to choose the statement, A or B, in each pair that best expressed how he felt about psychology. An attitude score for each student was obtained by counting the number of times he chose the A or more favorable statement in the 36 pairs.

Kuder-Richardson (1937) estimates of reliability were obtained from two samples of 175 and 174 students. For the first sample, the reliability coefficient was .87 and for the second it was .88. These reliability coefficients are comparable to those usually reported for attitude scales scored by the median method.

When the median method of scoring is used, a test-retest reliability coefficient can be obtained by having the same group of subjects indicate their agreement or disagreement with the statements twice, with a time interval separating the two administrations of the scale. Scores obtained at the time of the first administration can then be correlated with those obtained at the second. It may also be possible, under certain circumstances, to obtain two sets of statements with respect to the same psychological object such that the statements in each set have approximately the same scale values. These two sets of statements might then be regarded as comparable forms of the same attitude scale. If the two forms of the scale are given to the same group of subjects, two scores can be obtained, one on each form. By correlating the scores on the two forms, an estimate of the reliability of the scales can be obtained.

As in all attempts to measure attitudes by means of scales, the subject's position on the attitude continuum is unknown and must be estimated from his responses to the statements contained in the scale. In the case of the median method of scoring, the attitude score gives the position of the subject on the psychological continuum on which the statements themselves have been scaled. In the procedure used by Edwards, this is not the case. The score obtained by this method is regarded as a linear transformation of the subject's position on the psychological continuum on which the original statements were scaled.

References

EDWARDS, A. L. A technique for increasing the reproducibility of cumulative attitude scales. *J. appl. Psychol.*, 1956, 40, 263–265.

KUDER, G. F., and RICHARDSON, M. W. The theory of the estimation of test reliability. *Psychometrika*, 1937, 2, 151–160.

Chapter 10

THE METHOD OF EQUAL-APPEARING INTERVALS

ALLEN L. EDWARDS
University of Washington

The Sorting Procedure

In the method of equal-appearing intervals, as originally described by Thurstone and Chave (1929), each statement concerning the psychological object of interest is printed on a separate card and subjects are then asked to sort the statements on the cards into a number of intervals. Along with the cards containing the statements, each subject is given a set of 11 cards on which the letters A to K appear. These cards are arranged in order in front of the subjects with the A card to the extreme left and the K card to the extreme right. The A card is described as representing the card on which the statements that seem to express the most unfavorable feelings about the psychological object are to be placed. Statements that seem to express the most favorable feelings about the psychological object are to be placed on the K card. The middle or F card is described as the "neutral" card on which statements that express neither favorable nor unfavorable feelings about the psychological object are to be placed. Varying degrees of increasing favorableness ex-

Reprinted from *Techniques of Attitude Scale Construction* by Allen L. Edwards, pp. 83–94. Copyright, © 1957. Reprinted with permission of the author and Appleton-Century-Crofts, Educational Division, Meredith Corporation.

pressed by the statements are represented by the cards lettered G to K and varying degrees of unfavorableness by the cards E to A. It may thus be observed that the psychological continuum from least to most favorable is regarded as continuous with the psychological continuum from least to most unfavorable and the F or "netural" interval is, in essence, a zero point, as illustrated in Figure 10.1.

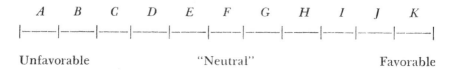

Unfavorable   "Neutral"   Favorable

Figure 10.1. The Thurstone equal-appearing interval continuum.

Each subject is asked to judge the degree of favorableness or unfavorableness of feeling expressed by each statement in terms of the 11 intervals represented by the cards. It was found by Thurstone and Chave that subjects required about 45 minutes to judge the 130 statements that these investigators used in developing their scale to measure attitude toward the church. Comparative judgments of the kind described were obtained from 300 subjects.

Thurstone and Chave also believed that the sorting or judging of the statements would be done similarly by those judges who had favorable and those who had unfavorable attitudes toward the psychological object under consideration.

Only the middle and the two extreme cards on which the statements were to be sorted were defined for the subjects. Thurstone and Chave believed it was essential that the other cards not be so defined in order that the intervals between successive cards would represent *equal-appearing intervals* or degrees of favorableness-unfavorableness for each subject. If the intervals are judged equal by the subjects, the successive integers from 1 to 11 can then be assigned to the lettered cards A to K, and, in essence, the subject has then rated each statement on an 11-point rating scale. The 11-point scale then becomes the psychological continuum on which the statements have been judged and all that is required is that some typical value be found for the distribution of judgments obtained for each statement. This typical or average value can then be taken as the scale value of the statement on the 11-point psychological continuum. As their measure of the average value of the distribution of judgments, Thurstone and Chave used the median of the distribution for a given statement. Before describing methods for finding the medians, however, another point should be considered.

As in the method of paired comparisons, some subjects in making equal-appearing interval judgments may undertake the task carelessly and with little interest. Still other subjects may misunderstand the directions and thus not be aware of the nature of the judgments desired. They may respond, for example, in terms of their *own agreement* or *disagreement* with the statements rather than in terms of the *judged degree of favorableness-unfavorableness*. A criterion used by Thurstone and Chave, for eliminating those subjects who apparently performed the judging task with carelessness or who otherwise failed to respond to the instructions for making the judgments, was to reject the judgments obtained from any subject who placed 30 or more statements on any one of the 11 cards.[1] They report that of 341 subjects making the judgments, 41 were eliminated by this criterion.

Calculation of Scale and Q Values

The data obtained from a large number of judges can be arranged in the form shown in Table 10.1. Three rows are used for each statement. The first gives the frequency with which the statement was placed in each of the 11 categories. The second gives these frequencies as proportions. The

Table 10.1. Summary Table for Judgments Obtained by the Method of Equal-Appearing Intervals

STATEMENTS		SORTING CATEGORIES											SCALE VALUE	Q VALUE
		A	B	C	D	E	F	G	H	I	J	K		
		1	2	3	4	5	6	7	8	9	10	11		
1	f	2	2	6	2	6	62	64	26	18	8	4		
	p	.01	.01	.03	.01	.03	.31	.32	.13	.09	.04	.02	6.8	1.7
	cp	.01	.02	.05	.06	.09	.40	.72	.85	.94	.98	1.00		
2	f	0	0	0	10	40	28	50	26	28	14	4		
	p	.00	.00	.00	.05	.20	.14	.25	.13	.14	.07	.02	6.9	2.8
	cp	.00	.00	.00	.05	.25	.39	.64	.77	.91	.98	1.00		
3	f	0	0	0	2	8	6	26	44	56	44	14		
	p	.00	.00	.00	.01	.04	.03	.13	.22	.28	.22	.07	8.7	2.0
	cp	.00	.00	.00	.01	.05	.08	.21	.43	.71	.93	1.00		

1. Edwards and Kilpatrick (1948) report that they also examine the sortings for each subject and eliminate those subjects who show obvious reversals of the continuum. This can be quickly done by looking at the judgments made for two or three key statements believed to fall at each of the two extremes of the continuum.

proportions are obtained by dividing each frequency by N, the total number of judges or, more simply, by multiplying each of the frequencies by the reciprocal of N. The third row gives the cumulative proportions, that is, the proportion of judgments in a given category plus the sum of all of the proportions below that category.

If the median of the distribution of judgments for each statement is taken as the scale value of the statement, then the scale values can be found from the data arranged in the manner of Table 10.1 by means of the following formula

$$S = l + \left(\frac{.50 - \Sigma p_b}{p_w} \right) i \qquad [1]$$

where $S =$ the median or scale value of the statement

$l =$ the lower limit of the interval in which the median falls

$\Sigma p_b =$ the sum of the proportions below the interval in which the median falls

$p_w =$ the proportion within the interval in which the median falls

$i =$ the width of the interval and is assumed to be equal to 1.0

Substituting in the above formula to find the scale value for the first statement in Table 10.1, we have[2]

$$S = 6.5 + \left(\frac{.50 - .40}{.32} \right) 1.0 = 6.8$$

The other scale values shown in Table 10.1 are found in the same manner.

Thurstone and Chave used the interquartile range or Q as a measure of the variation of the distribution of judgments for a given statement. The interquartile range contains the middle 50 per cent of the judgments. To determine the value of Q, we need to find two other point measures, the 75th centile and the 25th centile. The 25th centile can be obtained from the following formula

$$C_{25} = l + \left(\frac{.25 - \Sigma p_b}{p_w} \right) i \qquad [2]$$

where $C_{25} =$ the 25th centile

$l =$ the lower limit of the interval in which the 25th centile falls

$\Sigma p_b =$ the sum of the proportions below the interval in which the 25th centile falls

$p_w =$ the proportion within the interval in which the 25th centile falls

$i =$ the width of the interval and is assumed to be equal to 1.0

2. The interval represented by the number assigned to a given card or category is assumed to range from .5 of a unit below to .5 of a unit above the assigned number. Thus the lower limit of the interval represented by the card assigned the number 7 is 6.5 and the upper limit is 7.5.

The 75th centile will be given by

$$C_{75} = l + \left(\frac{.75 - \Sigma p_b}{p_w} \right) i \qquad [3]$$

where C_{75} = the 75th centile

Σp_b = the sum of the proportions below the interval in which the 75th centile falls

p_w = the proportion within the interval in which the 75th centile falls

i = the width of the interval and is assumed to be equal to 1.0

For the first statement in Table 10.1, we have

$$C_{25} = 5.5 + \left(\frac{.25 - .09}{.31} \right) 1.0 = 6.0$$

and

$$C_{75} = 7.5 + \left(\frac{.75 - .72}{.13} \right) 1.0 = 7.7$$

Then the interquartile range or Q will be given by taking the difference between C_{75} and C_{25}. Thus

$$Q = C_{75} - C_{25} \qquad [4]$$

or for the first statement in Table 10.1, we have

$$Q = 7.7 - 6.0 = 1.7$$

The interquartile range is a measure of the spread of the middle 50 per cent of the judgments. When there is good agreement among the subjects in judging the degree of favorableness or unfavorableness of a statement, Q will be small compared with the value obtained when there is relatively little agreement among the subjects. A large Q value, indicating disagreement among the judges as to the degree of the attribute possessed by a statement, is therefore taken as an indication that there is something wrong with the statement. Thurstone and Chave regard large Q values primarily as an indication that a statement is ambiguous. Large Q values may result from the fact that the statement is interpreted in more than one way by the subjects when making their judgments or from any of the other conditions producing large discriminal dispersions discussed in connection with the method of paired comparisons.

Scale and Q values for statements can also be found graphically. Figure 10.2 shows the cumulative proportion graph for the first statement in Table 10.1 and Figure 10.3 shows the cumulative proportion graph for the second statement. Dropping a perpendicular from the graph to the baseline at the value of cp equal to .50 will give the value of the

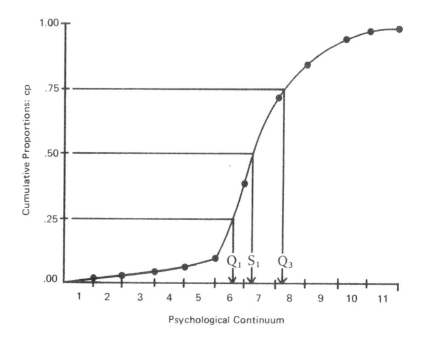

Figure 10.2. Cumulative proportion graph for Statement 1 of Table 10.1.

median or the scale value of the statement. Dropping perpendiculars to the baseline at values of cp equal to .25 and .75 will give the values of C_{25} and C_{75} and the distance between these two on the baseline will give the value of Q. Using graph paper ruled 10 to the inch, the scale and Q values of the statements can be found quite accurately by the graphic method.

A device used by Edwards and Kilpatrick (1948) facilitates the finding of scale and Q values by the graphic method. They prepared a master chart with the Y axis corresponding to the cumulative proportions and with the scale intervals on the X axis. This chart was taped to a ground-glass plate that fitted over the top of a small wooden box. Inside of the box was a 100 watt bulb. By placing tracing paper over the chart, the cumulative proportion graphs for the individual statements could be quickly drawn and the scale values and 75th and 25th centiles readily found.

Jurgensen (1943) has used a nomograph to obtain scale and Q values. He describes how the nomograph may be constructed and states it can be prepared in less than 10 minutes for any given number of judges. He

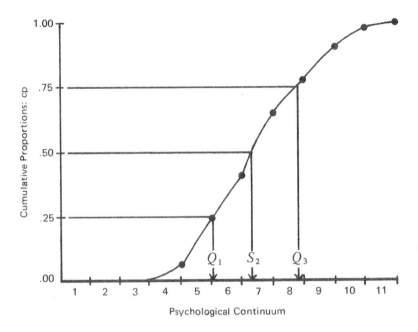

Figure 10.3. Cumulative proportion graph for Statement 2 of Table 10.1.

estimates that scale and Q values for statements can be found using the nomograph in less than one-fourth the time required by direct calculation.

The Attitude Scale

It may be observed that the scale values of the two statements whose cumulative proportion graphs are shown in Figures 10.2 and 10.3 are quite comparable. For the statement shown in Figure 10.2, for example, the scale value is 6.8 and for the one shown in Figure 10.3, the scale value is 6.9. The Q values for the two statements, however, differ considerably. For the first statement, the Q value is 1.7 and for the second it is 2.8. When there is a high degree of agreement among the judges, the cumulative proportion graph will, in general, have a steep slope and Q will be relatively small compared with the value obtained when the judgments are spread over the entire scale and the slope of the cumulative proportion graph is more gradual.

In general, what is desired in constructing an attitude scale by the method of equal -appearing intervals is approximately 20 to 22 statements such that the scale values of the statements on the psychological continuum are relatively equally spaced and such that the Q values are relatively small. Thus both S and Q are used as criteria for the selection of statements to be included in the attitude scale. If a choice is to be made among several statements with approximately the name S values, preference is given to the one with the lowest Q value, that is, to the one believed to be least ambiguous.

Assume that 22 statements have been selected from the much larger group for which we have scale and Q values. These statements can be arranged in random order and presented to subjects with instructions to indicate those that they are willing to accept or agree with and those that they reject or disagree with.[3] Taking only the statements with which the subject has agreed, an attitude score is obtained from the scale values of these statements that is regarded as an indication of the location of the subject on the psychological continuum on which the statements have been scaled. The attitude score is based upon the arithmetic mean or median of the scale values of the statements agreed with. If the subject has agreed with an odd number of statements, and if the median method of scoring is used, then the score is simply the scale value of the middle statement when they are arranged in rank order of their scale values. For example, if a subject has agreed with 5 statements with scale values of 3.2, 4.5, 5.6, 7.2, and 8.9, his score would be the scale value of the middle statement or 5.6. Using the arithmetic mean as the score would result in a value of 5.8 being assigned to the subject. If an even number of statements are agreed with and the median method of scoring is used, then the midpoint of the scale distance between the two middle statements is taken as the score. For example, if the subject has agreed with 4 statements with scale values of 4.5, 5.6, 7.2, and 8.9, his score would be $5.6 + (7.2 - 5.6)/2 = 6.4$. The arithmetic mean of these scale values would give a score of 6.6.

It has been customary among those working with the method of equal-appearing intervals to construct two comparable forms of the attitude scale. This is done by selecting from the initial group of statements for which scale and Q values have been obtained, in addition to the first set, a second set of 20 to 22 statements such that they also have scale values fairly equally spaced along the psychological continuum and with fairly low Q values. If both forms of the attitude scale are then given to the same group of subjects, the scores for the subjects on the two forms can

3. Research by Sigerfoos (1936) indicates that the arrangement of the statements in order of their scale values results in scores for subjects quite comparable to the scores obtained with a random arrangement of the statements. If this finding is, in general, true for equal-appearing interval scales, then the rank order arrangement would somewhat facilitate the scoring.

be correlated and this correlation taken as a measure of the reliability of the scales. Reliability coefficients typically reported for the correlation between two forms of the same equal-appearing interval scale are above .85.

References

EDWARDS, A. L. and KILPATRICK, F. P. A technique for the construction of attitude scales. *J. appl. Psychol.*, 1948, 32, 374–384.

JURGENSEN, C. E. A nomograph for rapid determination of medians. *Psychometrika*, 1943, 8, 265–269.

SIGERFOOS, C. C. The validation and application of a scale of attitude toward vocations. In H. H. Remmers (Ed.), *Further studies in attitudes, Series II, Studies in Higher Education, Bulletin of Purdue University*. Lafayette, Indiana: Purdue Univ., 1936, pp. 177–191.

THURSTONE, L. L., and CHAVE, E. J. *The measurement of attitude*. Chicago: Univ. Chicago Press, 1929.

Chapter 11

THE METHOD OF SUCCESSIVE INTERVALS

BERT GREEN

Johns Hopkins University

The method of successive intervals, first reported by Saffir (1937), was designed by Thurstone to overcome some of the difficulties inherent in the method of equal-appearing intervals. The method of successive intervals is recommended when there are too many items to be scaled feasibly by the method of paired comparisons (say 20 or more).

Each judge is required to sort the items into a fixed number of categories (usually 7, 9, or 11) spaced along a continuum of favorableness. The categories have a rank order on the continuum, but they are not necessarily equally spaced. The items in a given category are simply required to be less favorable than those in the adjacent category to the left and more favorable than those in the adjacent category to the right. When a number of judges have sorted the items, we can tabulate a judgment distribution for each item. This distribution shows the proportion of judges who placed the item in each category.

The mathematical model treats the categories as contiguous segments of the attitude continuum, separated by category boundaries. In the model it is assumed that the category boundaries can be located on the

Reprinted with permission of the author and the publisher from the *Handbook of Social Psychology*, Vol. 1, edited by Gardner Lindzey, pp. 347–50. Addison-Wesley, Reading, Massachusetts, 1954.

continuum in such a way that all the item judgment distributions are normal. The formal model is stated in terms of the cumulative judgment distributions. Let the M categories be numbered in order from least to most favorable. Let p_{jg} be the proportion of judges who placed item j in category g or any category of lower rank order. Let X_{jg} be the unit normal deviate corresponding to p_{jg}. In the model, t_g is the boundary between category g and category $g + 1$. S_j is the mean of the judgment distribution for item j, and hence the scale value for item j, and σ_j is the standard deviation of item j. The model is illustrated in Figure 11.1. It states that

$$X_{jg} = \frac{t_g - S_j}{\sigma_j}. \qquad [1]$$

A graphical procedure for obtaining the values of t_g, S_j, and σ_j is presented by Saffir (1937). Mosier (1940) described a short cut that is useful when a large number of items is involved. A least-squares solution has been developed by Gulliksen (1952), but its application is quite complicated.

Fortunately, it is usually adequate to assume that all standard deviations, σ_j, are equal, in which case they are set equal to unity. This model states that

$$X_{jg} = t_g - S_j. \qquad [2]$$

Procedures for obtaining the scale values and the category boundaries have been described by Saffir (1937), Attneave (1949), who uses the term "method of graded dichotomies," Garner and Hake (1951), who use the term "equal discriminability scale," and Edwards (1952). A least-squares method similar to these methods is described below.

The method of successive intervals is based on the assumption that

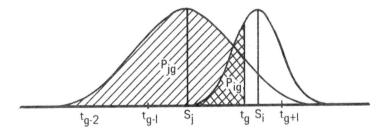

Figure 11.1. The successive intervals model. Two item judgment distributions with different standard deviations are shown. The abscissa is the latent attitude variable, and is measured in arbitrary units.

the item judgment distributions are normal on the derived scale. It should be noted that this is not an assumption about the distribution of attitude scores of those who endorse the item, but rather an assumption about the distribution of judgments concerning the proper location of the item on the scale. This assumption must be checked in order to ascertain whether the model fits the data. The check may be accomplished graphically by plotting the observed p_{jg} versus the estimated t_g on normal probability paper. Straight lines should result, and if equal standard deviations have been assumed, these lines should have the same slope. Wide departures from these conditions are viewed with alarm. Edwards and Thurstone (1952) suggest obtaining the differences between the observed proportions and the corresponding proportions computed from the estimated parameters of the model. Although no statistical criterion is presented, the authors give some representative average discrepancies.

The first step in obtaining the scale values and category boundaries is to convert the observed proportions p_{jg} into X_{jg} by the unit normal deviate transformation, using a table of the normal distribution. The X_{jg} are arranged in a data matrix with a row for each item, and a column for each category. This is illustrated by an artificial example with 6 items and 5 categories in Table 11.1. All X_{jg} that are greater than $+ 2.0$ or less than -2.0 are rejected. If no values are rejected, the average value of each column is the category boundary represented by that column, while the average value of each row, when subtracted from the over-all average of the table, gives the scale value of the corresponding item.

If some values are rejected, then for adjacent columns we subtract the entries in one column from the corresponding entries in the other column, and the differences are recorded in a column of a second matrix. Of course, wherever a value has been rejected, no difference is possible. The average of these differences is the difference between the category boundaries represented by the two columns. By assigning $t_1 = 0$, the remaining category boundaries can be computed, as shown in Table 11.2. Then a fourth matrix is constructed by subtracting the computed value of t_g from each entry in the gth row of the X_{jg} matrix, as shown in Table 11.2(d). The average value of a row of this matrix is the scale value of the corresponding item.

Since the scale factor and zero point are arbitrary, the scale factor has been set by letting all standard deviations equal unity. In the case of the complete matrix, the zero point was fixed by letting the average scale value be zero. In the case of the incomplete matrix, the zero point was fixed by setting $t_1 = 0$. Any other arbitrary scale factor and zero point may be used.

*Table 11.1. Successive Intervals Computations—Complete Matrix
(5 Categories, 6 Items)*

(a) **Matrix of** p_{ij}

		Category			
		1	2	3	4	5

		1	2	3	4	5
	1	.40	.55	.86	.96	1.00
	2	.28	.36	.71	.90	1.00
Item j	3	.15	.30	.65	.81	1.00
	4	.16	.25	.54	.82	1.00
	5	.09	.21	.50	.74	1.00
	6	.03	.08	.37	.57	1.00

(b) **Matrix of** X_{jg}

		Category boundary g				Row sum	Row average	Scale value
		1	2	3	4			
	1	— .25	.13	1.08	1.75	2.71	.68	— .78
	2	— .58	— .36	.55	1.28	.89	.22	— .32
Item j	3	—1.04	— .52	.39	.88	— .29	— .07	— .03
	4	— .99	— .67	.10	.92	— .64	— .16	.06
	5	—1.34	— .81	.00	.64	—1.51	— .38	.28
	6	—1.88	—1.41	— .33	.18	—3.44	— .86	.76
Column sum		—6.08	—3.64	1.79	5.65	—2.28		
Category boundary		—1.01	— .61	.30	.94		— .10	

The methods of paired comparisons and successive intervals yield interval scales in which the metric is obtained from the variability of judgments about the items. If items were always judged consistently in paired comparisons, so that the proportions were all either zero or unity, the mathematical model of paired comparisons could not be used. Identical sortings from all judges would render the method of successive intervals equally useless. Happily, judges seldom agree. There is usually sufficient "error" variance to allow a scale to be constructed. The metric in these scales is essentially a function of the degree of confusion, or overlap, between items. The difference of the scale values of two items may be used to make a precise probability statement about the degree of overlap. The scales are based on variability, but it is a remarkably predictable variability.

Table 11.2. Successive Intervals Computations—Incomplete Matrix
(5 Categories, 6 Items)

		(a) Matrix of p_{jg} Category					(b) Matrix of X_{jg} Category boundary g			
		1	2	3	4	5	1	2	3	4
	1	.44	.70	.94	.99	1.00	— .15	.52	1.56	—
	2	.16	.45	.81	.93	1.00	— .99	— .13	.88	1.48
Item j	3	.04	.24	.52	.78	1.00	—1.75	— .71	.05	.77
	4	.04	.16	.40	.69	1.00	—1.75	— .99	— .25	.50
	5	.01	.08	.32	.57	1.00	—	—1.41	— .47	.18
	6	.00	.02	.14	.36	1.00	—	—	—1.08	— .36

		(c) Matrix of differences			(d) Matrix of $t_g - X_{jg}$ Category boundary				Row sum	Scale value
		2–1	3–2	4–3	1	2	3	4		
	1	.67	1.04	—	.15	.31	.17	—	.63	.21
	2	.86	1.01	.60	.99	.96	.85	.94	3.74	.94
Item j	3	1.04	.76	.72	1.75	1.54	1.68	1.65	6.62	1.66
	4	.76	.74	.75	1.75	1.82	1.98	1.92	7.47	1.87
	5	—	.94	.65	—	2.24	2.20	2.24	6.68	2.23
	6	—	—	.72	—	—	2.81	2.78	5.59	2.80

Column sum	3.33	4.49	3.44
Column average	.83	.90	.69

$t_1 = 0$, $t_2 = .83$, $t_3 = 1.73$, $t_4 = 2.42$

Scoring the Respondents

When a scale of items has been established by one of the judgment methods, a rule for scoring the respondents must be adopted. The usual scoring procedure is to give each person a score which is the median of the scale values of the items which he endorses. The mean scale value could be used, but it is felt that the median is less sensitive to the particular set of scale values in the scale. Lorge (1939) found very little difference between mean scores and median scores. By either scoring method, the respondents are placed on the scale of favorableness toward the issue being studied.

Item Selection

Usually, many more items are scaled than will be used on the final questionnaire. Ambiguous or irrelevant items can be discarded when the items are picked for the final form of the scale. If the judgment distribution for an item has a large standard deviation, the item is presumed to be ambiguous, since it apparently means different things to different judges. In the extreme case of ambiguity, the judgment distribution might be spread uniformly over the categories. Items that are ambiguous according to this subjective criterion should be eliminated before the successive intervals scale is computed, so that the assumption of equal standard deviations becomes tenable. The simple estimation procedures described above can then be used.

Irrelevant items cannot easily be detected on the basis of judgments. If an item is almost always judged to be in the middle category, in the methods of successive, or equal-appearing, intervals, then it may be irrelevant, but it may also be a good neutral item. Thurstone (Thurstone and Chave, 1929) proposed a criterion of irrelevance which can be used when the questionnaire has been administered to a group of respondents. For each item in turn, the individuals who responded positively to a given item are selected; the popularity of each of the other items is determined for this group of individuals. These popularities are compared with the scale values of the items. Items with scale values near the given item should be very popular, while items far removed on the scale should have low popularity for this group. If the popularities are independent of the scale values, then some factor other than the latent attitude variable is determining the responses to this item.

Specifically, the relevance of item i is determined by plotting the ratio p_{ij}/p_j versus S_j. The divisor p_j is used to correct for the different popularities of items in the total group.

A similar technique can be used to get a rough picture of the operating characteristics of the items. In this case, individuals of nearly equal attitude scores are grouped together. The proportion of individuals who endorse the item can be determined for each group, and can be plotted as a function of the mean attitude score of the group. An acceptable operating characteristic will have a maximum near the scale value of the item, and will decrease toward zero on either side of this maximum.

Items having unacceptable operating characteristics, or giving indications of irrelevance, should be rejected. Although these criteria are subjective, they allow the investigator to compare items, and then to choose the most relevant items and the items with the best operating characteristics.

When the poor items have been discarded, the investigator may still have more items than he needs for the final scale. In this case, he should selected a set of items with scale values spread evenly along the continuum. If particularly good discrimination is desired in some portion of the continuum, then relatively more items may be selected with scale values in that range. If enough items are available, parallel forms of the scale can be produced by using items with matched scale values.

References

ATTNEAVE, F. A method of graded dichotomies for the scaling of judgments. Psychol. Rev., 1949, 56, 334–340.

EDWARDS, A. L. The scaling of stimuli by the method of successive intervals. J. appl. Psychol., 1952, 36, 118–122.

EDWARDS, A. L., and THURSTONE, L. L. An internal consistency check for the method of successive intervals. Psychometrika, 1952, 17, 169–180.

GARNER, W. R., and HAKE, H. W. The amount of information in absolute judgments. Psychol. Rev., 1951, 58, 446–459.

GULLIKSEN, H. A least-squares solution for successive intervals. Amer. Psychologist, 1952, 7, 408. (Abstract)

LORGE, I. The Thurstone attitude scales: I. Reliability and consistency of rejection and acceptance. J. soc. Psychol., 1939, 10, 187–198.

MOSIER, C. I. A modification of the method of successive intervals. Psychometrika, 1940, 5, 101–107.

SAFFIR, M. A. A comparative study of scales constructed by three psychophysical methods. Psychometrika, 1937, 2, 179–198.

THURSTONE, L. L., and CHAVE, E. J. The measurement of attitude. Chicago: Univer. of Chicago Press, 1929.

PART III

Scalogram Analysis

Scalogram analysis, or Guttman scaling, has been a very popular scaling procedure, particularly in sociologically oriented social psychology. It has, in fact, served to define what is meant by scaling for many people, perhaps because it is the scaling method most typically presented and described in introductory methods books. The method has lent itself to this popularity because it is easily understood, theoretically interesting, and esthetically pleasing. It is also method of scaling that has served to stimulate and redirect research interests. In addition, the process has stimulated the development of a variety of scaling equipment and techniques ranging from the scalogram board described by E. Suchman in *Measurement and Prediction,* through the Gradgram, to the numerous counter-sorter procedures and computer programs.

Scalogram analysis as a method of scaling encounters a variety of problems that, unfortunately, have emerged only in the more recent literature. Indifference to problems of probability by the early advocates of Guttman scaling led to a progressive disenchantment with the method on the part of many researchers and methodologists. Researchers began to notice the extent to which Guttman scaling frequently capitalized on probability to attain the requisites of a scale. A continuing concern with solutions to these problems has led to an increasingly sophisticated examination of these issues (See Chapters 15, 16, 17, and 18).

The method of scalogram analysis has been greatly enriched by auxiliary considerations such as Louis Guttman's delineation of the components of scalable attitudes and Paul Lazarsfeld's attention to the process of latent structure analysis. Guttman scaling is a generally useful scaling method—one that is not limited to attitude measurement. For example, Guttman scales have been developed to order state and city laws; voting patterns of office holders; and cultural, national, and community characteristics.

The papers included in this section has been selected to introduce the basic methods and problems of Guttman scaling. They serve also to describe the development of concern with the probability of securing reproducible scales by chance alone and to present some of the proposed solutions for this problem.

The scalogram method was most completely described and discussed in the book *Measurement and Prediction* based upon the World War II studies of the American soldier. Chapter 12, by Samuel A. Stouffer, taken from *Measurement and Prediction*, presents an overview of the basic idea of cumulative scaling. This is followed by Louis Guttman's presentation of the procedures for scalogram analysis (Chapter 13). The coefficient of reproducibility, a measure of the amount of deviation from ideal scale pattern, is also described, along with the other criteria for scalability. Another explicit consideration of assessing reproducibility is presented in Chapter 14, by Benjamin W. White and Eli Saltz. They consider and discuss Guttman's coefficient of reproducibility, Jackson's Plus Percentage Ratio, Loevinger's Index of Homogeneity, Green's Summary Statistics Method, and the Phi Coefficient.

In Chapter 15 Karl F. Schuessler emphasizes that Guttman scaling lacks a statistical test for a goodness of fit of the model. Schuessler provides three tests that appraise the fit between the model and any particular set of data. Roland J. Chilton (Chapter 16) deals with the notion of chance reproducibility. He introduces an extension of one of Schuessler's tests that allows for the comparison of chance and observed reproducibility and permits the testing of the significance of the difference between these two values. The test rests upon an assumption of a normal distribution of reproducibilities. Chilton subsequently used a computer to generate thousands of sets of scaled data. An examination of the distribution of these reproducibilities supported the assumption which in turn indicates the utility of the index based upon the notion of chance reproducibility. In Chapter 17, Chilton reviews and compares various statistical tests for scalogram analysis. The works of Goodman, Green, Sagi, and Schuessler are summarized and compared. This selection points up still further the importance of taking into account chance reproducibility.

Finally, Carmi Schooler, in Chapter 18, underscores again the difficulties in Guttman scaling and adds a note of "extreme caution." A thorough consideration of the selections presented in this section should familiarize the reader with the process and procedures of Guttman scaling. It should also make one aware of the problems associated with the method. The method should be employed with informed caution and with particular attention to matters relating to capitalizing upon probability.

Chapter 12

AN OVERVIEW OF THE CONTRIBUTIONS
TO SCALING AND SCALE THEORY

SAMUEL A. STOUFFER

Introduction

Vital to the development of social psychology and sociology as scientific disciplines is the definition and classification of the objects of study.

The requires rigorous yet economical methods for handling data which are initially *qualitative,* not quantitative. The objective of much of the Research Branch methodological endeavors, is to deal *with theoretical models of ordered structures or scales and with technical procedures for testing the applicability of a particular model to a particular set of qualitative data.*

Perhaps the most drastic departure from earlier approaches is represented in the initial thinking of Louis Guttman. In 1940, just before the war, Guttman contributed a series of studies on the logic of measurement and prediction to a monograph of the Social Science Research

Reprinted by permission of Princeton University Press from *Measurement and Prediction* by Samuel A. Stouffer *et al.,* Vol. IV of *Studies in Social Psychology in World War II* (copyright 1950 by Princeton University Press), pp. 3, 4–6, 9–18.

While the author has had the benefit of critical readings of drafts of this chapter by the authors of Chapters 2 to 11, namely, Louis Guttman, Edward A. Suchman and Paul F. Lazarsfeld, as well as by other experts in the field, he must assume responsibility for the summary statements in the present chapter. He has sought as faithfully as possible to represent the points of view of these authors and to reconcile divergencies of opinion, though not always succeeding completely, perhaps, to each author's satisfaction.

Council.[1] This work contained the basic principle of ideas which he was to apply in the Research Branch a year later and which were to be greatly expanded theoretically and adapted for quick practical use.

Guttman offered a model which dispenses with the concept of a latent or underlying continuum to which the response to a particular item is to be related. He considered an attitude area "scalable" if responses to a set of items in that area arranged themselves in certain specified ways. In particular, it must be possible to order the items such that, ideally, *persons who answer a given question favorably all have higher ranks than persons who answer the same question unfavorably*. From a respondent's rank or scale score we know exactly which items he endorsed. Thus we can say that the response to any item provides a *definition* of the respondent's attitude.

Guttman and his associates in the Research Branch developed simple and practical techniques for testing hypotheses as to whether attitude areas were scalable by this definition. Not all attitude areas satisfied the criteria of goodness of fit of the model, but it was possible to find many areas which seemed to be scalable and several hundred such scales were worked out during Research Branch experience. Also important, theoretically, is the fact that the ranking of people by the Guttman model appears to represent only one of a set of *principal components*. In fact, if there are m scale types or rank groups, in the case of perfect scalability, there are m principal components which Guttman found to have a definite law of formation. The first component is a monotonic increasing function of the ranks—a straight line in the special case where the frequencies are the same for each rank group. The second component will always be a U-shaped or J-shaped curve with one bend in it. This has been interpreted by Guttman as a measure of *intensity*, leading, in the ideal case, to the determination of an objective zero point, at the point where the curve of intensity is a minimum.

The perfect Guttman scale is not likely to be found in practice, but satisfactory approximations to it are not so rare as some critics of the early papers describing it have implied. In analyzing data which failed to fit the model, Guttman has distinguished between two classes of misfits. One, which he calls the *quasi scale,* has the property that the errors are at random. He proves that the correlation of the quasi scale with an outside criterion is the same as the multiple correlation between responses to the individual items forming that scale and the outside criterion. This important result leads to great operational economies and justifies the use of sets of items from an area not scalable in his strictest sense. If the

---

1. Louis Guttman, "The Quantification of a Class of Attributes: A Theory and Method for Scale Construction," in P. Horst et al., *The Prediction of Personal Adjustment* (Social Science Research Council. New York, 1941) , pp. 319–348.

condition of random errors is not satisfied we are confronted with what are called *nonscale types*. These nonscale types set problems for further analysis, by indicating the presence of more than one variable. Intensive study in any given situation may lead to the dissection of a supposed attitude area into two or more subareas each of which may turn out to be scalable, in Guttman's sense, when analyzed separately.

1. *The Scalogram Theory of Establishing Rank Order*

The approach which was developed in the Research Branch under the guiding hand of Louis Guttman has been named scalogram analysis. Several attitude areas were analyzed during the war to see whether items from these areas could be accepted as scalable. As experience proceeded, some of the earlier criteria were revised, and many of the areas once deemed sufficiently scalable do not satisfy the more rigorous criteria subsequently imposed. But the general logical outline, which is quite a simple one, has stood the test of a wide variety of applications.

The scalogram hypothesis is that the items have an order such that, ideally, *persons who answer a given question favorably all have higher ranks on the scale than persons who answer the same question unfavorably*. From a respondent's rank or scale score we know exactly which items he endorsed. Thus, ideally, scales derived from scalogram analysis have the property that the responses to the individual items are reproducible from the scale scores.

In practice, this ideal is not perfectly attained. In the following section, chapter 13, standards are set forth for accepting data as constituting such a scale. There have been some misunderstandings in the profession as to these standards, and as to the reasoning as well as the empirical experience on which they are based.

The standards have been and doubtless will be criticized from two quite opposite points of view: (a) that they are too lenient, permitting areas to be accepted as scalable when they are not, and (b) that they are too stringent, limiting too severely the number of areas which can be accepted as scalable. There is an arbitrary element of judgment involved in arriving at a standard, just as there is in arriving at a standard such as the 5 percent level for a test of significance in sampling theory. It is quite possible that some of the standards here proposed may be changed with increasing experience. Research Branch experience has shown that when the proposed criteria are adopted, many areas have been found to be scalable; rank scores based on even a small number of items from such an area necessarily have high test-retest reliability (according to a theorem of scalogram analysis) and have been found in practice to correlate satisfactorily with external variables.

The items used in a scalogram analysis must have a special *cumulative property*. The general idea may be illustrated by a hypothetical scale of stature comprised of responses to three items:

1. Are you over 6 feet tall?	———Yes	———No
2. Are you over 5 feet 6 inches tall?	———Yes	———No
3. Are you over 5 feet tall?	———Yes	———No

If a person checks item 1 "Yes," he must, unless he is careless, also check items 2 and 3 "Yes." If he checks item 1 "No" and item 2 "Yes" he must also check 3 "Yes." Hence, if we give a score of 2 to a man who has endorsed two items we know exactly *which two items* he endorsed. He could not say "Yes" to item 1, "No" to item 2, and "Yes" to item 3. The four admissible response patterns to the three items are shown below:

Rank order of respondents	Score	Says yes to item			Says no to item		
		1	*2*	*3*	*1*	*2*	*3*
1	3	X	X	X			
2	2		X	X	X		
3	1			X	X	X	
4	0				X	X	X

This simple diagram is called a scalogram—hence the name scalogram analysis for the procedure.

It is essential to note that the items are *cumulative*. A man who answers "Yes" to item 1 has the stature of a man who answers "Yes" to item 2 and and "No" to item 1, *plus additional stature*. Thus the following items would not yield a scale in a scalogram analysis. They are not cumulative.

1. Are you over six feet tall?	———Yes	———No
2. Are you between 5 feet 6″ and 6 feet tall?	———Yes	———No
3. Are you between 5 feet and 5 feet 6″ tall?	———Yes	_____No
4. Are you under 5 feet tall?	———Yes	———No

Here it is possible to make only one affirmative response and we must use additional knowledge of the fact that a man who says "Yes" to item 1 is taller than a man who says "Yes" to item 2 if we are to order the items. The scale picture would look as follows:

Rank order of respondents	Says yes to item				Says no to item			
	1	*2*	*3*	*4*	*1*	*2*	*3*	*4*
1	X					X	X	X
2		X			X		X	X
3			X		X	X		X
4				X	X	X	X	

In this example, the order of the items is implicit in the structure of the questions. Where this is not the case, a device sometimes used is to have a set of judges rank the items in advance. The consistency with which the judges agree in their initial ranking can be evaluated. But since only one answer is permissible, there is no test of internal consistency of the subsequent *respondents*. Such a test could be devised by, for example, asking each respondent to check "Yes" to the *two* items closest to his position. This would reduce the number of rank groups, but would provide objective evidence of order such that initial judges would no longer be needed, although they might still be useful. The scale picture would be:

Rank order of respondents	Says yes to item				Says no to item			
	1	*2*	*3*	*4*	*1*	*2*	*3*	*4*
1	X	X					X	X
2		X	X		X			X
3		X	X	X	X			

It would seem to be quite possible to use as a model this type of "interval" or "position" scale. In fact, Frederick Mosteller has developed the theory of such a scale, in connection with a study of scaling sponsored by the RAND Corporation at the Harvard Laboratory of Social Relations. It will be noted that the cumulative type of item which fits the scalogram model will not fit this model, and vice versa.

One can handle the stature items 1, 2, 3, and 4 above in still other ways. For example, the respondent might be asked to rank the four stature intervals in terms of their closeness to his height. If he is 5 feet, 2 inches tall, the order would be interval 3, rank 1; interval 4, rank 2; interval 2, rank 3; interval 1, rank 4. For all respondents, we would get the following scale picture (the man whose stature is 5 feet, 2 inches, is in rank group 5, for example):

Rank order of respondents	Rank assigned to item number			
	1	*2*	*3*	*4*
1	1	2	3	4
2	2	1	3	4
3	3	1	2	4
4	4	2	1	3
5	4	3	1	2
6	4	3	2	1

The same result could be derived from a set of $n(n-1)/2$ paired comparisons, which would have the additional advantage of providing a check on the internal consistency of an *individual* as well as the internal consistency of a *group* of *individuals*. Questions would be of the type.

Which of these is closer to your stature? (Check one)
――――Over 6 feet
――――Between 5 feet 6″ and 6 feet

All of these examples illustrate how a model can be set up without *explicitly* postulating *a latent* continuum. The ordering of manfest responses *if perfectly consistent* will reveal the appropriate structure.[2] The scalogram model, requiring cumulative items, has at the present time been studied far more thoroughly than these others. Therefore, our attention will now be focused on it. We must keep clearly in mind the fact that the cumulative character of the items satisfying a scalogram analysis puts a restriction upon the type of data which can be used with this model. Yet the restriction is not, in practice, as severe as may at first appear.

It is often possible to find items which have an intrinsic cumulative character. The prototype is perhaps the social distance scale, with such items as the following:

1. Would you want a relative of yours to marry a Negro?
2. Would you invite a Negro to dinner at your home?
3. Would you allow a Negro to vote?

An illustration from Research Branch data of this type of intrinsic scale may be taken from a study of riflemen overseas who had recently experienced combat. It was desired to see if there was a scale order in the way in which respondents reported experiencing fear. The following question was asked:

Soldiers who have been under fire report different *physical reactions to the dangers of battle*. Some of these are given in the following list. How often have you had these reactions when you were under fire? Check one answer after

2. Thurstone's law of comparative judgment postulates and underlying continuum and seeks to derive, usually with the aid of the method of paired comparisons, not only a rank order but a metric. In psychophysical measurement, such as the classical example of lifted weights, the assumption is made that each individual respondent has the same position and that variations in judgment either within the same individual or between a group of individuals represent discriminal error. This discriminal error provides a unit of measurement. But in the present example, each individual can be perfectly consistent in all his responses, yet because each individual can have a different position (i.e., different stature), the group variability will not represent *discriminal* error. Detailed analysis of this phenomenon has been made by Clyde H. Coombs, in connection with his work on the RAND study at the Harvard Laboratory of Social Relations. Among other things, he has shown that a rank order table like that above, or a paired comparisons table, does reveal under certain conditions whether some rank intervals have greater magnitude than others.

each of the reactions listed to show how often you had the reaction. Please do it carefully.

[There followed 10 items, each with a four-step check list. For example:

Shaking or trembling all over

—————Often

—————Sometimes

—————Once

—————Never]

A scale picture indicated that nine of these items formed a very satisfactory scale when ordered as follows:

1. Urinating in pants
2. Losing control of the bowels
3. Vomiting
4. Feeling of weakness or feeling faint
5. Feeling of stiffness
6. Feeling sick at the stomach
7. Shaking or trembling all over
8. Sinking feeling of the stomach
9. Violent pounding of the heart

One item—"cold sweat"—did not fit into the scale; that is, some people who experienced less frequent fear symptoms than cold sweat also experienced cold sweat, but this was also true of some people who experienced more frequent symptoms than cold sweat. Hence "cold sweat" was shown to involve a factor or factors additional to the scale variable.

The fact that the nine items satisfied the scalogram criteria means that if a man did not report vomiting, for example, he also did not report urinating in his pants or losing control of his bowels. If he did report vomiting, he also reported all of the other experiences (4 through 9 on the list) which were generally more frequent that vomiting. In this case he would have a score of 7 and if the scale were perfect we would know exactly *which* seven experiences he reported. Actually, of course, the scale was not quite perfect. By an easy procedure, it was possible to compute a *coefficient of reproducibility*, which was .92. This means that if we knew any scale score, such as 7, and if we guessed the exact items which respondents with this scale score endorsed and did not endorse, we would guess 92 out of 100 of the items correctly and 8 out of 100 incorrectly. The scale picture also met other criteria with respect to individual item reproducibility, wide range of marginals, and randomness of error.

It may be that only a limited range of psychological and social phenomena have the intrinsic cumulative characteristic described above. An area which should eventually be quite rich in data of this type should be that of level of difficulty. For example, three problems in calculus:

1. Integrate $\dfrac{dy}{dx} = xy\,(y - a)$

2. Integrate $\dfrac{dy}{dx} = x^2$

3. Differentiate $y = x^2$

It is unlikely that a person who can do problem 1 would fail on problems 2 and 3, and it is also unlikely that a person who failed on problem 2 could do problem 1. In other words, if A has a higher score than B, A can do all that B can do, *plus* something more.

The possibility should be faced, however, that large areas of psychological or social behavior may not yield items which approximate the scale pattern required by scalogram analysis. For example, a short catalogue of psychosomatic symptoms which the Research Branch used in constructing a psychiatric screening test simply did not order itself in a manner satisfying the rigorous criteria set up. That is, a man who said he was bothered with shortness of breath might or might not complain of spells of dizziness, and vice versa. As we shall see later, it is still possible to construct a scale—called a *quasi scale*—which lacks the property of reproducibility but which does have the valuable property of *yielding the same correlation with an outside criterion as does the multiple correlation of the individual items with that criterion.*

It is frequently possible, however, to order items cumulatively (and hence to order respondents) by constructing items with multiple check lists and choosing cutting points by combining categories such that the error of reproducibility is minimized.

Consider, for example, three items which happen to have uniform format. (Such uniformity is not at all essential—the number of response categories can vary arbitrarily from item to item and the wording of the categories can be varied):

1. How many of your officers take a personal interest in their men?
   1————All of them
   2————Most of them
   3————About half of them
   4————Some of them
   5————Few or none of them

2. How many of your officers will go through anything they ask their men to go through?
   1————All of them
   2————Most of them

3————About half of them
4————Some of them
5————Few or none of them

3. How many of your officers are the kind you would want to serve under in combat?
1————All of them
2————Most of them
3————About half of them
4————Some of them
5————Few or none of them

If each item is dichotomized by taking the two top categories as "favorable" responses, the three items will not ordinarily form a scale by scalogram analysis. In a sense, in spite of manifest difference in content, these three particular items are more or less synonymous. Hence, the frequencies probably will not be cumulative. However, if for item 1 we take response category 1 as "positive," for item 2 response categories 1 and 2, and for item 3 response categories 1, 2, and 3, we perhaps can build a cumulative set of items. This is, of course, just one of various ways in which a cumulative set of responses might be built from these items. The procedure for selecting cutting points which maximize scalability has been reduced to a simple routine. It is *not arbitrary,* but is based upon a study of the *simultaneous distribution* of all the original responses.[3] From a practical standpoint, the operational procedures developed in the Research Branch for swiftly evaluating a large number of questions simultaneously may rank as one of the major contributions of the Branch to social science technique. Work which would have required hundreds of hours of elaborate machine analysis can now be done in a few hours by one semi-skilled clerk.

There can be no doubt that the empirical choice of cutting points on more or less synonymous items in order to achieve a set of cumulative responses is rather crude. It needs to be studied and criticized. But it does supply an ordering of respondents with a high degree of reliability and the ultimate test will be its utility in comparison with other techniques.

3. Two people can check different response categories and still have the same attitude, because they differ in verbal habits. For example, one person may have a general tendency toward extreme statements and say "all of them" to a given question, whereas another person may have the same attitude toward officers but express it on the same question by answering "most of them." If these two responses are treated as separate categories they could exhibit substantial error of reproducibility, which tends to vanish when the categories are combined (that is, treated as though they were the same response). The scalogram analysis shows how best to combine categories in order to reduce errors of reproducibility.

Relatively few of the scales described in this volume can be strictly described as based on near synonymous items of the kind illustrated above. For example, we can examine a scale of general attitudes toward the Army. Twelve items comprise this scale, and some of the items rather closely resemble each other in format. But they are certainly not synonymous. Consider four of them:

In general, do you think you yourself have gotten a square deal in the Army?

————Yes, in most ways I have

————In some ways yes, in other ways, no

————No, on the whole I haven't gotten a square deal

In general, how well do you think the Army is run?

————It is run very well

————It is run pretty well

————It is not run so well

————It is run very poorly

————Undecided

Do you think the Army has tried its best to look out for the welfare of enlisted men?

————Yes, it has tried its best

————It has tried some, but not enough

————It has hardly tried at all

Do you think when you are discharged you will go back to civilian life with a favorable or unfavorable attitude toward the Army?

————Very favorable

————Fairly favorable

————About 50–50

————Fairly unfavorable

————Very unfavorable

These items, along with eight other items also of a general character but varying in format, were found to cumulate, after the choice of appropriate cutting points, such that they satisfied the scalogram criteria.

The greater the variety of questions, of course, the wider the generalization which can be made about the coverage of a particular scale, assuming that criteria of scalability are satisfied.

It is recommended that a relatively large number of items, preferably as many as ten or twelve, be used in the initial testing of the hypothesis of scalability, and that, if possible, some of these items be trichotomized rather than dichotomized. This will ordinarily make it quite difficult to achieve high reproducibility, but protects against spurious results which might be obtained by chance with only three or four dichotomous items.

Critics of the use of only three or four items seem to have overlooked an important distinction, namely, that while it is desirable to use as many as ten or a dozen items in initial testing of scalability, a smaller number of items can be safely selected from the scalable list for practical research purposes. If a dozen items reveal a scale pattern, then a smaller number selected from these dozen will show the same pattern (though with fewer ranks). Because they belong to the same scale, according to scalogram criteria, a smaller subset of items finally selected for use—possibly comprising only three or four—will of necessity correlate very highly with other subsets of items in the same scale. Items comprising such a subset will have higher reliability than similiar items from most other types of scales, since each response is a *definition* of the respondent's position on a single continuum and is minimally corrupted with other affective material which correlates with some items and not with others.

In fact, one of the important consequences of finding an attitude to be scalable is that one is then justified in selecting three or four items which can be used to order respondents in a limited number of ranks. It may be possible eventually to use a pretest for selecting *a single item* for practical use—such as is the staple of much conventional market research or public opinion polling. But now the item is selected *with full knowledge* as to its place in the attitude structure. Intensity analysis in particular, as discussed in the last section of this chapter, may help determine objectively which item is unbiased with respect to the attitude as a whole, in the sense of dividing people into two groups, those favorable and those unfavorable on a given issue.

In attitude research of the future, an important desideratum will be to obtain simultaneous measures on a *large number of continua* from a given respondent. Scalable attitudes lend themselves particularly well to this in practice. Why? Because each attitude can be represented by only a few items, so that a large complex of attitudes can be observed in a single study. Thus multiple factor analysis might be employed to construct a typology of these continua, and we may well be on the road to an analysis of socio-psychological problems comparable to the road which has led to progress in the study of human abilities.

When the hypothesis of the scale structure is found not to be borne out by the observed data, the attitude items may still have an ordered structure, as has been said, namely, that of the *quasi scale*. The conditions for existence of a quasi scale need more study than is provided in this volume. Here we are focusing attention on the Guttman procedure. There is, of course, the possibility that entirely different models might have fitted the data.

Chapter 13

THE BASIS FOR SCALOGRAM ANALYSIS

Louis L. Guttman
Hebrew University, Israel

One of the fundamental problems facing research workers in the field of attitude and public opinion measurement is to determine if the questions asked on a given issue have a single meaning for the respondents. Obviously, if a question means different things to different respondents, then there is no way that the respondents can be ranked in order of favorableness. Questions may appear to express but a single thought and yet not provide the same kind of stimulus to different people. The responses even to the simplest question can differ in kind as well as in degree.

That two people can give the same response to the same question and yet have different attitudes has long been of primary concern to public opinion pollsters. This is particularly the case when more than one question is asked about the same topic, and the replies to the different questions appear to be inconsistent. Consistency of response is a problem that plagued attitude research long before public opinion polling captured the fancy of the public. How can one tell if there is enough consistency in the responses of a population to a series of questions to indicate that

Reprinted with permission of the author and of Princeton University Press from *Measurement and Prediction* by Samuel A. Stouffer *et al.*, Vol. IV of *Studies in Social Psychology in World War II* (copyright 1950 by Princeton University Press), pp. 60–90.

A bibliography of published articles on scalogram analysis is given at the end of this chapter.

only a single factor is being measured? Are all respondents interpreting the questions to mean the same things? Are differences in responses due only to differences in degree of feeling and not to differences in kind? Is it meaningful to score the people from high to low with respect to a given set of items?

This problem of consistency underlies a great deal of research in the social and psychological sciences dealing with large classes of qualitative observations. For example, research in marriage is concerned with a class of qualitative behavior called marital adjustment which includes an indefinitely large number of interactions between husband and wife. Public opinion research is concerned with large classes of behavior like expressions of opinion by Americans about a military treaty with the British. Educational psychology deals with large classes of behavior like achievement tests. Other problems in social and psychological research where the consistency underlying a series of items is of fundamental importance include the study of aptitudes involved in fitting people into jobs, the measurement of human intelligence and abilities, the study of neurotic behavior and other aspects of personality, the appraisal of social status—in short, any problem involving the assigning of numerical values to qualitative observations in an attempt to evolve a single rank ordering. It is often desired in such areas to be able to summarize data by saying, for example, that one marital couple is better adjusted than another marital couple, or that one person has a better opinion of the British than has another person, or that one student has a greater knowledge of arithmetic than has another student.

While the data to be presented in the present chapter will deal almost entirely with social attitudes or opinions, the approach is appropriate for all of the types of problems mentioned. This rather new approach seems to afford an adequate basis for the quantification of many types of qualitative data.

The approach to scale analysis to be presented here has been used successfully during the war in investigating morale and other problems in the United States Army. While some interesting mathematics lie in the background, no knowledge of this mathematics is required in actually analyzing data. Simple routines have been established which require no knowledge of statistics, which take less time than the different manipulations now used by various investigators (such as critical ratios, biserial correlations, factor analyses, etc.), and which give a complete picture of the data not afforded by any of these other techniques. The word "picture" might be interpreted here literally, for the results of the analysis are presented and easily assimilated *in the form of a "scalogram," which gives the configuration of the qualitative data.*

Theory of Scale Analysis

A main condition for an attitude scale has often been pointed out by psychologists. For example, Murphy, Murphy, and Newcomb say that "no scale can really be called a scale unless one can tell from a given attitude that an individual will maintain every attitude falling to the right or to the left of that point."[1] A similar consideration is the starting point for our theory of scale analysis. Instead of focusing on the ranking of items, however, we focus on the ranking of individuals. The ranking of items apparently is restricted to dichotomous items, where a person either endorses or does not endorse a statement. In such a case, it is possible to consider a ranking of endorsements, so that if a person endorses a more extreme statement, he should endorse all less extreme statements if the statements are to be considered a scale. But if the items have more than two categories, such a consideration breaks down; "agree" to one item might be equivalent to, or even less "favorable" than, "undecided" to another item, so that there remains a problem of how to rank items and response categories.

The ranking of people provides a more general approach to the problem of scaling, since it turns out to be equivalent to the ranking of items when all items are dichotomous, and it also includes the case where items have more than two answer categories. We shall call a set of items of common content a scale if a person with a higher rank than another person is just as high or higher on every item than the other person. This involves no problem of ranking the categories of one item against those of the other items, but only needs a ranking of the categories *within* each item.

An equivalent definition of a scale for our approach is that, within each item, if one response category is higher than another, then *all* people in the higher category must have higher scale ranks than those in the lower category.

A third equivalent definition of a scale is the one upon which our practical scalogram analysis procedures are directly based. It requires that each person's responses should be reproducible from his rank alone. A more technical statement of the condition is that each item shall be a simple function of the persons' ranks. The meaning of this is expanded below, where it is easily seen that the three definitions of a scale just given here are all equivalent. This third definition, while perhaps the least intuitively obvious, has proved to be the most convenient formulation for practical procedures. Of course, when it is fulfilled, the first

1. G. Murphy, L. B. Murphy, and T. M. Newcomb, *Experimental Social Psychology* (Harper & Bros., New York, 1937), p. 897. We shall use the word "item" where they use "attitude."

two definitions are also fulfilled; and when the items are dichotomous, then the psychologists' condition first mentioned is also fulfilled.

Murphy, Murphy, and Newcomb further note that, "As a matter of fact there is every reason to believe that none of the rather complex social attitudes which we are primarily discussing will ever conform to such rigorous measurement."[2] Perhaps such a belief may account for the fact that the mass of current attitude research pays little or no attention to this fundamental rationale. The common tendency has been to plunge into analysis of data without having a clear idea as to when a single dimension exists and when it does not. For example, *bivariate* techniques—like critical ratios and biserial correlations—are commonly used to find items that "discriminate" and to determine "weights," without testing whether or not the *multi*variate distribution of the items is actually indicative of a single dimension.

One of the main purposes of this chapter will be to propose a rigorous definition of a scale. This definition applies not only to the study of general attitudes, but also to the study of public opinion on specific issues. Furthermore, specific examples will be given to show that *consistent scales satisfying the rigorous requirements proposed above have been obtained in actual practice.* The proposed method provides a simple analysis of a series of questions which will enable one *to determine quickly whether or not the basic condition for a scale is satisfied by the data.*

The notions of variable, function, and simple function. First, a word about what is meant by a variable, whether qualitative or quantitative. We use the term in its conventional logical or mathematical sense, as denoting a set of values. These values may be numerical (quantitative) or nonnumerical (qualitative). We shall use the term "attribute" interchangeably with "qualitative variable." The values of an attribute (or of a quantitative variable, too, for that matter) may be called its *subcategories*, or simply *categories.*

A variable y is said to be a single-valued function of a variable x if to each value of x there corresponds a single value of y.

In particular, suppose y is an attribute, say like the attribute about expression of liking for the British, and takes on three values. We may denote by y_1 the statement, "I like the British"; by y_2, the statement, "I don't like the British"; and by y_3, "I don't know whether or not I like the British." If x is a quantitative variable which takes on at least three values, and if we can divide the x values into three intervals which will have a one-to-one correspondence with the values of y, then we shall say the attribute y is a *simple* function of x. For example, suppose x takes on the ten values 0, 1, 2, 3, 4, 5, 6, 7, 8, 9. Then the correspondence table might be as follows:

2. *Ibid.*

x	0	1	2	3	4	5	6	7	8	9
y	y_1	y_1	y_1	y_3	y_3	y_2	y_2	y_2	y_2	y_2

The three intervals for x are: 0–2, 3–4, and 5–9, to which correspond the values y_1, y_3, and y_2 respectively. Every person who has an x-value between 0 and 2 has y_1 as his y-value; every person who has an x-value of 3 or 4 has y_3 as his y-value; and every person with an x-value between 5 and 9 has y_2 as his y-value.

We might show this graphically by plotting the x-values on a straight line, and cutting it into intervals:

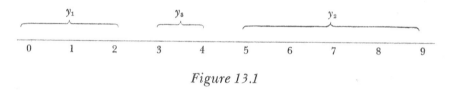

Figure 13.1

For statistical variables, another representation is in terms of a bar chart of frequencies, and this is what is used for convenience below.

THE DEFINITION OF SCALE

For a given population of objects, the multivariate frequency distribution of a universe of attributes will be called a *scale* if it is possible to derive from the distribution a quantitative variable with which to characterize the objects such that each attribute is a simple function of that quantitative variable. Such a quantitative variable is called a scale variable.

Perfect scales are not to be expected in practice. The deviation from perfection is measured by a *coefficient of reproducibility*, which is simply the empirical relative frequency with which the values of the attributes do correspond to the proper intervals of a quantitative variable. In practice, 90 percent perfect scales or better have been used as efficient approximations to perfect scales.

A value of a scale variable will be called a *scale score*, or simply a *score*. The ordering of objects according to the numerical order of their scale scores will be called their scale order.

Obviously, any quantitative variable that is an increasing (or decreasing) function of a scale variable is also a scale variable. For example, in the illustration above, consider x to be a scale variable. Any constant could be subtracted from or added to each of the x scores, and y would remain a simple function of the transformed x. Thus, the scores 0, 1, 2, 3, 4, 5, 6, 7, 8, 9 could be replaced by respective scores —5, —4, —3, —2,

—1, 0, 1, 2, 3, 4. Or the *x* scores could be multiplied by any constant, or their square roots or logarithms could be taken—any transformation, continuous or discontinuous, could be used, as long as the rank order correlation between the original *x* and the transformed variable remained perfect. All such transformations will yield scale variables, each of which is equally good at reproducing the attributes.

Therefore, the problem of metric is of no particular importance here for scaling. For certain problems like predicting outside variables from the universe of attributes, it may be convenient to adopt a particular metric like a least squares metric, which has convenient properties for helping analyze multiple correlations. However, it must be stressed that such a choice of metric is a matter of convenience; any metric will predict an outside variable as accurately as will any other.

In practice, the rank order has been used as a scale variable. (It is in fact a least squares metric for a rectangular distribution of scale scores.)

While rank order is sufficient for the *mechanical* aspects of testing for scalability and of external prediction, it is not adequate for certain problems of *psychological* description and generalization.

An Example of a Scale of Dichotomies

As may be expected, the universe of attributes must form a rather specialized configuration if it is to be scalable. Before describing a more general case, let us give a little example. Suppose that a statistics test is composed of the following problems:

Consider a population of voters in which 60 percent are Democrats and 40 percent are Republicans.

1. What is the probability that one person chosen at random will be a Democrat?
2. What is the probability that two people chosen at random will both be Democrats?
3. What is the probability that out of ten people chosen at random, *at least* three will be Democrats?

If this test were given to the population of members of the American Sociological Society, we would perhaps find it to form a scale for that population. The responses to each of these questions might be reported in dichotomous form as "right" or "wrong." There are $2 \times 2 \times 2 = 8$ possible types for three dichotomies. Actually, for this population of sociologists we would probably find only four of the eight types occurring. There would be (a) the type which would get all three questions right, (b) the type which would get the first and second questions right, (c) the type which would get only the first question right, and (d) the type which would get none of the questions right. Let us assume that this is what would actually happen. That is, we shall assume the other four types, such as the type getting the first and third questions right but the second question wrong, would not occur. In such a case, it is possible to assign to the population a set of numerical values like 3, 2, 1, 0. Each member of the population will have one of these values assigned to him. This numerical value will be called the person's score. From a person's score we would then know precisely to which problems he knows the answers and to which he does not know the answers. Thus a score of 2 does not mean simply that the person got two questions right, but that he got two particular questions right, namely, the first and second. A person's behavior on the problems is reproducible from his score. More specifically, each question is a *simple function* of the score, as will be shown below.

THE MEANING OF "MORE" AND "LESS"

Notice that there is a very definite meaning to saying that one person knows "more" statistics than another with respect to this sample. For example, a score of 3 means more than a score of 2 because the person with a score of 3 knows *everything a person with a score of 2 does, and more.*

There is also a definite meaning to saying that getting a question right indicates "more" knowledge than getting the same question wrong, the importance of which may not be too obvious. People who get a question right all have higher scale scores than do people who get the question wrong. As a matter of fact, we need no knowledge of which is a "right" answer and which is a "wrong" answer beforehand to establish a proper order among the individuals. For convenience, suppose the questions were given in a "true-false" form, with suggested answers (1) .50, (2) .36, (3) .42 for the respective questions.[3] Each person records either a T or an F after each question, according as he believes the suggested answers to be True or False. If the responses of the population form a scale, then we do not have to know which are the correct answers in order to rank the respondents (only we will not know whether we are ranking them from high to low or from low to high). By the scale analysis, which essentially is based on sorting out the joint occurrences of the three items simultaneously, we would find only four types of persons occurring. One type would be $F_1 T_2 F_3$, where the subscripts indicate the questions; that is, this type says F to question 1, T to question 2, and F to question 3. The other three types would be $F_1 T_2 T_3$, $F_1 F_2 T_3$, and $T_1 F_2 T_3$. These types

---

3. We shall assume that no one gets an answer right by guessing. Scale analysis can actually help one pick out responses that were correct merely by guessing from an analysis of the pattern of errors. But for this, much more than three items are necessary.

can be shown in a chart (a "scalogram") where there is one row for each type of person and one column for each answer category of each question.

The scale analysis would establish an order among the rows and among the columns which would finally look like this:

		Question				
Type score	F_3	T_3	F_1	T_2	F_2	T_1
3	x	x	x			
2		x	x	x		
1			x	x	x	
0				x	x	x

Figure 13.2

Or, alternatively, both rows and columns might be completely reversed in order. Each response to a question is indicated by an x. Each row has three marks because each question is answered (either correctly or incorrectly). The *parallelogram* pattern in the chart[4] is necessary and sufficient for a set of *dichotomous* attributes to be expressible as simple functions of a single quantitative variable.

From this chart we can deduce that F_1, T_2, and F_3 are all correct answers, or are all incorrect answers. That is, if we were now told that F_1 is a correct answer, we would immediately know that T_2 and F_3 are also correct answers. This means that we can order the men according to their knowledge even if we do not know which are the correct answers and which are the incorrect answers, only we do not know whether we are ordering them from highest to lowest or from lowest to highest. Except for direction, the ordering is a purely formal consequence of the configuration of the behavior of the population with respect to the items. The importance of this fact will become more apparent in more complicated cases where the attributes are not dichotomous but have more than two categories. As will be shown later, the scale analysis automatically decides, for example, where an "undecided" response to a public opinion poll question belongs, whether it is above "yes," below "no," in between, equivalent to "yes," or equivalent to "no." A priori judgments of content order are not essential to scale analysis.

THE BAR CHART REPRESENTATION

Another way of picturing the dichotomous scale of the sample of three items would be as follows: suppose that 80 percent of the population

---

4. Such a chart, where one column is used for each *category* of each attribute, we call a *scalogram*. The scalogram boards used in practical procedures are simply devices for shifting rows and columns to find a scale pattern, if it exists.

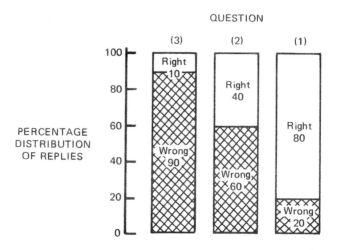

Figure 13.3

got the first question right, 40 percent got the second question right, and 10 percent got the third question right. The univariate distributions of the three respective items could be shown by the bar charts in Figure 13.3.

The bars show the percentage distributions for the respective questions. The multivariate distribution for the three questions, *given that they form a scale for the population,* can also be indicated *on the same chart,* since all those who are included in the group getting a harder question right are also included in the group getting an easier question right. Thus, we could draw the bar chart over again, but connect the bars with dashed lines in the fashion shown in Figure 13.4.

Here again we can see how the three questions are simple functions of the scores. From the marginal frequencies of the separate items, *together with the fact* that the items form a scale, we are enabled to deduce that 10 percent of the people got a score of 3. The 10 percent who got the hardest question right are included in those who got the easier questions right. This is indicated by the dashed line on the right, between the scores 2 and 3, which carries the same 10 percent of the people (those with a score of 3) through the three bars. The 40 percent who got the second question right include the 10 percent who got the hardest question right and 30 percent out of those who got the hardest question wrong, but all 40 percent got the easiest question right. This leaves us 30 percent who got just the first and second questions right. And so on. Thus we can think of an ordering of the persons along a vertical continuum, and each dichotomy can be thought of as resulting in one additional

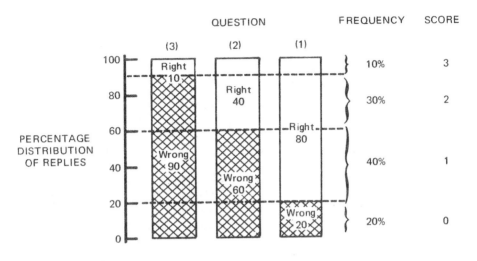

Figure 13.4

cut on that continuum. All those above the cutting point get the question right, and all those below the cutting point get the question wrong. Thus, there is a one-to-one correspondence between the categories of an item and segments of the continuum. *Or we can say that each attribute is a simple function of the rank order along the continuum.*

If the "right" and "wrong" answer categories are separated, as in Figure 13.2, this bar chart representation assumes the pattern of a parallelogram. Two basic steps are involved in this procedure, as will be described in detail in the next chapter. First, the questions are ranked in order of "difficulty" with the "hardest" question, i.e., the one that fewest persons got right, placed first and with the other questions following in decreasing order of "difficulty." Second, the people are ranked in order of "knowledge" with the "most informed" persons, i.e., those who got all questions right, placed first, the other individuals following in decreasing order of "knowledge." These two steps are the basic procedure for the scalogram board technique of scale analysis. The resulting parallelogram, assuming a scale is present, is shown in Figure 13.5.

ZERO-ORDER CORRELATIONS BETWEEN ITEMS

It is because all the items in the sample can be expressed as simple functions of the same ordering of persons that they form a scale. Each item is perfectly correlated with or reproducible from the ordering along the continuum. However, the point correlations between the items are not

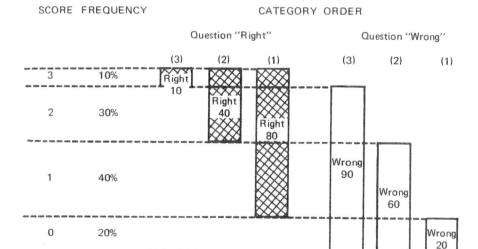

Figure 13.5

at all perfect. For example, the fourfold table between the second and third items is as follows:

Question 2

		Right	Wrong	
	Right	10	0	10
Question 3				
	Wrong	30	60	90
		40	60	100

The point correlation between the two items is .41. As a matter of fact, the point correlation between two dichotomous items may be anything from practically zero to unity, and yet they may both be perfect simple functions of the same quantitative variable.[5]

5. A tetrachoric coefficient for the fourfold table above, assuming a bivariate normal distribution, would be unity. However, this is *not* the correlation between the items. It does not tell how well one can predict one item from the other. The tetrachoric coefficient expresses instead the correlation between two quantitative variables of which the items are functions, provided the assumptions of normality are true. The reason the terachoric is unity in this case is that the quantitative variables of which the items are functions are one and the same variable, namely, the scale variable. Notice, however, that the distribution of the scale variable according to the rank order is not at all normal. One of the contributions of scaling theory is to do away with untested and unnecessary hypotheses about normal distributions.

An important feature of this fourfold table is the zero frequency on the upper right hand corner cell. Nobody who got the third question right got the second question wrong. Such a zero cell must always occur in a fourfold table between two dichotomous items which are simple functions of the same quantitative variable. This zero cell, furthermore, must occur in the column or row which contains the lowest frequency and in that cell which represents a "positive" answer on one question and a "negative" answer on the other question. Given only the marginal distributions of any two scale questions, it is possible to compute the frequencies in each cell of a correlation table of the two questions. In fact, *it is possible to construct the multivariate distribution of all the scale questions from a knowledge of their straight distributions alone.*[6] This can be done regardless of how many answer categories are retained for each question. An example of a scale using trichotomous items will be given later.

This requirement concerning zero-order correlation tables between items suggests a procedure for scale analysis based on these tables. However, zero-order relationships do not tell the whole story about the entire multivariate distribution. Actually, it is simpler to study the complete distribution by means of the scalogram board technique, or related techniques, than to study all the bivariate tables. It is helpful, however, to learn what a scale pattern signifies by exploring some of the consequences in terms of bivariate tables. While the zero-order correlation tables are not good to use as a practical technique for testing items for scalability, they are a good pedagogical device for understanding scales.

ZERO CELLS FOR DICHOTOMIES

The presence of a zero cell in the proper place is a necessary but not sufficient condition for the existence of a scale. If a correlation table of two attitude questions does *not* reveal this zero cell, then one can be certain that these two questions are *not* members of a single attitude continuum, i.e., do not have but one meaning to the respondents. However, the existence of a zero cell in itself is insufficient proof of a single variable, even if the cell is in the proper position. First, the two questions must be judged to belong to the same content universe by virtue of their content. Second, a sample of two questions is too subject to sampling error (with respect to the universe of content) to provide an adequate test; the occurrence of the zero cells in their proper position should be found in the multivariate distribution of several questions. The desired pattern

6. The importance of this predictability for correlation analysis of attitude questions will be discussed later.

for more than two questions will become more clear to the reader as additional examples are given.

Let us translate this simple example into a problem of public opinion analysis. Suppose the issue to be studied is public opinion toward the continued maintenance of a large Army now that the war is won. Three questions dealing with this topic, together with hypothetical percentage distributions of replies, are given below.

1. In your opinion, is it necessary or unnecessary for the protection of the United States to have a strong Army?
    (a) It is necessary   70%
    (b) It is not necessary   30%

2. Do you think the United States should or should not increase the present size of the Army?
    (a) It should   50%
    (b) It should not   50%

3. If all other countries agree to disarm, do you think the United States should or should not maintain a large Army?
    (a) It should   20%
    (b) It should not   80%

For the present, we will not deal with the problem of interpreting which of these percentages—20 percent, 50 percent or 70 percent—represents the division of public opinion upon the issue of the size of the Army.

Following the procedure outlined in relation to the statistical knowledge example given previously, we can test the hypothesis that these three questions form a scale by seeing whether the three questions are simple functions of the scale scores. One way of doing this would be to draw the bar chart diagram shown in Figure 13.5 and to compare the obtained frequency of the four scale types with the theoretical or expected frequency.

To draw this diagram of expected frequencies, we first place the questions in descending order from the question calling for the most extreme expression of "favorableness" toward a large Army, i.e., that question to which fewest persons reply "in favor of" a large Army, to the question calling for the least extreme expression of "favorableness" toward a large Army, i.e., that question to which the largest number of persons reply "in favor of" a large Army.[7] This ordering of the questions shows us immediately how many scale types should occur and what their expected frequencies should be, if we have a scale. Scoring each of the subjects and placing them in rank order of "favorableness" toward a large Army, i.e.,

---

7. The terms "favorable" and "unfavorable" are used in the sense of *more* or *less* favorable only. To divide such a continuum into "favorable" and "unfavorable" calls for the determination of a zero point.

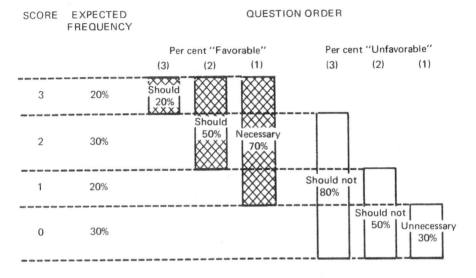

Figure 13.6

the number of questions upon which they express opinions "in favor of" a large Army, should produce the parallelogram shown in Figure 13.6.

Thus, in order for these three questions to form a scale, all persons who felt that the United States *should* maintain a large Army even if all other countries disarm must also feel that the United States *should* increase the present size of the Army and that it was *necessary* for the protection of the United States to have a strong Army. Examination of the fourfold table between any two of the questions would have to reveal a zero cell in the corner representing the "favorable"-"unfavorable" cell of the lowest marginal frequency. For example, the frequencies of the cells in the cross tabulation between questions 2 and 3 are completely predictable from the marginals, as follows:

Question 2

		"Favor- able"	"Unfavor- able"	
	"Favorable"	20	0	20
Question 3	"Unfavorable"	30	50	80
		50	50	100

To conform to the scale pattern, there should be *no* respondents who feel that the United States *should* maintain a large Army even if all other countries disarm, but who think that the United States *should not* increase the present size of the Army. On the other hand, a respondent who feels that the United States *should not* maintain a large Army if all other countries disarm, may or may not favor an increase in the present size of the Army.

Given a scale pattern, the same predictability or reproducibility of intercorrelations between any questions is possible from a knowledge of the straight distributions. In addition, a knowledge of the scale score enables one to predict or reproduce the responses of any individual to each of the questions asked. The main condition for a scale is satisfied: a person with a higher rank than another is just as high or higher on each item.

EXAMPLE OF A SCALE OF TRICHOTOMIES

The same technique of scale analysis outlined for dichotomous items applies equally well to items with any number of answer categories. In fact, this technique enables one to determine whether the different answer categories should be treated separately as representing meaningfully different replies or whether they should be combined. For example, in response to a question, "How important is it for the United States to have a large Army?" how should the answer categories "fairly important" and "not so important" be treated? Should they be left as separate categories, or should they be combined into a kind of neutral category, or should "fairly important" be combined with "very important" while "not so important" is combined with "not at all important?" Scale analysis, using the technique outlined, will tell one quite automatically how these combinations should be made.

Let us see what a scale composed of trichotomous items would look like. Suppose three questions from the same attitude universe were asked, each of which had three answer categories, the answers to which distributed as follows:

		Question	
Answer categories	1	2	3
a	25%	20%	40%
b	20	60	30
c	55	20	30
	100%	100%	100%

The number of scale types for these three trichotomous items is seven, whereas the number of possible types for a nonscalable area is twenty-seven. This reduction in scale types is indicative of the highly restrictive nature of the scale pattern. (The number of scale types equals the *sum* of all item categories, less the number of questions, plus one, while the number of all possible types equals the *product* of all the item categories.) The characteristics of the scale types can be determined simply by joining the answer categories of the different questions, as below in Figure 13.7.

Arranging the questions in order of the frequency of "favorable" responses (for all three answer categories) and arranging the subjects in order of scale scores (assume simple weights of two for *a*, one for *b*, and zero for *c*), the following bar chart parallelogram emerges for the ideal scale pattern (Figure 13.8).

This pattern is completely predictable on the basis of the marginal distribution of answers to each of the questions. From a person's scale score it is possible to reproduce exactly how he answered each of the questions. From the marginal distribution of responses to any of the questions it is possible to construct any of the internal correlations between questions. The pattern of intercorrelation for questions 1 and 3, for example, would have to be as follows:

Question 1

		a	b	c	
	a	25	15	0	40
Question 3	b	0	5	25	30
	c	0	0	30	30
		25	20	55	100

The basic condition to be satisfied is that persons who answer a question "favorably" all have higher scale scores than persons who answer the same question "unfavorably." This constitutes a rigorous definition of a scale. It provides a simple, objective technique for testing the existence of a single variable, that is, for determining whether the questions have the same meaning for all respondents.

The Measurement of Error

The amount by which a scale deviates from the ideal scale pattern is measured by a *coefficient of reproducibility*. This coefficient is simply a measure of the relative degree with which the obtained multivariate distribution corresponds to the expected multivariate distribution of a

			Scale type	Frequency	Scale score
a	a	a	aaa	20%	6
			aba	5%	5
b			bba	15%	4
		b	bbb	5%	3
	b		cbb	25%	2
c			cbc	10%	1
	c	c	ccc	20%	0
Item 1	Item 2	Item 3			

Figure 13.7

perfect scale. It is secured by counting up the number of responses which would have been predicted wrongly for each person on the basis of his scale score, dividing these errors by the total number of responses and subtracting the resulting fraction from 1. As will be seen in the actual examples to be presented, the occurrence of errors is easily determined by visual inspection of the scale pattern. An acceptable approxi-

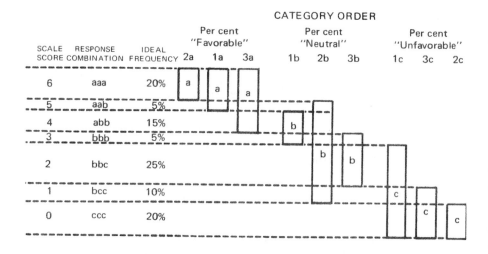

Figure 13.8

mation to a perfect scale has been arbitrarily set at 90 percent reproducibility. Thus, if a scale consisted of five items tested on 100 people, the total number of responses would be $5 \times 100 = 500$. To secure a coefficient of reproducibility of at least .90, there could be at most 50 errors for the entire sample of 100 respondents on all five questions.

The coefficient of reproducibility of the universe can be observed only with sampling error if the scale pattern is not perfect. There may be error due to the sampling of items and error due to the sampling of people. The problem of item sampling is considered later. With respect to sampling deviations due to people, the following remarks may be made.

Reproducibility is computed by counting up the errors for each person on each item. If the sample consists of 100 persons and five items, the percent reproducibility is based on 500 observations. If the errors of reproducibility are random, and if the population reproducibility is at least .90, then the standard error of a sample proportion of reproducibility is at most .013, which allows a deviation in the proportion of at most .040 at the three standard error level of confidence.

Such a calculation of a standard error is, of course, voided if the errors are not random, which may often be the case. Empirical experiments on samples of 100 cases each on five items, however, have consistently shown less than a variation of .04 when the combined sample reproducibility has been over .90. Occasionally an item has behaved a little differently for comparable samples of people. Raising requirements for reproducibility, and using items with more than two answer categories, reduces the possibility for such minor variations.

In some cases, a sample of 200 or more people may be necessary to get a clear picture of the situation.

REPRODUCIBILITY IS NOT A SUFFICIENT CRITERION FOR
SCALABILITY

Reproducibility by itself is not a sufficient test of scalability. It is the principal test, but there are at least four other features that should be taken into account: (a) range of marginal distributions, (b) pattern of errors, (c) number of items in the scale, (d) number of response categories in each item.

(a) *Range of marginal distributions:* The reproducibility of any individual item can never be less than the percentage of respondents falling into a single answer category of that item, regardless of whether or not a scale exists. For example, if a dichotomous item has 80 percent of the people in one category and 20 percent in the other, there cannot be less than 80 percent reproducibility in reproducing that item from a rank order obtained from all the items, regardless of the scalability of the set of items as a whole. Thus, if a sample comprises only items with extreme kinds of dichotomizations, reproducibility will be automatically high for that sample, regardless of the scalability of the universe. Therefore, to test a universe for scalability, attempts should be made to include in the sample as wide a range of marginal distributions as possible, and specifically to attempt to include items with marginals around 50–50.

(b) *Pattern of errors:* If an area is scalable with but 10 percent error (and not artificially so because of extreme marginals), this implies that there is but one dominant variable in the area along which to order the persons. The errors of reproducibility may be caused either (a) by one or two other variables of lesser magnitude that may be in the area, or (b) by many small variables.

The existence of one or two additional small variables as opposed to many small variables in the area is indicated by *nonscale types* in the scale pattern which occur with sufficient frequency to be noticed, but not with enough frequency to impair substantially the reproducibility of the area from only the dominant variable. If such definite nonscale types exist, then the multiple correlation of an outside variable with the whole area would *not* be quite equivalent to the simple correlation with rank order on the dominant variable, and would be attained only by taking the nonscale types into account.

On the other hand, if error of reproducibility is random, then the multiple correlation of any outside variable on the area will be precisely equal to the simple correlation with the rank order on the area. This property, it is important to note, holds *no matter how low the reproducibility is.* Some areas which are not scalable are called *quasi scales*; their reproducibility may not be high but their errors occur in a sort of gradient. This means that although they lack the essential property of a scale—rank order cannot reproduce persons' characteristics on the items in the area very well—nevertheless the rank order is perfectly efficient for relating any outside variable to the area.

(c) *Number of items:* The more items included in a scale, the greater is the assurance that the entire universe of which these items are a sample is scalable. If the items are dichotomous (or dichotomized from more than two categories as a result of the scale analysis), it is probably desirable that at least ten items be used, with perhaps a lesser number being satisfactory if the marginal frequencies of several items are in the range of 30 percent to 70 percent. Just four or five items, with marginal frequencies outside such a range, would not give much assurance as to how scalable the universe was, no matter how scalable the sample might be. In practice, ten or more items can be used on a pretest to determine whether or not a universe is scalable but fewer items can be used in

the larger study—if the universe is shown to be scalable by the pretest—
to obtain the number of ranks necessary for the amount of discrimination
between people required by the study.

(d) *Number of response categories:* The more response categories for
items included in a scale, the greater is the assurance that the entire universe is scalable. A caution to be observed in combining response categories to reduce error is to make sure that the reduction in error is not
just a consequence of obtaining new extreme marginal frequencies (e.g.,
90–10) that do not permit much error (see *a* above). Equally important
is the fact that keeping answer categories separate, while it will usually
increase the amount of error, decreases the possibility of a scale pattern
appearing if in fact the universe is nonscalable. For example, four dichotomous items with high reproducibility do not provide as dependable
an inference concerning the scalability of an area as would four trichotomous items which were equally as reproducible. It is especially important to keep as many response categories as possible when the total number
of items is small. The more categories that can remain uncombined, the
more reliable is the inference that the universe from which they come is
scalable.

In many cases, of course, sufficient reproducibility may exist without
combining categories. In such a case categories may be combined anyhow
for convenience in final scoring. Combining categories in such a case does
not disturb rank order except that two adjacent ranks are merged.

The Universe of Attributes[8]

A basic concept of the theory of scales is that of the universe of attributes. In social research, a universe is usually a large class of behavior
such as described in the introduction. The universe is the concept whose
scalability is being investigated, like marital adjustment, opinion of
British fighting ability, knowledge of arithmetic, etc. The universe consists of all the attributes that define the concept. Another way of describing the universe is to say it consists of all the attributes of interest to the
investigation which have a common content, so that they are classified
under a single heading which indicates that content.

An important consideration of the present theory of scales becomes
that of the sampling of items. In studying any attitude or opinion, there
is an unlimited number of questions or question wordings which could

---

8. The words *population* and *universe* are ordinarily used interchangeably in statistical literature. For scales, it is necessary to refer both to a complete set of objects
and to a complete set of attributes, so it will be convenient to reserve *population* for the
former, and *universe* for the latter. In social research, the objects are usually people, so
that *population* is appropriate for them.

be used. Any question asked in an attitude or opinion survey is ordinarily but a single sample of indefinitely many ways the question could be put. It is well known that changing the wording of the questions, changing the order of presentation of questions, changing order of check lists of answers, etc., can yield apparently different results in the responses. Conceivably, one could ask questions which would secure "favorable" replies ranging from 0 to 100 percent, depending upon the extremeness of the statement that the respondents are asked to approve or disapprove. It is, therefore, essential to inquire into the nature of the *universe of all possible questions of the same content,* and to determine what inferences can be made about that universe that will not depend on the particular sample of questions used.

Scalogram theory shows that if the universe contains but a single variable, that is, if all questions have but a single content ordering, then the same rank order of the individuals upon this content will be obtained regardless of which sample of questions is selected from the universe. The problem of sampling of items thus has a simple solution for the case of a scalable universe.

An important property of a scalable universe is that the ordering of persons based on a sample of items will be essentially the same as that based on the universe. If the universe is scalable, the addition of further items merely breaks up each type given by the sample into more differentiated types. But it would not interchange the order of the types already in the sample. For example, in Figure 13.8 above, type 5 would always have a higher rank order than type 4. People in type 5 might be ordered within the type into more subcategories; people within type 4 might be ordered into more subcategories; but all subcategories within 5 would remain of higher rank than all those in type 4. This may be seen in reverse, for example, by deleting one of the questions or by combining answer categories so as to make a trichotomous question dichotomous, and noticing that all that is accomplished is to collapse the number of types to a smaller number so that two neighboring types may now become indistinguishable; but any types two steps apart would still remain in the same order with respect to each other.

Hence, we are assured that if a person ranks higher than another person in a sample of items, he will rank higher in the universe of items. This is an important property of scales, that *from a sample of attributes we can draw inferences about the universe of attributes.*

One of the criteria for selecting a sample of items is to choose a sample with enough categories to provide a desired amount of differentiation between individuals.[9] Thus if individuals are desired to be differentiated,

9. The number of possible scale types may be determined as follows: add unity to the total number of categories in all questions (after combination) and subtract the number of questions.

say, into only ten groups, items should be chosen which will yield ten types.[10] The shape of the distribution of the rank orders in a sample of items will of course depend upon the marginal frequencies of the items selected. One sample of items may yield a distribution of one shape; another sample may yield a completely different shape. This need not be a matter of concern, since our primary interest lies in the ordering of people, not the relative frequency of each position.

It might be asked how one can know the universe forms a scale if all one knows is a sample from the universe.

At present it seems quite clear that in general the probability of finding a sample of items to form a scale by chance for a sample of individuals is quite negligible, even if there are as few as six dichotomous items in the sample and as few as one hundred individuals.[11] It seems quite safe to infer in general that if a sample of items is selected without knowledge of their empirical interrelationships and is found to form a scale for any sizable random sample of individuals, then the universe from which the items are selected is scalable for the entire population of individuals. This problem has already been discussed under "Measurement of Error."

THE RELATIVITY OF SCALES

A universe may form a scale for a population at a given time and may not at a later time. Such a change in time would tend to indicate that the change is one of *kind*, rather than *degree*. A new meaning has been added to the previous single variable. For example, the items in a scale of expression of desire of American soldiers to go back to school after the war may not prove to be scalable if they were asked once more at the close of the war.

Conversely, a universe may not be scalable at one time, but scalable at another. This would indicate a change in the structuralization of the attitude from many dimensions to one dimension, or that an "unstructured" attitude has become "structured."

A universe may form a scale for one population of individuals, but not for another. Or, the items may form scales for two populations in dif-

10. We are of course not considering problems of reliability in the sense of repeated observations of the same attributes. For convenience, we are tacitly assuming perfect reliability.

11. To work out the complete probability theory would require two things: first, a definition of a sampling process for selecting items, and, second, a definition of what is meant by a scale not existing. A definition of the sampling process is difficult because items are ordinarily developed intuitively. Stating a null hypothesis that a scale does not exist leads to many possible analytical formulations, for different limiting conditions may be imposed upon the multivariate distribution of the items. For example, should the marginal frequencies be considered fixed in all samples, should the bivariate frequencies be considered fixed, etc.? These are questions which may become clearer as the theory of scaling develops, and in return may clarify our conceptions of what observation of social phenomena implies.

ferent manners. For example, a sample of items of satisfaction with Army life which formed a scale for combat outfits in the Air Force did not form a scale for men in the technical schools of the Air Force. The structure of camp life for these two groups was too different for the same items to have the same meaning in both situations.

A universe may not form a scale for the total population, but still form a scale for subgroups of that population. The essential definition of a scale is that of "single-meaning," and while a series of questions may contain different meanings to a cross section of the population, they may contain only a single meaning for some subgroup of that population. However if a scale is obtained for a cross section of the population, then that same scale pattern necessarily holds for all major subgroups.

If a universe is scalable for one population but not for another population, we cannot compare the two populations in degree and say that one is higher or lower on the average than another with respect to the universe. They differ in more than one dimension, or in kind of attitude rather than in degree of "favorableness" on the same attitude. It is only if two groups or two individuals fall into the same scale that they can be ordered from higher to lower. A similar consideration holds for comparisons in time. An important contribution of the present theory of scaling is to bring out this emphasis quite sharply.

CONTENT ALONE DEFINES THE UNIVERSE

Before the structure of a universe can be analyzed, the universe itself must be defined. Let us take an example from opinion research where it is desired to observe the population of individuals in a standardized manner by a check list of questions. The behavior of interest to the investigation is responses of individuals to such questions. Suppose the universe of items consists of all possible questions which could be asked in such a list concerning the fighting ability of the British. Such questions might be: "Do you think the British Army is as tough as the German Army?" "Do you think the RAF is superior to the Luftwaffe?" etc. (We do not pause here for problems of wording, interpretation, and the like. The reader is urged rather to focus on the general outline we are trying to establish.) There may be an indefinitely large number of such questions which belong in the universe; and in a particular investigation, ordinarily only a sample of the universe is used.

An attribute belongs to the universe by virtue of its content. The investigator indicates the content of interest by the title he chooses for the universe, and all attributes with that content belong in the universe. There will, of course, arise borderline cases in practice where it will be hard to decide whether or not an item belongs in the universe. The evaluation of the content thus far remains a matter that may be decided by consensus of judges or by some other means. It may well be that the formal analysis for scalability may help clarify uncertain areas of content. However, we have found it most useful at present to utilize informal experience and consensus to the fullest extent in defining the universe.

It should be pointed out that an area of qualitative data which has been carefully thought through and judged to comprise a homogeneous universe of content does not necessarily form a scale. The concepts of universe and of scale are distinct and separate. If a universe is not a scale, it cannot be represented by a single rank order. In some cases, scale analysis may suggest that there are two or more subareas in the universe which might be scalable separately. Then the universe could be represented by *several* scale variables, by giving each person a rank order on each of the scalable subareas. It may happen that a sample of items is analyzed for a group of people and is not found to be scalable; but one or more subsets of the items seem to be scalable separately. Finding scalable subsets of items may sometimes imply that the original universe of content can be divided into subuniverses, at least one—or perhaps all—of which are scalable separately. To test the hypothesis that a scalable subset is part of a scalable subuniverse, it is necessary to show that the content of this subuniverse is ascertainable by inspection, and is distinguishable by inspection from that of the rest of the universe. A practical procedure to test this hypothesis might be as follows: construct new items of two types of content, one type which should belong in the original universe but should not belong to the scalable subuniverse. If the new items designed for the apparently scalable subuniverse do scale, and scale together with the old subset; and if the new items designed not to be in this subuniverse do not scale with the subscale; then the hypothesis is sustained that a subuniverse has been defined and has been found scalable. Each hypothesized subuniverse might be tested this way.

It often happens, when about a dozen questions in an area are pretested, that all but one or two are found to be scalable. There are at least two possibilities as to why some items do not scale while others in the same area do: (a) the universe is not scalable as a whole, but contains a scalable subuniverse; (b) the items have been imperfectly constructed. This latter reason is so easy and glib that it is best to avoid it as much as possible. If the vast majority of a sample of items do scale, then it may be plausible to blame faulty construction for the nonscalability of one or two items. This hypothesis can be tested by rebuilding the apparently faulty items and retesting them. In practice, if the vast majority of a sample of items do scale, the one or two items that do not may often be ignored. If their construction is faulty, there are sufficient items without them to establish rank order for the people; and if they really represent a nonscalable part of the original universe, this part may be so small as not to be essential to the study.

SCALE ANALYSIS DOES NOT DEFINE CONTENT

The question may be asked: to what extent, if any, can scale analysis be used to arbitrate differences of opinion with respect to content, that is, as to whether or not a given item belongs in an area? The answer is quite simple. Scale analysis as such gives no judgment on content;[12] *it presumes that the universe of content is already defined,* and merely tests whether or not the area is representable by a single variable. It might serve as an *auxiliary* argument with respect to content in the special case where there is controversy over but one or two items of a large sample of items in which the remaining items are scalable. In such a case, if the items in doubt do scale with the others, this may be taken as additional evidence supporting the contention that they belong in the area. It should be emphasized that this kind of inference is but *auxiliary*—there must be a cogent initial argument based on content if an item is to be classified in an area. Sheer scalability is not sufficient; an item may happen to scale with an area, and yet not have the content defining the area—it may be a correlate rather than part of the definition.

An important emphasis of our present approach is that a criterion for an attribute to belong in the universe is *not* the magnitude of the correlations of that item with other attributes known to belong in the universe. Attributes of the same type of content may have any size of intercorrelations, varying from practically zero to unity.[13]

METHODS OF OBSERVATION

Let us assume that somehow we have a universe of attributes and a population of individuals defined. Next, observations are made as to the behavior of the population with respect to the universe. (In practice this will often be done only with samples. A sample of individuals from the population will have their behavior observed on a sample of attributes from the universe.) How the observations are to be made is of no theoretical concern here. In opinion research and other fields, questionnaires and schedules have been used. But any technique of observation which yields the data of interest to the investigation may be used. Such

12. More generally, *no* correlation analysis as such determines content; it studies only formal relationships. If x relates perfectly with y, that does not mean x is identical with y. If it is known that the correlation between x and y is .6, that alone does not help to name the content of either x or y.

13. That correlations are no criterion for content has been quite well known. See, for example, R. F. Sletto, *Construction of Personality Scales by the Criterion of Internal Consistency* (Sociological Press, Hanover, N.H., 1937).

techniques for the social and psychological sciences might be case histories, interviews, introspection, and any other technique from which observations may be recorded. The important thing is not how the observations were obtained, but that the observations be of central interest to the investigation.

Use of a questionnaire implies that the investigator is interested in a certain type of universe of verbal behavior. Participant observation may imply that the investigator is interested in a certain type of nonverbal behavior. Such distinct universes may each be investigated separately. It may often be of interest to see how well one universe correlates with another, but such a correlation cannot be investigated until each universe is defined and observed in its own right.

A good deal of attention has been given by various research workers to the types of observations recorded. Thurstone has approached the problem of attitude measurement by asking essentially, "What are your opinions?"[14] Other research workers have attempted to get at attitudes by asking different questions. For example, Bogardus asks, "What social relations are you willing to tolerate with these people?"[15] while Pace asks, "What would you do under these circumstances?"[16] Ford studied an attitude area by asking about past behavior and concludes, "One may say that the scale for estimating . . . experiences are really attitude scales in disguise."[17] And so in the literature of attitude analysis we find much space devoted to the construction of areas by means of different types of items and check lists. All of these different question forms represent only differences in techniques of observing the attitude universe. There is no a priori reason to assume that if the attitude universe is, say, attitude toward the British, any one way of securing observations of this universe is better than any other. In fact, if the universe is scalable, all types of questions may be used indiscriminately with the same result as far as the rank order of individuals in that universe is concerned. If, however, one series of items (for example, questions about past experiences with the British) does not scale with another series of items concerning the British (for example, behavior in hypothetical situations), then these two series may be regarded as two scalable subuniverses of a larger but nonscalable universe.

14. L. L. Thurstone and E. J. Chave, *The Measurement of Attitude* (University of Chicago Press, Chicago, 1929).

15. See an application of this in Stuart C. Dodd, "A Social Distance Test in the Near East," *American Journal of Sociology*, Vol. 41, No. 2 (September 1935), p. 195.

16. C. R. Pace, "A Situations Test to Measure Social-Political-Economic Attitudes," *Journal of Social Psychology*, Vol. 10, No. 3 (August 1939), pp. 331–344.

17. R. N. Ford, "Scaling Experience by a Multiple-Response Technique: A Study of White-Negro Contacts," *American Sociological Review*, Vol. 6, No. 1 (February 1941), p. 21.

OBSERVATIONS MADE IN PUBLIC OPINION POLLS

It is with regard to the manner of observation that public opinion polls have, perhaps, been most often differentiated from attitude scales. A question concerning public opinion, it is argued, is more specific and controversial than an attitude question. This specificity has led public opinion pollsters to place a great deal of stress upon the "unbiased" construction of a single meaningful question. Since the validity of the poll question depends upon its manifest content, a great deal of concern is shown over the "meaning" of the single question. It is felt that practically every word must be tested for understandability, bias, single meaning, etc., while check-list answer categories must be presented in reverse form or combined in different ways, etc. Actually, it is this susceptibility of the single poll question to slight changes in question wording that reinforces the position of the present approach that *attitude questions and opinion questions represent simply different forms of observation.*

Any single opinion question is only a sample of one from the whole universe of questions in the opinion area. With regard to differences in "specificity" between attitudes and opinions, it is our position that any area of behavior can usually be regarded as a subuniverse of some larger area, and can itself often be divided further into subareas. In other words, there is no means of determining when an attitude is specific enough to be called an opinion. Our definition of both involves the determination of verbal or nonverbal behavior in some area, which may be relatively specific or relatively general. In either case, of attitudes or of opinions, validity will depend upon the adequate sampling of the entire universe of questions in the area of interest. Scale analysis provides an objective test of whether or not any particular opinion poll question contains but a single dimension of meaning that is common to all similar questions that could have been asked in its place.

The examples of scales to be given later happen to comprise observations made by means of questionnaires. It should not be inferred, however, that scaling refers only to that technique. *Scale analysis is a formal analysis, and hence applies to any universe of qualitative data of any science, obtained by any manner of observation.*

Summary

1. The need for scale analysis arises out of the fundamental problem of attitude scaling and opinion polling of how to determine the dimensions of meaning which the questions asked have for the respondents. Scale analysis affords a rigorous test for the existence of single-meaning

for an area and provides a rank order of individuals for such areas as are found scalable.

2. Basic to the present scale theory is the concept of sampling the attitude or opinion universe. An unlimited number of questions could be asked in any area; the problem is one of selecting a sample of questions which are representative of all possible questions that might have been asked. Scale analysis affords a test of this sampling.

3. Scale analysis applies equally well to the study of attitudes as to the study of opinions. More generally, any technique of observation—questionnaires, interviews, participant observation, etc.—of any form of verbal or nonverbal behavior which is qualitative, yields data which may be studied by scale analysis.

4. Scale analysis tests the hypothesis that a group of people can be arranged in an internally meaningful rank order with respect to an area of qualitative data. A rank order of people is meaningful if, from the person's rank order, one knows precisely his responses to each of the questions or acts included in the scale.

5. More precisely, the multivariate frequency distribution of a universe of attributes or items for a population of objects or people is a scale if it is possible to derive from the distribution a quantitative variable with which to characterize the objects such that each attribute in the universe is a simple function of that quantitative variable.

6. The scalogram board technique for determining the existence of a scale involves two basic steps: (1) the questions and answer categories are ranked in a preliminary order of extermeness with the "most extreme" category, i.e., the one which is endorsed by fewest people, placed first and the other categories following in decreasing order of "extremeness," and (2) the people are ranked in order of "favorableness" with the "most favorable" persons, i.e., those who answer all questions "favorably," placed first and the other individuals following in decreasing order of "favorableness." The resulting pattern, if a scale is present, will be a parallelogram.

7. There is an unambiguous meaning to the order of scale scores and the order of categories within each item. An object with a higher score than another object is characterized by higher, or at least equivalent, values on each attribute. Similarly, one category of an item is higher than another if it characterizes persons all of whom are higher on the scale.

8. From the multivariate distribution of a sample of attibutes for a sample of objects, inferences can be drawn concerning the complete distribution of the universe for the population.

a. The hypothesis that the complete distribution is scalable can be adequately tested with a sample distribution.

b. The rank order among objects according to a sample scale is essentially that in the complete scale.

9. Perfect scales are not found in practice.

a. The degree of approximation to perfection is measured by a *coefficient of reproducibility,* which is the empirical relative frequency with which values of the attributes do correspond to intervals of a scale variable.

b. In practice, 90 percent perfect scales or better have been used as efficient approximations to perfect scales.

10. The predictability of any outside variable from the scale scores is the same as the predictability from the multivariate distribution with the attributes. The zero-order correlation with the scale score is equivalent to the multiple correlation with the universe. Hence, *scale scores provide an invariant quantification of the attributes for predicting any outside variable whatsoever.*

11. Scales are relative to time and to populations.

a. For a given population of objects, a universe may be scalable at one time but not at another.

b. A universe may be scalable for one population but not for another, or it may not be scalable for an entire population, but scalable for a subpopulation. However, if a universe is scalable for an entire population, it will be scalable for all major subpopulations.

c. Comparisons with respect to degree can be made only if the same scaling obtains in both cases being compared.

References

1. CLARK, K. E., and KRIEDT, P. H., "An Application of Guttman's New Scaling Techniques to an Attitude Questionnaire," *Educational and Psychological Measurement,* Vol. 8, No. 2 (Summer 1948), pp. 215–223.

2. EDWARDS, A. L., and KILPATRICK, F. P., "Scale Analysis and the Measurement of Social Attitudes," *Psychometrika,* Vol. 13, No. 2 (June 1948), pp. 99–114.

3. EDWARDS, A. L., and KILPATRICK, F. P., "A Technique for the Construction of Attitude Scales," *Journal of Applied Psychology,* Vol. 32, No. 4 (August 1948), pp. 374–384.

4. FESTINGER, L., "The Treatment of Qualitative Data by 'Scale Analysis,'" *Psychological Bulletin,* Vol. 44, No. 2 (March 1947), pp. 149–161.

5. GOODENOUGH, W. H., "A Technique for Scale Analysis," *Educational and Psychological Measurement,* Vol. 4, No. 3 (1944), pp. 179–190.

6. GUTTMAN, L., "The Quantification of a Class of Attributes: A Theory and Method of Scale Construction," in P. Horst et al., *The Prediction of Personal Adjustment* (Social Science Research Council, New York, 1941), pp. 319–348.

7. GUTTMAN, L., "A Basis for Scaling Qualitative Data," *American Sociological Review*, Vol. 9, No. 2 (April 1944), pp. 139–150.

8. GUTTMAN, L., "The Cornell Technique for Scale and Intensity Analysis," *Educational and Psychological Measurement*, Vol. 7, No. 2 (Summer 1947), pp. 247–280.

9. GUTTMAN, L., "Suggestions for Further Research in Scale and Intensity Analysis of Attitudes and Opinions," *International Journal of Opinion and Attitude Research*, Vol. 1, No. 1 (March 1947), pp. 30–35.

10. GUTTMAN, L., and SUCHMAN, E. A., "Intensity and a Zero Point for Attitude Analysis," *American Sociological Review*, Vol. 12, No. 1 (February 1947), pp. 57–67.

11. GUTTMAN, L., "On Festinger's Evaluation of Scale Analysis," *Psychological Bulletin*, Vol. 44, No. 5 (September 1947), pp. 451–465.

12. NOLAND, E. W., "Worker Attitude and Industrial Absenteeism: A Statistical Appraisal," *American Sociological Review*, Vol. 10, No. 4 (August 1945), pp. 503–510.

13. SHAPIRO, G., "Myrdal's Definitions of the 'South': A Methodological Note," *American Sociological Review*, Vol. 13, No. 5 (October 1948), pp. 619–621.

14. SUCHMAN, E. A. and GUTTMAN, L., "A Solution to the Problem of Question 'Bias,'" *Public Opinion Quarterly*, Vol. 11, No. 3 (Fall 1947), pp. 445–455.

Chapter 14

MEASUREMENT OF REPRODUCIBILITY

Benjamin W. White
San Francisco State College
and Eli Saltz
Wayne State University

Much of our knowledge of human behavior is based upon data obtained through the administration of multiple-choice tests to groups of subjects. Such instruments are used in many ways: selection, attitude measurement, ability measurement, and clinical diagnosis, to name only a few. Particularly since the publication of Guttman's model for measuring a test's reproducibility (7), there has been increasing concern over one aspect of the responses of groups of subjects to groups of items—the extent to which the patterns of subjects' responses can be predicted from their total scores. While these considerations have been of great interest to social and clinical psychologists, they have also proved pertinent to constructors of ability tests. It is the purpose of this article (*a*) to examine some of the techniques which have been devised to assess a test's "reproducibility," "homogeneity," or internal consistency, (*b*) to evaluate these techniques against certain criteria, and (*c*) to suggest possible logical relationships of these techniques to the concept of reliability.

Reprinted with permission of the author and the publisher from the *Psychological Bulletin*, Vol. 54, No. 2, 1957, pp. 81–99. Copyright 1957 by the American Psychological Association.

In the ensuing discussion the word *test* will be used to describe any technique whereby two or more subjects respond to two or more stimuli in such a way that the responses of all subjects to each item can be dichotomized. It is assumed that every subject responds to every such item. It is further assumed that the experimenter assigns a value of unity to all responses on one side of the dichotomy and a value of zero to the rest. A "total score" for a subject is computed by adding the weights assigned to his responses thus dichotomized. With this system, a subject's total score is the number of responses he has made which fall into the unity-weighted class.

Often such scores are presumed to yield an ordering of the subjects on some hypothetical linear continuum, ability, or trait. For some time social scientists have been aware that this process of assigning a simple order to people on the basis of their responses to a number of test items is a legitimate representation of their test behavior only when their responses possess certain characteristics. There are many ways of stating this, but for the purposes of this discussion, it will be most convenient to use the following: a total score, computed by counting the number of test responses which have been classified in one of two ways, will yield a perfect mapping of the entire pattern of responses of all subjects when, and only when, the interitem covariances are maximal.

For purposes of illustration, consider a six-item test. On such a test, total scores can take seven possible values from 0 to 6. When interitem covariance is maximal, there is only one way in which a subject can make any given total score. Naturally he can make a total score of 0 only by "failing" all six items, and a score of 6 only by "passing" all six items. He can make a score of 1 only by passing the easiest item. By "easiest" is meant the item which was passed by more subjects than any other. Similarly he can make a score of 2 only by passing the two easiest items. In other words, given the information that the interitem covariances are maximal, the order of difficulty of the items, and a subject's total score, one can tell exactly which items the subject got wrong and right. On such a test there are only seven ways in which people respond to the items, and each of these corresponds with one of the seven possible total scores.

At the other extreme, consider a six-item test whose items are independent, i.e., exhibit zero covariances. Such a test could yield 2^6 or 64 different response patterns. There would be 15 different ways in which a person could get a total score of 2, for example. In this situation, given knowledge of the total score, the order of difficulty of the items, and the fact of zero covariance between items, one would not be able to reconstruct a subject's pattern of responses to the test, unless the total score happened to be 0 or 6. Representation of the test behavior of the subjects with the conventional total score would result in a considerable loss of information.

Various indices have been developed which will permit the tester to ascertain the degree to which the total scores of a given test yield a complete mapping of the responses of all subjects to all the items (reproducibility). These indices differ not only in their computational formulas, but in their underlying assumptions, though all start with the same primary data: the dichotomized responses of a group of subjects to a group of test items. Four criteria are suggested against which each index may be evaluated.

1. *Does it yield a theoretical maximum value which is the same for any test?*

2. *Does it yield a theoretical minimum value which is the same for any test?*

3. *Does it permit evaluation of the null hypothesis that the obtained reproducibility index is not significantly different from chance?*

4. *Does it permit evaluation of each item in the test as well as of the test as a whole?*

The rationales for these criteria are reasonably straightforward. If maximum or minimum possible values differ from test to test, it is difficult to evaluate one test against another. For example, two tests having reproducibility quotients of .90 are differently evaluated when it is discovered that the minimum theoretical reproducibility of one is .60, and of the other .90. If the quotient does not have a known sampling distribution, there is the possibility that the obtained quotient does not differ significantly from chance. And finally, if the items can not be evaluated it is difficult to improve reproducibility by omission or inclusion of specific items.

In the light of these criteria, we propose to discuss several techniques which have been devised to yield an index of reproducibility. In order to demonstrate the computations involved in each technique, we shall use the responses of ten subjects to a six-item test, illustrated in Table 14.1.

In this matrix the rows represent subjects and the columns test items. The marginal entries at the bottom of the matrix indicate the number of subjects who "passed" a given item, and the marginals in the last column of the matrix represent the number of items each subject "passed."

Guttman's Reproducibility

Guttman (7) originated the term reproducibility. The term means essentially the degree to which one can reproduce a subject's entire response

Table 14.1. Responses of Ten Subjects to a Six-Item Test Where Rows and Columns Are Unordered

Subject	Item 1	2	3	4	5	6	Total Score
A	0	0	0	1	0	0	1
B	1	0	1	1	1	1	5
C	0	0	1	0	0	0	1
D	1	1	1	0	1	1	5
E	1	0	1	1	1	0	4
F	0	0	1	0	1	0	2
G	0	1	1	0	0	0	2
H	1	0	1	0	1	0	3
I	1	0	1	0	1	1	4
J	1	0	0	1	1	0	3
Item Difficulty	6	2	8	4	7	3	

pattern from a knowledge of his total score and the order of difficulty of the items. Originally Guttman's technique of obtaining the index of reproducibility involved mechanical operations on a matrix of N subjects and K test items similar to Table 14.1. A device, the scalogram board, permits interchange of rows and columns of this matrix in a particular manner so that the unity entries are maximally concentrated above the main diagonal of the matrix. Such rearrangement of the response matrix in Table 14.1 is shown in Table 14.2.

Table 14.2. Responses of Ten Subjects to a Six-Item Test Where Rows and Columns Are Ordered

Subject	Item 2	6	4	1	5	3	Total Score
D	1	1	0	1	1	1	5
E	0	1	1	1	1	1	5
B	0	0	1	1	1	1	4
I	0	1	0	1	1	1	4
H	0	0	0	1	1	1	3
J	0	0	1	1	1	0	3
G	1	0	0	0	0	1	2
F	0	0	0	0	1	1	2
C	0	0	0	0	0	1	1
A	0	0	1	0	0	0	1
Item Difficulty	2	3	4	6	7	8	

It should be noted that if there are any ties in total score or in the number of subjects passing items, the arrangement of the matrix may not be unique. In this example the order of the columns is unique since there are no ties in the number of subjects passing items, but the order of rows is not, since there are two subjects at each total score level. In such cases further permutations of rows and columns are made until errors are minimized. The index of reproducibility is a function of the number of errors, i.e., unity entries which are below the diagonal and zero entries which are above it. This diagonal is not necessarily exactly coincident with the main diagonal, and Guttman has several rules to be followed in its determination. Since Guttman's original procedure is unwieldly, we shall in this illustration use a procedure developed by Jackson (10) for arriving at cutting points for each item. For all practical purposes, Jackson's R_t quotient is identical with Guttman's.[1] Jackson's method is illustrated in Table 14.3.

This is the same matrix shown in Table 14.2, except that the unity and zero entries under each item have been placed in separate columns. In order to draw cutting points, one simply draws a line across each column at the place where the number of zero entries above the line and the number of unit entries below the line (errors) are minimized. These cutting points are seen as descending steps in the table. In the first column there is one entry of unity which falls below the cutting line and this has been put in parentheses. If the cutting line had been drawn directly below this unity entry, the five zero entries above it would be counted as errors and put in parentheses. In this illustration there is a unique cutting point for five of the six items, i.e., a line which yields an absolute minimum number of errors. In Item 5 however, the line could be either where it is drawn, or two rows higher. Either solution yields 1 error. The lower one was chosen because it yielded an additional cutting point for the scale,[2] whereas the higher cutting point would have been identical for that of Item 1.

1. It should be noted that many people have suggested modifications in the calculations of Guttman's R_t (3, 6, 11, 17, 18). These refinements of procedure are, by and large, identical in their logical properties with Guttman's quotient, and were so intended by their authors. Consequently no space is given to them in this article.

2. In Jackson's method, the cutting points are used to determine minimum number of errors. Once the minimum number of errors has been determined the exact locations of the cutting points no longer enter into the computation of reproducibility. Consequently, for Jackson's method it doesn't matter which of the two cutting points is used for Item 5, since both result in one error. However, Guttman's original procedure made use of the specific cutting point used. Guttman assigned the cutting points to the row marginals (the total scores) and then rescored every S on the basis of the cutting points. All Ss below the lowest cutting point would be scored as having failed all the items. In Table 14.3, for example, Item 3 has the lowest cutting point; subject A is below this cutting point and so he would be rescored as having failed all the items.

Table 14.3. Jackson's Method of Computing Reproducibility (R), Minimum Reproducibility (MR), and Plus Percentage Ratio (PPR)

Subject	2 +	2 −	6 +	6 −	4 +	4 −	1 +	1 −	5 +	5 −	3 +	3 −
						Item						
D	1		1			(0)	1		1		1	
E		0	1		1		1		1		1	
B		0		0	1		1		1		1	
I		0	(1)			0	1		1		1	
H		0		0		0	1		1		1	
J		0		0	(1)		1		1			(0)
G	(1)			0		0		0	(0)		1	
F		0		0		0		0	1		1	
C		0		0		0		0		0	1	
A		0		0	(1)			0		0		0
# right (*P*)	2		3		4		6		7		8	
# wrong (*Q*)		8		7		6		4		3		2
Errors	1		1		3		0		1		1	
R_i	.90		.90		.70		1.00		.90		.90	
MR_i	.80		.70		.60		.60		.70		.80	
PP_i	.10		.20		.10		.40		.20		.10	
PPR_i	.50		.67		.25		1.00		.67		.50	

Note.—Rights are listed under +. Wrongs are listed under −.
Total errors = 7; $R_t = 88\%$; $MR_t = 70\%$; $PP_t = 18\%$; $PPR_t = .61$.

After the cutting points have been assigned, the errors in each column are counted. From these it is possible to compute the reproducibility for each item (R_i) by dividing the number of errors (E) by the number of subjects (N) and subtracting the quotient from 1.

$$R_i = 1 - \frac{E}{N}. \qquad [1]$$

The reproducibility for the entire test (R_t) may be computed by summing the errors for all items

$$\left(\sum_{i=1}^{k} E \right)$$

All *S*s between the first and second cutting points would be rescored as having passed one item. And so forth. The Guttman reproducibility index indicates the percentage of actual reproducibility as compared with the maximum reproducibility obtained by these rescoring processes. Therefore, if two items are given the same cutting point, the number of different classes or "cutting point scores" will be decreased—that is, the number of discriminations made by the scale is diminished.

dividing this by the number of subjects (n) times the number of items (k), and subtracting the quotient from 1.

$$R_t = 1 - \frac{\sum\limits_{i=1}^{k} E}{NK}. \qquad [2]$$

For this example, the reproducibility of the test is 88.3 percent, somewhat below the 90 percent figure which Guttman uses as one criterion of scalability.

The Guttman index of reproducibility meets our first criterion in that it has an absolute maximum of 100 percent for any test with more than one item, and our fourth criterion in that one can compute the index for each item as well as for the test as a whole. However, it suffers a serious shortcoming in having no unique minimal value. As Jackson (10) and others (1, 2, 13, 14) have pointed out, the index of reproducibility is drastically affected by the difficulty levels of the items in a test. The reason for this is that the difficulty of an item (percentage of persons passing) places a limit on the likelihood of an error: passing a difficult item, and failing an easy one. The reproducibility figure can approach its absolute lower limit of 50 percent only when all the items have a difficulty level of 50 percent, a trivial case in which 100 percent reproducibility could be obtained only if one-half the subjects passed all the items while the other half failed all the items. With even slight departures from this strict condition, the lower limit of the reproducibility index rises sharply. In our illustrative example minimum reproducibility is 70 percent. This fact makes it exceedingly difficult to evaluate an obtained index of reproducibility. With short scales and wide spread in item difficulties, Guttman's figure of 90 percent may on occasion be very little higher than the minimum reproducibility of the scale.

Jackson's Plus Percentage Ratio (PPR)

In order to circumvent this drawback of Guttman's reproducibility index, Jackson (10) has developed another statistic which he calls the Plus Percentage Ratio (PPR). Unlike the Guttman index, PPR has the same absolute minimum for all tests. Referring again to Table 14.3, note the minimum reproducibility figures in the row labelled MR_i. Here the minimum reproducibility figure for each item (MR_i) is obtained by dividing the number of subjects who got a given item right (# right), or wrong (# wrong), whichever figure is the larger, by the number of subjects (N).

$$MR_i = \frac{\text{\# rights or \# wrongs (whichever is larger)}}{N}. \qquad [3]$$

The minimum reproducibility for the entire test (MR_t) is computed by taking for each item the number of rights

$$\left(\sum_{i=1}^{k} \text{\# rights} \right)$$

or the number of wrongs

$$\left(\sum_{i=1}^{k} \text{\# wrongs} \right),$$

whichever number is larger, summing the numbers so obtained over all items and dividing this sum by the product of the number of items (K) and the number of subjects (N).

$$MR_t = \frac{\sum_{i=1}^{k} \text{\# rights, or \# wrongs (whichever is larger)}}{KN}. \qquad [4]$$

In the next to the last row of Table 14.3, the "Plus %" (PP_i) figures are listed. Here the differences between the obtained reproducibility and the minimum reproducibility ($R_i - MR_i$) for each item are entered. These figures indicate how much better obtained reproducibility is than the minimum for that item. In the last row, the "Plus % Ratios" (PPR_i) are entered for each item. These figures may be obtained by dividing the Plus % figure for a given item by one minus the minimum reproducibility (MR_i) for that item.

$$PPR_i = \frac{R_i - MR_i}{1 - MR_i}. \qquad [5]$$

The Plus Percentage Ratio for the total test (PPR_t) is similarly computed by dividing the difference between R_t and MR_t by one minus MR_t.

$$PPR_t = \frac{R_t - MR_t}{1 - MR_t}. \qquad [6]$$

The Plus % Ratio has a distinct advantage over the index of reproducibility in that it has both an absolute maximum of one and an absolute minimum of zero for any test of more than one item. For the test illustrated here the PPR_t is .61. As Jackson points out, testers should be prepared for the fact that this index will almost inevitably be lower than the Guttman index of reproducibility, often considerably lower. The index has not often been used on well-known tests, so it is difficult to say what an acceptable level should be. Jackson tentatively suggests 70 percent. It remains to be seen whether this figure is a reasonable one in mental testing or attitude scaling. The *PPR* in any event has much to recommend it since it circumvents one of the most serious criticisms which has been leveled at Guttman's reproducibility index.

Loevinger's Index of Homogeneity (H)

HOMOGENEITY OF A TEST (H_t)

A rather different approach to the measurement of the reproducibility of mental tests has been put forth by Loevinger, who uses the following as a definition of homogeneity (13, p. 29).

> The definitions of perfectly homogeneous and perfectly heterogeneous tests can be restated in terms of probability. In a perfectly homogeneous test, when the items are arranged in the order of increasing difficulty, if any item is known to be passed, the probability is unity of passing all previous items. In a perfectly heterogeneous test, the probability of an individual passing a given item A is the same whether or not he is known already to have passed another item B.

It can be seen that this definition comes quite close to the Guttman notion of reproducibility, and in fact the perfectly reproducible and the perfectly homogeneous test are identical.

With the test items arranged in order of increasing difficulty, Loevinger computes the quantity S by finding, for *all pairs of items*, the proportion of subjects who have passed both items (P_{ij}). From this is subtracted the theoretical proportion who would have passed both items had they been independent ($P_i P_j$). These differences are summed over the $k(k - 1)$ pairs of items (i.e., each item is paired with every other item in the test).

$$S = \sum_{i=1}^{k-1} \sum_{j=i+1}^{k} P_{ij} - P_i P_j. \tag{7}$$

For a test made up of completely independent items, S would have a value of zero. S does not have an upper limit of unity when the test is perfectly homogeneous. The upper limit is fixed by the proportion of subjects passing the more difficult item in each pair (P_j).

$$S_{max} = \sum_{i=1}^{k-1} \sum_{j=i+1}^{k} P_j - P_i P_j. \tag{8}$$

The homogeneity of a test (H_t) is then given by the ratio of these two quantities

$$H_t = \frac{S}{S_{max}} . \qquad [9]$$

This procedure is exactly analogous to that used by Jackson in computing the Plus Percentage Ratio. This can be seen more easily if Loevinger's equation is rewritten as follows:

$$H_t = \frac{S}{S_{max}} = \frac{\displaystyle\sum_{i=1}^{k-1} \sum_{j=i+1}^{k} (1 - P_{ij}) - (1 - P_i P_j)}{\displaystyle\sum_{i=1}^{k-1} \sum_{j=i+1}^{k} 1 - (1 - P_i P_j)} . \qquad [10]$$

The first term in the parentheses of the numerator $(1 - P_{ij})$ indicates the proportion of subjects passing a harder item *and* failing an easier one subtracted from unity. This is very like the reproducibility coefficient which is given by the proportion of errors subtracted from unity. The second term in the numerator $(1 - P_i P_j)$ is the product of the proportion of subjects passing the harder item and the proportion failing the easy item, this product then subtracted from unity. The quantity $(1 - P_i P_j)$ is analogous to Jackson's minimum reproducibility. The denominator is seen to be the difference between unity (perfect reproducibility) and minimum reproducibility. The two methods differ only in the procedure for counting errors. Loevinger's technique involves the equivalent of an examination of all pairs of items $i \neq j$ and counting every occasion upon which the harder item is passed and the easier item failed. In the illustrative example, such a tabulation yields a total of 13 errors, whereas Jackson's error count is 7. This is the reason that Loevinger's H_t will usually be lower than Jackson's PPR_t. The former is .23, and the latter .61. In Jackson's system for counting errors, a deviant response is counted only once no matter where it occurs in the response pattern. For example, if items are arranged in order of decreasing difficulty, a response pattern of (1, 0, 0, 0) would be credited with one error, while in Loevinger's system, since the passed item was the hardest of the four, there would be three errors. The two methods also have somewhat different ways of computing minimum reproducibility, Jackson's yielding a figure of .70, and Loevinger's .72.

Loevinger points out that her formula for H_t is equivalent to

$$H = \frac{\sigma^2_x - \sigma^2_{het}}{\sigma^2_{hom} - \sigma^2_{het}} , \qquad [11]$$

where all the variances refer to total raw scores. The first term in the numerator (σ^2_x) is the variance of the obtained scores, the second numerator term (σ^2_{het}) is the variance of the total scores which would be obtained from items of the same difficulties which were completely independent, and the first term in the denominator (σ^2_{hom}) is the variance in total scores which would be obtained if the same items were perfectly correlated. The raw score variance of a test made up entirely of independent items is the familiar

$$\sum_{i=1}^{k} pq,$$

or the sum of the item variances. The raw score variance of a test made up wholly of perfectly correlated items is given by

$$\sigma^2_{hom} = \sum_{i=1}^{k} P_i Q_i + 2 \sum_{i=1}^{k-1} \sum_{j=i+1}^{k} P_j - P_i P_j. \qquad [12]$$

The first term on the right of this equation is the item variance employed above, and the second term is two times a sum which is seen to be identical to S_{max}.

This relationship is interesting since it shows that total score variance increases with reproducibility, being at a minimum when the item covariances are zero, and reaching an upper limit when item covariances are maximal.

Both Loevinger's H_t and Jackson's PPR have the advantage of being uninfluenced by the distribution of item difficulties which makes them preferable to the Guttman reproducibility index when it is given without further information. The procedures are objective and can be reduced to routine computations. When "errors" occur mainly on item pairs which are close together in difficulty level, the two procedures should yield practically identical indices, but if there are "errors" which occur in item pairs which are widely different in difficulty level, Loevinger's H_t will be lower than Jackson's PPR_t. Loevinger's technique has the aesthetic advantage of making full use of the information contained in the response matrix, but the practical drawback of being tedious to compute when the number of items is large since $k(k - 1)$ cross breaks have to be made to compute the P_{ij}s. However, Jackson's method is also laborious since it requires an initial posting of the entire response matrix.

The sampling distribution of H_t is unknown, and Loevinger advises that it should not be used as an estimate of homogeneity unless the sample of subjects exceeds 100.

HOMOGENEITY OF AN ITEM WITH A TEST (H_{it})

Loevinger's H_t yields an index for the test as a whole, but does not provide an index of the homogeneity of each item with the test. For this purpose, she suggests another index, (H_{it}), the logic of which is the same as that employed in H_t. In a perfectly homogeneous test, subjects passing a given item should have higher total scores than those failing the item. The starting point is a formula developed by Long (15).

$$\text{Long's Index} = 1 - \frac{2 \sum \text{``passes'' below ``fails''}}{PQ} . \quad [13]$$

In 13, the numerator is two times the number of subjects passing a given item who have total scores lower than those of subjects who failed the same item, and the denominator is the product of the number of passes on the item (P) and the number of fails (Q). Loevinger points out that difficulties arise when two subjects have identical total scores, one of whom has failed the item and the other has passed it. There is also the question of whether the response to the item should in this computation be included in the total score. In order to circumvent these difficulties with Long's index, Loevinger proposes the modification

$$H_{it} = 1 - \frac{2 \sum \text{``passes'' below or tied with ``fails''}}{PQ - \sum \text{``passes'' one above ``fails''}} . \quad [14]$$

It is clear that this index can take values from minus to plus unity, but it is not clear that a zero value is obtained when there is no relation between an item and the total test. The sampling properties of the index are unknown and will have to be investigated to establish the value to be expected for a chance relation. The obtained H_{it} values for the illustrative test may be seen in the last column of Table 14.6.

Green's Summary Statistics Method (I)

Green (4, 5) has recently developed a method for computing an index of consistency for a test (I) which has all the advantages of Jackson's PPR, and Loevinger's H_t, plus greater ease of computation. Like Jackson's PPR, I is given by

$$I = \frac{Rep - Rep_{ind}}{1.00 - Rep_{ind}} , \quad [15]$$

where Rep is the obtained reproducibility of the test, Rep_{ind} is the reproducibility which would be obtained with the same set of item difficulties and complete independence between items, and 1.00 is perfect reproducibility.

Green's method of computing errors is the same as that employed in 10 above, except that the summation is not over all pairs of items, $i \neq j$, but only over those item pairs whose members are adjacent in difficulty level. Green's reproducibility is given by

$$Rep = 1 - \frac{1}{NK} \sum_{i=1}^{k-1} n_{i, i+1} - \frac{1}{NK} \sum_{i=2}^{k-2} n_{\overline{i-1}, i, i+1, i+2}, \qquad [16]$$

where N is the number of subjects, K the number of items. Items are ranked in order of difficulty, the most difficult item receiving rank k, and the easiest item rank 1. The quantity $n_{i, i+1}$ is the number of subjects who both fail the ith item and pass the next most difficult item $(i + 1)$. There will be $k - 1$ such item pairs. The last quantity, $n_{\overline{i-1}, i, i+1, i+2}$ is the number of subjects who have failed both item $i - 1$ and i and passed both item $i + 1$ and $i + 2$. There will be $k - 3$ such terms in this summation.

The reproducibility that would be expected if the items had their observed difficulties, but were mutually independent is given by

$$Rep_{ind} = 1 - \frac{1}{N^2 K} \sum_{i=1}^{k-1} n_i n_{i+1} - \frac{1}{N^4 K} \sum_{i=2}^{k-2} n_i n_{i+1} n_{i+2} n_{\overline{i-1}}. \qquad [17]$$

These values for Rep and Rep_{ind} are then put in 15 to obtain I, which will be unity for a perfectly reproducible test and zero for a test whose items are completely independent. Green suggests that I should be .50 for a test before its items can be considered scalable. Since this method makes only a partial count of the "errors" in a response matrix, it produces a slight overestimate of reproducibility. In one empirical investigation (5) it was found that the average discrepancy between Green's reproducibility and the exact reproducibility of ten scales was .002.

Following a suggestion of Guttman (7), Green furnishes an approximation to the standard error of Rep.

$$\sigma_{\text{Rep}} \approx \sqrt{\frac{(1 - Rep)(Rep)}{NK}}. \qquad [18]$$

With this standard error it is possible to ascertain whether an obtained Rep is significantly larger than Rep_{ind}. Green warns, however, that when such a test yields borderline significance, one should be cautious in interpretation since both Rep and σ_{Rep} are approximations. A high significance level does not necessarily indicate that the items are homogeneous, merely that the item intercorrelations are significantly greater than zero.

*Table 14.4. Green's Method of Computing Reproducibility (Rep),
Chance Reproducibility (Rep$_{ind}$), and Index of Consistency (I)*

Subjects	Items						Total Scores
	2	6	4	1	5	3	
D	1	1	0	1	1	1	5
E	0	1	1	1	1	1	5
B	0	0	1	1	1	1	4
I	0	1	0	1	1	1	4
H	0	0	0	1	1	1	3
J	0	0	1	1	1	0	3
G	1	0	0	0	0	1	2
F	0	0	0	0	1	1	2
C	0	0	0	0	0	1	1
A	0	0	1	0	0	0	1
Rank Order of Difficulty	6	5	4	3	2	1	
n_i	2	3	4	6	7	8	
n_j	8	7	6	4	3	2	
$n_{j,i+1}$	—	1	2	1	0	1	
$n_{i-1,j,i+1.i+2}$	—	—	0	0	0	—	

$$Rep = 1 - \frac{1}{(10)\,(6)}(1+0+1+2+1) - \frac{1}{(10)\,(6)}(0+0+0) = .917$$

$$Rep_{ind} = \frac{1}{(10^2)\,(6)}(7 \cdot 2 + 6 \cdot 3 + 4 \cdot 4 + 3 \cdot 6 + 2 \cdot 7) - \frac{1}{10^4(6)}(4 \cdot 6 \cdot 3 \cdot 2 + 3 \cdot 4 \cdot 4 \cdot 3 + 2 \cdot 3 \cdot 6 \cdot 4) = .860$$

$$I = \frac{Rep - Rep_{ind}}{1.00 - Rep_{ind}} = \frac{.916 - .860}{1.000 - .860} = .407$$

For the illustrative test, the computation of *Rep*, *Rep$_{ind}$*, and *I* are shown in Table 14.4.

The obtained *Rep* is .917, as compared with Jackson's .88, and .78 by Formula 10. The index of consistency (*I*) is seen to be .41, as compared with Jackson's *PPR* of .61, and Loevinger's *H$_t$* of .23.

The Phi Coefficient (ϕ_{it})[3]

A measure of item reproducibility can be derived from the phi coefficient. This measure has the advantages of an absolute maximum of 1.00, an absolute minimum of 0.00, a known sampling distribution, and direct relationship to conventional test construction procedure.

The logic behind the procedure is simple. Take as an example an item which 30 percent of the subjects pass and 70 percent fail. If the item is perfectly reproducible in a perfectly reproducible test, the 30 percent of the subjects with the highest total scores should all pass the item; the 70 percent with the lowest total scores should all fail the item. Subjects can easily be ranked on total score and this distribution cut in the same ratio as the pass-fail ratio on any particular item being evaluated. It is then simple to determine the number of persons high on total score who pass the item, the number of high persons who fail the item, the number of low persons who fail the item and who pass the item. The data may be put in a fourfold table as in Table 14.5.

Table 14.5. Item-Total Score Phi Coefficient (ϕ_{it})

		Total Score*		Total
		Low	High	
Item Score	Pass Item i	A	B	$A+B$
	Fail Item i	C	D	$C+D$
	Total	$A+C$	$B+D$	N

* Total score distribution is broken so that number of subjects in low group is equal to number failing item i: $(C+D = A + C)$.

Obviously, one has only to determine the marginal sums (which are determined by the pass-fail ratio of the item) and one of the cell frequencies, since the rest can be computed by subtraction from the marginals.

Splitting subjects on the basis of total score in the same ratio as the pass-fail split on an item may produce a problem if several subjects are tied for total score across the cutting points. The tied subjects should be randomly distributed between the high and low groups so that the total scores are split in the same ratio as the pass-fail ratio. Take as a simple example the case in which 100 subjects have answered a questionnaire in such a manner that the pass-fail ratio on a particular item is 30/70.

3. The writers find that Cronbach (1, p. 324) has anticipated them in this suggested manner of estimating reproducibility.

To evaluate this item, the subjects must be split on total score so that the highest 30 percent constitute one group and the lowest 70 percent constitute the second group. If three persons are tied for rank 30 in total score, two will be arbitrarily considered ranks 29 and 30 respectively, and will be placed in the high group. The third person will be assigned rank 31, and, despite the fact that his score is the same as that of two subjects in the high group, he will be placed in the low group. If the total number of subjects is reasonably large, and if the number of subjects having the *critical tied score* is not a large percentage of the total number of subjects, this will not distort the resulting phi.

Since the marginals for the total score have been determined in a manner that forces them to be equal for the marginal for the particular item the usual phi formula can be simplified to

$$\phi_{it} = \frac{BC - AD}{(A + B)(C + D)}, \qquad [19]$$

where the quantities A, B, C, and D correspond to cell entries in Table 14.5.

The null hypothesis for such a phi coefficient is, in every case, that the obtained phi is not significantly greater than zero. This can be tested by a chi square or a Fisher exact test on the fourfold table.

If the investigator desires to "purify" his test, he must choose a cutting point and select all the items with phi coefficients above this cutting point to constitute his reproducible scale. New total scores can then be computed on the basis of the selected items, and phi coefficients recalculated to give an estimate of the reproducibility of the new scale. The coefficients for some of the items not included in the new total score may be so high that these items can be included in the scale, while those for some of the included items may drop to a level which makes it advisable to exclude them.

Unlike some indices of reproducibility, this index is not affected by extremes of item difficulty. This is true because phi is not an index of the frequency in one cell, but is determined by the intercorrelation between cells. The method has a disadvantage, a purely aesthetic one, but one that may prejudice some workers against it; the phi coefficients so obtained are not likely to yield many values in the .80's or .90's. The phi coefficients computed for the items in the illustrative test are shown in Table 14.6, where they may be compared with those computed by Jackson's PPR_i and Loevinger's H_{it}.

Though this method of computing a phi coefficient between a test item and the total score has the advantages of a known sampling distribution, absolute maximum and minimum values, and freedom from re-

Table 14.6. Item-Total Score Phi Coefficients
for Illustrative Six-Item Test

Item	ϕ_{it}	PPR_i	H_{it}
1	1.000	1.000	1.000
2	.375	.500	.714
3	.375	.500	.333
4	.167	.250	.619
5	.524	.667	.867
6	.524	.667	.889

strictive distribution assumptions, it does not furnish an index for the test as a whole. It is possible however to derive one by an averaging of the obtained phi coefficients. Such an approach is shown in Formula 20, which Cronbach says is analogous to Guttman's formula for reproducibility.

$$R = \frac{1}{K} \sum_{i=1}^{k} 1 - 2p_i q_i (1 - \phi_{it}). \qquad [20]$$

Cronbach explains (1, p. 324):

> The correlation of any two-choice item with a total score on a test may be expressed as a phi coefficient, and this is common in conventional item analysis. Guttman dichotomizes the test scores at a cutting point selected by inspection of the data. We will get similar results if we dichotomize scores at that point which cuts off the same proportion of cases as pass the item under study. [Our ϕ_{it} will be less in some cases than it would be if determined by Guttman's inspection procedure.] Simple substitution in Guttman's definition . . . leads to [Formula 20 above] where the approximation is introduced by the difference in ways of dichotomizing. The actual R obtained by Guttman will be larger than that from [this formula].

In our example the value turns out to be .80 as compared with the reproducibility figure of .88.

This composite index for the entire test will have a maximum value of 1.00 and a chance value of

$$\frac{1}{K} \sum 1 - 2p_i q_i,$$

which approaches .50 as the average item difficulty approaches 50 percent. The sampling distribution of this statistic, to our knowledge, is not known.

Discussion

This concludes the exposition of the major methods which have been put forward to give an index of the reproducibility of tests. Of those which yield indices for the test as a whole, several meet serious objections which have been leveled at Guttman's scalogram analysis. The techniques of Jackson, Loevinger, and Green are all objective, and result in measures which are not affected by the distribution of item difficulties. All have the same underlying rationale, but differ slightly in the way in which "errors" are counted. Loevinger's H_t is the most conservative of the three since all possible errors are counted; Jackson's PPR_t is the least conservative, and Green's I will usually fall between the two. The principal and not inconsiderable advantage of Green's technique is ease of computation, an important factor when the number of subjects and test items is large. Green's technique is the only one discussed that gives an estimate of significance for the reproducibility of the entire test.

Of the methods of computing the homogeneity of an item with the total test, the phi coefficient seems the most desirable because computation is easy and because the significance level of the obtained statistic can be determined exactly. Almost any of the commonly used item-analysis statistics—point biserial, biserial, or Flanagan correlation coefficient—may of course be interpreted as an index of item reproducibility, since in a reproducible test any person passing a given item will pass more other items than a person failing that item. They differ from the phi coefficient mainly in the number of assumptions they impose upon the data. Those employing conventional item-analysis statistics have been quite willing to assume an interval scale and a distribution function, usually normal, while those working within the framework of the concept of reproducibility have in general foresworn the unit of measurement and have thus confined themselves to distribution-free statistics.

All the reproducibility indices rest upon the same assumption that in a reproducible or homogeneous test, one can reproduce the entire response pattern of passes and fails, given the total number of items correct, and the item difficulties. All the methods employ the same data in the response matrix. All agree that in the response matrix of the perfectly reproducible test there will be no instances in which a subject passes an item more difficult than one he has failed. This is equivalent to saying either that all interitem covariances are maximal, or that the variance in total scores is maximal. Conversely, the test with lowest reproducibility will exhibit zero interitem covariances, and minimal variance in total scores.

REPRODUCIBILITY AND FACTOR ANALYSIS

It is obvious that the phi coefficient method of determining the homogeneity of an item with total test is very similar to the procedure in classical test construction for "purifying" a test.

A common procedure for evaluating an item in conventional test construction is to compare the number of subjects passing the item among the 27 percent of the sample making the highest total scores as opposed to the 27 percent making the lowest total scores. A "good" item is one that discriminates between these highs and lows. Consequently, the items which would be chosen as producing the most reproducible scale in the phi procedure for obtaining reproducibility would also be selected as the most discriminating in conventional test statistics. This point is important when considering the relationship between reproducibility and factor analysis.

Several authors have been concerned with the question of the relationship between reproducibility and factor analysis. Loevinger (14) has stated that factor analysis and reproducibility are unrelated. Humphreys (9) appears to agree with Loevinger on this point and attacks reproducibility for not being as satisfactory a tool for research as factor analysis. He feels that reproducibility will lead to a confusing multiplicity of tests, while a factor analytic approach will not. Humphreys uses the hypothetical case of the problems involved in constructing a mechanical information test. The criterion of reproducibility, he fears, would require the construction of separate tests for the cross saw, the brace and bit, the pipe wrench, etc. On the other hand, all these tests would probably appear on a single common factor that would be orthogonal to other factors.

The writers disagree with both Loevinger and Humphreys, feeling that reproducibility and factor analysis are closely related. This relationship can be made obvious by consideration of the Wherry-Gaylord iterative analysis (19). This is a method for discovering homogeneous groupings of items in a test. It involves correlating each item with the total score. Items with the highest correlations are selected and the test rescored on the basis of these items. All the items are then correlated with the new total scores. This procedure is continued until a stable group of items is extracted. These items constitute a single factor. The remaining items can be rescored and additional factors extracted. The first factor removed would be the general factor. As can be seen, the phi method of obtaining reproducibility corresponds very closely to the Wherry-Gaylord extraction of the general factor. The principal differences are that the Wherry-Gaylord does not require that the finally selected items have a range of item difficulties, and does not cut total scores at the same ratio as the item-difficulty levels. Evidence reported by Wherry, Campbell, and Perloff (20) suggests that the Wherry-Gaylord general factor will correspond to the general factor obtainable in a Thurstone multiple factor analysis. The present writers found similar evidence in an analysis of a morale scale. After the morale scale had been subjected to a Thurstone multiple factor analysis, it was administered to a new group of subjects and subjected to a phi reproducibility scaling. The resulting scale was almost identical in item content with the Thurstone general factor.

While it appears to be true that a highly reproducible scale will tend to measure a single factor (since the phi analysis will isolate the general factor in the test items), not all single factor tests will be highly reproducible scales. This is because a reproducible scale must have a range of difficulty levels if all persons are not to be forced into two categories: either all items passed or all items failed. The following example points up the reason this is true. If all items were at the 50-percent difficulty level, and if the test were perfectly reproducible, the 50 percent of the subjects with the highest total scores would score correct on all items; the 50 percent of the subjects with the lowest total scores would score incorrect on all items. This restriction is not necessary for all single-factor tests. Single-factor tests can have all items at the same difficulty level and still have a wide range of total scores due to the almost inevitable presence of error variance in the items. Reproducibility is impossible in such a case. Despite this lack of reproducibility, the single-factor test might be quite adequate since, if two persons score high on a single-factor test it is because they are high in the factor, and the differential patterns of their responses must be irrelevant for prediction since the differential patterns must be a result of error variance and do not represent stable patterns. If the differential patterns were differentially predictive, the test could not be a single-factor test. In those situations, therefore, where it is desirable to have all items at the same difficulty level, reproducibility is usually not a useful approach. The exception to this rule is the case in which a single discrimination is desired—e.g., pass vs. fail. In this case all items should have pass percents which are proportional to the pass percent desired for the whole test (16).

In many practical test-construction situations, where the logic of the situation is not incompatible with reproducibility, it appears to the writers that obtaining a general-factor test through phi reproducibility is simpler than through a Thurstone multiple-factor analysis. In addition to the relative ease of computation, the set of items so obtained should form not only a single-factor test, but also a reproducible scale.

REPRODUCIBILITY AND RELIABILITY

It is obvious that the techniques for computing so-called reliability co-efficients from a single test administration employ exactly the same data which have been used to compute the indices of reproducibility described above. Cronbach (1) has already pointed out the intimate relation of Guttman's reproducibility to the Kuder-Richardson Formula 20, which he has rechristened *alpha*. The key term in *alpha* is the ratio of two variances, $\Sigma\, pq/\sigma^2_x$. As Loevinger points out (13, p. 31 $\Sigma\, pq$ gives the raw score variance which would be obtained from a test whose items were completely independent, (σ^2_{het}); and σ^2_x is the obtained raw score variance. Loevinger's Formula 11 has these same quantities in it, plus a third representing the raw-score variance of a test whose items were perfectly correlated. It should be noted that the lower limit of *alpha* is always zero, but the upper limit is dependent upon the distribution of item difficulties. The obtained *alpha* for our illustrative test is .47, and the upper limit of *alpha* for this set of item difficulties is .88.

In order to make *alpha* independent of the distribution of item difficulties, Horst (8) has developed a formula which turns out to be identical with Loevinger's 11, except for a correction term composed of the ratio of the maximal to obtained score variance. Since this ratio has a lower limit of 1.00, figures obtained by Horst's method will necessarily be larger than Loevinger's except in the perfect case. The Horst formula for the reliability coefficient corrected for dispersion of item difficulties is given below

$$r_{tt} = \frac{\sigma^2_x - \sum_{pq}}{\sigma^2_{max} - \sum_{pq}} \left(\frac{\sigma^2_{max}}{\sigma^2_x} \right). \qquad [21]$$

The striking similarity of the Loevinger Formula 11 and the Horst Formula 21 cause one to suspect that the difference between single-trial reliability and homogeneity or reproducibility is more apparent than real.

The critical difference between the "reproducibility" and the "reliability" camps of test construction is seen most clearly in the ways they interpret their indices. When a test shows perfect reproducibility, it will also show perfect reliability by any of the formulas described so far. In order for this unlikely event to occur, several conditions must be met: all the items must be homogeneous in content, all subjects must be similarly constituted in the trait, attitude, or ability being tapped; and this trait, attitude, or ability must remain stable during the testing period. *Any* departure from these conditions will cause *any* of these measures to fail, and there is no way to tell on the basis of the response matrix alone what is amiss. An astute dropping of rows or columns from the matrix (subjects and/or items) will, of course, make things look better. In any event, a low figure indicates that considerable information will be lost in attempting to order subjects on a single linear continuum on the basis of their total scores. It is here that techniques such as Lazarsfeld's latent structure analysis (12) may be used to determine the minimal number of dimensions (classes) needed to account for the information contained in a response matrix. With this technique, a subject, instead of being given a total score, is assigned a probability of belonging in each of several classes. No unidimensionality is assumed, so there is no question of item or subject elimination to force unidimensionality, a procedure routinely employed by those addicted to Guttman scaling and classical test construction. When any such item-elimination procedure is used in test construction, a reliability or reproducibility figure computed on the final sample of items cannot be evaluated until the new version of the test has been administered to another sample of subjects. A low reproducibility figure is generally taken as an indication of item heterogeneity in a test, while a low reliability figure of the Kuder-Richardson variety is usually seen as an indication of the presence of considerable error variance. In the absence of other information, either interpretation is equally plausible, or suspect, since, as was pointed out above, the indices employ the same information from the response matrix.

ITEMS AND SUBJECTS

There is no reason why the techniques of computing reproducibility or single trial reliability cannot be reversed to yield coefficients about the homogeneity of subjects, instead of test items. It is surprising that this has not been done more often, especially in the area of attitude measurement. Lack of reproducibility in a response matrix is just as likely to be due to heterogeneity in the population tested, as to heterogeneity in the test items. For most of the indices described above, computation of subject homogeneity would merely involve switching row and column marginals in the formulas. Such a technique would seem to be a promising one for the identification of deviants.

WHY REPRODUCIBILITY OR SINGLE-TRIAL RELIABILITY?

Having come this far, it is high time we asked why a test with high reproducibility or single-trial reliability is a good thing. Social scientists are all too prone to assume that it is, and to think no further about it. As Cronbach (1) has pointed out, reproducibility is in a sense a measure of the redundancy in a test. For many purposes, this is undesirable. Whenever test results are used to predict a dichotomous criterion such as hire/not hire, pass/fail, butcher/candlestickmaker, psychotic/normal— in short to classify subjects—it can be argued that the last thing in the world a test should have a high internal consistency. The real need is a set of items highly related to the criterion but not to each other. This is, of course, a restatement of the multiple-correlation approach to prediction. Ideally each item would represent a different pure factor. In such a situation, interest lies not in ordering subjects on some linear hypothetical trait, attitude, or ability continuum, but in an efficient dichotomization of the subjects or an ordering on the basis of the probability of membership in a class. To the extent that the test items are redundant, valuable testing time is wasted. It is a mistake to think such a test is "measuring" something, in the usual sense of that word. That a test can differentiate between neurotics and normals is no indication that "neuroticism" is a trait on which people can be ordered in some simple fashion. Much confusion in clinical literature is based on this fallacy. Unless the instrument exhibits high homogeneity-reproducibility– single-trial reliability, there is no reason to assume that the score on the test can yield an ordering of the subjects on some unidimensional continuum—which can be given a label.

It is the person doing "basic" research who is apt to be more interested in ordering subjects on a unidimensional continuum. For him, the question of the internal consistency of his multiple-item test or questionnaire is of immediate concern. He may start with the unshakable conviction that the trait he has in mind *is* unidimensional, in which case he will engage in an often lengthy process of test construction, weeding out items until he achieves an instrument with internal consistency at a satisfactorily high level. This type of worker usually longs for an infinite population of items and subjects. When this longing is fulfilled, or even approximated, he can usually come up with a selection of items which, when administered to an appropriate population, will yield a response matrix of the desired internal consistency. He may even regard this achievement as support for his initial assumption about the unidimensionality of the trait, though the logic of such a conclusion is somewhat less than perfect, considering the amount of information thrown away in order to make things come out so neatly.

On the other hand, he may begin with a more modest aim: to find out, for a given set of items, the minimum number of parameters needed to account for the obtained responses of subjects to these items. If he finds that the response matrix shows high reproducibility or high single-trial reliability, he is apt to be pleased because life is so simple; but if he does not find his data so neatly arranged, he is likely to resign himself to fairly laborious procedures in order to find out the dimensionality of the data he has collected rather than to attribute any departure from unidimensionality to error variance.

The important point is that all the techniques mentioned here, whether they are regarded as indices of reproducibility, homogeneity, or single-trial reliability, are based upon the same raw data in the response matrix; and all are more or less interchangeable with a little algebraic manipulation, though, as we have seen, they yield different numbers. How the number is interpreted depends not upon which one of these formulas is employed, since they are all basically equivalent, but upon what assumptions are made. One can assume that the items are homogeneous and that the subjects are similarly constituted in the trait being measured, in which case one uses the index as a measure of intra-individual trait stability. On the other hand, one can assume trait stability and subject homogeneity, in which case the index is said to reflect the homogeneity of the items. As was mentioned above, one may equally well assume trait stability and item homogeneity and employ the index as a measure of the homogeneity of the subjects. Any pair of assumptions appears to be about as plausible as any other. The important point is that from a single response matrix there is no way of telling what assumptions are reasonable. An obtained index, be it Jackson's PPR_t, Loevinger's H_t, Green's I, Cronbach's *alpha*, or Horst's r_{tt}, will be less than 1.00 when any or all of these conditions are not met. The plausibility of the assumptions can be ascertained only by recourse to further data, and the kind of data required will be different for testing each assumption. An estimate of intraindividual trait stability, for example, demands retesting the same subjects with the same items, but such retest data will be of little value in arriving at estimates of subject or item heterogeneity.

The one thing these indices of reproducibility or single-trial reliability will reflect without equivocation is the amount of information thrown away by representing the subject's performance on the test by a total score based on the number of items passed. They indicate, in other words, how adequately a unidimensional model fits the obtained data.

Proponents of homogeneity or reproducibility have been criticized because their criteria for a "good" test are unrealistically strict. It is true that perfect reproducibility will occur when, and only when: (*a*) the factors determining subjects' responses to the test do not change during the testing period, (*b*) the factors determining subjects' responses to the test are the same for all subjects, and (*c*) all the items in the test are identical in the factors determining the responses they elicit. It is also true that perfect single-trial reliability will be obtained only under the same circumstances. These are stringent conditions, and they are seldom, if ever, met. Human beings are just not that simple, but the fault is hardly Guttman's. There is nothing wrong in continuing to assume that many human abilities, attitudes, and traits are unidimensional continua, but we should be fully aware that this is at best a useful first approximation, and that an appreciable proportion of the information in our raw data will thereby be sacrificed on the altar of error variance.

References

1. CRONBACH, L. J. Coefficient alpha and the internal structure of tests. *Psychometrika,* 1951, 16, 297–334.
2. FESTINGER, L. The treatment of qualitative data by "scale analysis." *Psychol. Bull.,* 1947, 44, 146–161.
3. FORD, R. N. A rapid scoring procedure for scaling attitude questions. *Publ. Opin. Quart.,* 1950, 14, 507–532.
4. GREEN, B. F. Attitude measurement. In G. Lindzey (Ed.), *Handbook of social psychology.* Cambridge: Addison-Wesley, 1954.
5. GREEN, B. F. A method of scalogram analysis using summary statistics. *Psychometrika,* 1956, 21, 79–88.
6. GUTTMAN, L. The Cornell technique for scale and intensity analysis. *Educ. psychol. Measmt,* 1947, 7, 247–279.
7. GUTTMAN, L. The basis for scalogram analysis. In S. A. Stouffer et al., *Measurement and prediction.* Princeton: Princeton Univer. Press, 1950.
8. HORST, P. Correcting the Kuder-Richardson reliability for dispersion of item difficulties. *Psychol. Bull.,* 1953, 50, 371–374.
9. HUMPHREYS, L. G. Test homogeneity and its measurement. *Amer. Psychologist,* 1949, 4, 245. (Abstract)
10. JACKSON, J. M. A simple and more rigorous technique for scale analysis. In *A Manual of scale analysis.* Part II. Montreal: McGill Univer., 1949. (Mimeographed.)
11. KAHN, L. H., & BODINE, A. J. Guttman scale analysis by means of IBM equipment. *Educ. psychol. Measmt,* 1951, 11, 298–314.
12. LAZARSFELD, P. F. The logic and mathematical foundation of latent structure analysis. In S. A. Stouffer et al., *Measurement and prediction.* Princeton: Princeton Univer. Press, 1950.
13. LOEVINGER, JANE. A systematic approach to the construction and evaluation of tests of ability. *Psychol. Monogr.,* 1947, 61, No. 4 (Whole No. 285).
14. LOEVINGER, JANE. The technic of homogeneous tests compared with some aspects of "scale analysis" and factor analysis. *Psychol. Bull.,* 1948, 45, 507–529.
15. LONG, J. A. Improved overlapping methods for determining the validities of test items. *J. exp. Educ.,* 1934, 2, 264–268.
16. LORD, F. M. Some perspectives on "the attenuation paradox in test theory." *Psychol. Bull.,* 1955, 52, 505–510.
17. MARDER, E. Linear segments: a technique for scalogram analysis. *Publ. Opin. Quart.,* 1952, 16, 417–431.
18. NOLAND, E. W. Worker attitude and industrial absenteeism: a statistical appraisal. *Amer. sociol. Rev.,* 1945, 10, 503–510.
19. WHERRY, J. J. & GAYLORD, R. H. The concept of test and item reliability in relation to factor pattern. *Psychometrika,* 1943, 8, 247–269.
20. WHERRY, R. J., CAMPBELL, J. T., & PERLOFF, R. An empirical verification of the Wherry-Gaylord iterative factor analysis procedure. *Psychometrika,* 1951, 16, 67–74.

Chapter 15

A NOTE ON STATISTICAL SIGNIFICANCE OF SCALOGRAM

KARL F. SCHUESSLER
Indiana University

In his recent book on measurement, W. S. Torgerson states that one limitation of Guttman's method of scaling resides in its incapacity to test statistically whether the cumulative scale model is a good fit (2). The Guttman scale model (1), it will be recalled, consists of an ordered set of

$$\left[(\sum_{1}^{m} c_i - m) + 1 \right]$$ combinations from a total of $(c_1 c_2 \ldots . c_i \ldots . c_m)$ pos-

sible combinations, where m is the number of items, and c_i is the number of categories of the ith item. It is by definition a single-valued function in which each scale score, x, corresponds to only one combination, y.

While there is truth to Torgerson's statement, it is perhaps unnecessarily categorical and may act to divert attention from scale analysis before its potentialities as an ordering device have been properly considered by the investigator. For it is possible to construe the matrix of observations (scalogram) in such a way that statistical hypotheses may be

Reprinted by permission of the author and the American Sociological Association from *Sociometry*, Vol. 24, No. 3, pp. 312–18, 1961.

The author is indebted to Professor Louis Guttman of Hebrew University, Jerusalem, Israel, who kindly read the original manuscript of this note and made suggestions for its refinement.

tested, and inferences drawn in respect to the presence of basic dimensions which could account for the configuration of observed responses. This paper sets forth three such possibilities, primarily as suggestions to the research worker who is uncertain about the statistical significance of his scale findings and who wishes as a minimum to rule out chance as an explanation.

Although the methods to be unfolded are general, as a convenience we shall have recourse throughout to the hypothetical responses of 100 individuals to three dichotomized attitude items, with the following marginals:

Item	Response* +	Response* −
I	.80	.20
II	.60	.40
III	.35	.65

* Plus sign represents positive response; minus sign, a negative response.

and the following distribution of persons by response patterns:

Number of Positive Responses	Response Pattern			Frequency
(x)	I	II	III	
3	+	+	+	30
2	+	+	−	25
2	+	−	+	3
2	−	+	+	1
1	+	−	−	22
1	−	+	−	4
1	−	−	+	1
0	−	−	−	14
				$n = 100$

Finally, for ease of exposition, we shall employ, along with x, m, and c_i, the following notation:

n = number of persons in sample

n_{ij} = number of persons giving jth response to ith item

p_{ij} = proportion of persons giving jth response to ith item

Test I

Essentially, this procedure is a comparison between the number of scale types expected on the hypothesis that responses to successive items are statistically independent and the number of scale types observed in a sample of n persons. If this difference is sufficiently improbable by chi-square, then the hypothesis of independence may be waived and an alternative hypothesis accepted, whether or not explicitly formulated. Test I, then, takes as its point of departure the proportion of scale types expected by chance, against which the observed proportion of scale types is evaluated for possible significance.

As a preliminary step, we must designate scale types, since it is their combined frequency that is to be checked for statistical significance. In general, we designate as scale types those arrangements which have the largest chance expectation within their respective score intervals, and which together permit a perfect fit. In the event of ties within a given score interval, we designate whichever arrangement has the largest observed frequency, since that arrangement will yield a better fit than any other combination in the same score class.

To illustrate the identification of scale types by this method, we calculate below the chance frequencies corresponding to the response patterns given above:

x	Combination	p_{1j}	p_{2j}	p_{3j}	$(p_{1j}p_{2j}p_{3j})$	Expected Frequency
*3	+ + +	.80	.60	.35	.168	16.8
*2	+ + −	.80	.60	.65	.312	31.2
2	+ − +	.80	.40	.35	.112	11.2
2	− + +	.20	.60	.35	.042	4.2
*1	+ − −	.80	.40	.65	.208	20.8
1	− + −	.20	.60	.65	.078	7.8
1	− − +	.20	.40	.35	.028	2.8
*0	− − −	.20	.40	.65	.052	5.2
					1.000	100.0

* Denotes scale type.

Of those arrangements yielding a scale score of two, we treat + + − as the scale type, since it has the largest chance frequency (31.2); similarly, of the arrangements producing a score of one, we designate + − − as the scale type, since its chance expectation of 20.8 is largest. Where a given score can result from only one arrangement, as in the case of extreme scores, no choice is possible, so that extreme scores, when they occur, must always be designated as scale types.

Once scale types have been identified, we merely add their individual probabilities, and treat the resulting sum as the probability of a scale type on a single trial. We next apply this probability to n to obtain the frequency of scale types expected on the hypothesis that the m items are statistically independent. Lastly, against this expected frequency we check the observed frequency for possible statistical significance. Combining scale types and their corresponding probabilities, we obtain:

$$Pr \text{ (Scale Type)} = .168 + .312 + .208 + .052$$
$$= .74$$

Accordingly, the expected frequency of scale types is $(.74)$ $(100) = 74$, which leads to $\chi^2 = 15.02$, 1 df. Now, since we are interested only in the possibility that the observed frequency of scale types exceeds the frequency expected by chance, and not in the possibility of its being deficient, we take the positive square root of 15.02 and calculate the probability of z (normal deviate) larger than $15.02 = 3.87$. Since the probability value in this case is very small, we abandon the initial hypothesis that the m items are uncorrelated and accept the possibility that determining factors are present.

Alternative Hypotheses

At this juncture, it would be possible to give consideration to alternative hypotheses of varying degrees of stringency. For instance, we might hypothesize that the probability of a scale type is equal to, or less than, .8 $(H: P \leq .8)$, and proceed to determine the probability of the observed frequency of scale types against that hypothesis. Since any frequency smaller than $.8n$ is consistent with this one-tailed hypothesis, we need consider only those observed frequencies which are larger than $.8n$, as, for example, $.91n = 91$, in our illustration. Upon testing 91 against the expected frequency of 80, we obtain $\chi^2 = 7.44$, $z = 2.72$, and $Pr < .01$, a significance value which would prompt rejection of $H: P \leq .8$, and acceptance of the alternative that the probability of a scale type is greater than .8.

By giving due consideration to such alternatives to the chance hypothesis, we meet in part Torgerson's comment that scale analysis cannot be checked for goodness of fit. While it is not possible to test the hypothesis of perfect scalability by probabilistic methods, still we may test weaker alternatives, including the chance hypothesis, and by that technique assure ourtselves, as a minimum, that the chance hypothesis is probably false. More positively, we may assess the tendency in the sampled population to adhere to the scale model, although we can never prove that the population is perfectly scalable.

Test II

Like the first, this procedure requires that we initially calculate the chance frequencies of the respective response patterns, and designate scale types accordingly. But, instead of combining scale types into a single class, we now compare the observed and expected frequency of scale types within specific score intervals, along with the observed and expected frequency of non-scale types.[1] Obviously, this procedure is more sensitive than the first, since the comparisons within specific score intervals will indicate at what points the divergence from chance is negligible, and thereby lead to a more refined statistical interpretation of the matrix of responses. However, since χ^2 will necessarily be based on more than 1 df, it will be impossible to test by this statistic the one-tailed directional hypothesis that the observed scale frequencies are excessive. It will be possible to test only the hypothesis that scale frequencies are a chance phenomenon. Consequently, it will be necessary to determine visually whether the observed scale frequencies exceed, or fall short of, those expected by chance. This procedure is illustrated in the tabulation below:

x	O	E	$(O–E)$	$\dfrac{(O–E)^2}{E}$
3	30	16.8	13.2	10.37
2	25	31.2	—6.2	1.24
2	4	15.4	—11.4	8.44
1	22	20.8	1.2	.07
1	5	10.6	—5.6	2.96
0	14	5.2	8.8	14.90
	100	100.0	0.0	37.98

Inasmuch as $Pr\,(\chi^2 > 37.98/5df) < .01$, we reject the hypothesis that items are independent and tacitly accept, as in Test I, the hypothesis of response dependency.

Test III

As a third possibility, we may analyze the variation between and within items for each set of persons having the same score, a procedure which will usually result in $[(\Sigma c_i — m) — 1]$ separate analyses, or two less than the number of scale types. If persons with identical scores are consistently alike in their response to each item, then the total response variation will be completely accounted for by whatever variation exists between items and the intraclass correlation coefficient, δ, will be 1. At the other extreme, where total response variation is almost wholly accounted for by the within-item variation, the intraclass correlation coefficient will generally lie close to zero, thereby demonstrating a random distribution

---

1. Since each score yields two comparisons, excepting the extremes which yield one, the total number of comparisons will be $2(\Sigma c_i — m)$.

of values within classes. Evidently, we may employ the intraclass correlation coefficient to test the hypothesis that items are uncorrelated within each subset of identical scores, and to gauge the degree of score homogencity within classes.

To exemplify this type of analysis, let us return once again to the table above, in which 29 persons show scores of two, and 27 persons show scores of one. (Extreme scores are not subject to this analysis since there is no variation to be partitioned.) For each group, we set up a person-by-item matrix, the elements of which are arbitrary values—0 or 1 in our example—and proceed to the calculation of the intraclass coefficient (computation details available on request to the author). The 29 x 3 score matrix yields $\delta = .71$, and the 27 x 3 matrix yields $\delta = .61$. Both results are sufficiently improbable by the F-statistic to warrant rejection of the null hypothesis under test.

Comparison of Procedures

While the techniques presented above will normally yield identical decisions, as in our running illustration, they may at times lead to conflicting inferences (rejection *and* acceptance), owing to the fact that they answer to different but related questions and/or rest on different assumptions concerning the observed data.

Procedures I and II are essentially alike in testing an observed frequency against an expected frequency, but they differ in that I merges all scale types together before testing for what may be called gross significance, while II tests the observed frequency of scale types against the expected frequency within specific score intervals.

Both I and II assume that responses are purely qualtative, whereas III treats them as numerical values subject to arithmetic averaging. It follows that conflicting decisions between II and III, both of which examine the pattern of responses within score intervals, could result from the fact that in one instance observations are arbitrarily quantified, whereas in the other case they are merely enumerated. This difference points up a possible limitation of III, and a corresponding advantage of II: the former has recourse to artificial measures, while the latter only counts events.

Further, the intraclass correlation coefficient is a function of the variance ratio, and, consequently, to test its significance it is necessary to satisfy the conditions of the F-distribution. On the other hand, it is a conventional measure of correlation, and if such a measure be required, then its use would be dictated by that need, even though its statistical significance could not be exactly determined. But, all in all, the more conservative distribution-free method is less open to question.

Reproducibility and Chance Frequency of Scale Types

Since the procedures suggested in this paper intend primarily to rule out chance as an explanation of the scalogram, they in no way supplant Guttman's well-known criteria of a perfect scale. They would normally be applied in advance of Guttman's four principal criteria, namely: (a) 90 percent reproducibility, (b) balanced marginals, (c) less error than non-error within items, and (d) random scatter of error within items. At times, to be sure, the significance of the scalogram may appear obvious and a null test of chance unnecessary. But, whenever doubt exists within the mind of the worker, then null tests should be run as a matter of course. In particular, such checks are essential when the number of scale types is small $[(\Sigma c_i - m + 1) < 10]$ and/or the marginals are uniformly unbalanced (e.g., $p_{ij} > .80$). For in those situations, the frequency of scale types will necessarily be high as a matter of chance, and is likely therefore to be spuriously credited with significance when none actually exists.

It is evident from inspection of the foregoing examples that reproducibility need not be calculated in order to apply any of the foregoing tests. However, each procedure automatically provides a test of the hypothesis that reproducibility is a chance outcome, since reproducibility is a function of the ratio of scale types to non-scale types, increasing as that ratio increases. Hence, a demonstration of the significance of this ratio is also a demonstration of the significance of observed reproducibility. But since the number of scale types expected by chance is much simpler to calculate than chance reproducibility, we naturally have recourse to that simpler quantity.

Summary

It is the thesis of this paper that Guttman scalograms may be statistically interpreted and that such interpretations have a bearing on the decision to accept the hypothesis that the items constitute a scale or quasi-scale. While the Guttman hypothesis of perfect scalability is statistically untestable, it is possible to determine whether the observed configuration of responses might have arisen by chance. Additionally, alternative hypotheses may be tested in order to set a probable lower boundary to the proportion of scale types in the sampled population and thereby to gauge the tendency in the population to conform to the scale model. Altogether, the methods set forth in this paper enable the research worker to approach his scale findings within the framework of statistical inference, and correspondingly to make his decisions with greater or lesser confidence.

References

1. GUTTMAN, L., "The Quantification of a Class of Attributes," in Paul Horst (ed.), *The Prediction of Personal Adjustment,* Social Science Research Council, Bulletin 48, 1941, 319–348.
2. TORGERSON, W. S., *Theory and Methods of Scaling,* New York: John Wiley & Sons, 1958, 324.

Chapter 16

COMPUTER GENERATED DATA AND THE STATISTICAL SIGNIFICANCE OF SCALOGRAM

ROLAND J. CHILTON
University of Massachusetts

The three phases of scalogram analysis—measurement of deviation from pattern, elimination of chance as an explanation of pattern, and scoring—though rarely treated separately, are viewed in this note as distinct operations. From this perspective, the analysis consists of an initial phase during which the responses of a number of people to a set of questions are examined to determine the extent to which the pattern of responses approximates the cumulative pattern described by Guttman and others.[1] During this initial phase no attempt is made to improve the pattern; the items are simply ordered and the deviations from pattern counted. If an acceptable approximation of pattern emerges an attempt is then made to rule out chance as an explanation of the observed pattern.

It is important to emphasize that neither an acceptable approximation of pattern nor a rejection of chance factors as an explanation of an observed pattern may be interpreted as conclusive evidence that the items

Reprinted by permission of the author and the American Sociological Association from *Sociometry*, Vol. 29, No. 2, pp. 175–81, 1966.

Revision of a paper read at the annual meeting of the American Sociological Association, August, 1965.

1. Louis Guttman, "The Utility of Scalogram Analysis," in Samuel A. Stouffer *et al.*, *Measurement and Prediction*, Princeton: Princeton University Press, 1950.

do in fact scale in a Guttman sense. However, a large proportion of error responses (resulting in a low coefficient of reproducibility) makes it more difficult to conclude that the items present a cumulative pattern. More importantly, if chance factors might easily produce a similar pattern—even when the observed coefficient of reproducibility is greater than .90, there would be no reason to proceed on the notion that the items were somehow interrelated. Therefore, the central questions in scalogram analysis are these: Does the observed pattern of responses fit the theoretical cumulative pattern upon which the scoring is based? Could the observed pattern be the result of purely random factors?

Answers to the first question have generally been based on some modification of Guttman's well known criteria for an acceptable scale. To the extent that these criteria are intended to make it unlikely that an acceptable pattern would occur by chance alone, they provide indirect assurance that chance factors are not a problem. However, in practice these criteria are rarely met, and in some situations they may be too stringent. For these reasons, a simple statistical test would provide a less subjective basis for eliminating chance factors as an explanation of scalogram results. Tests developed independently by Goodman, Schuessler, and Sagi provide such a basis.[2]

The statistical procedure suggested in this note differs from Schuessler's and Sagi's tests in simplicity and utility. It requires only an observed reproducibility, a proportion of positive responses for each item, and the number of cases involved, in contrast to the analysis of the complete response matrix which is necessary in both Schuessler's and Sagi's tests.[3] Sagi's test is also limited to four item sets of data and to small numbers of cases—a severe limitation. Finally, the proposed test is applicable when Guttman's original coefficient of reproducibility is used. In this respect it differs from the tests suggested by Goodman which are developed for alternative measures of deviation from pattern.[4]

2. Karl F. Schuessler, "A Note on Statistical Significance of Scalogram," *Sociometry*, 24 (September, 1961), pp. 312–318; Leo A. Goodman, "Simple Statistical Methods for Scalogram Analysis," *Psychometrika*, 24 (March, 1959), pp. 29–42; Philip O. Sagi, "A Statistical Test for a Coefficient of Reproducibility," *Psychometrika*, 24 (March, 1959), pp. 19–27.

3. The simplicity of the test means that the results of a scalogram analysis may be checked for the importance of chance factors if the investigator presents this minimum of summary information and the asurance that the values were obtained *before* purification. If this were standard practice, a number of questionable uses of scalogram analysis might be avoided.

4. The measures of deviation from pattern reviewed by Goodman differ from Guttman's in that different procedures are followed in counting errors. Using Guttman's coefficient, an error is a positive response which must be substituted for a negative response or a negative response which must be substituted for a positive response in order to transform the observed pattern to the ideal cumulative pattern.

Proposed Test

The test presented here is essentially an extension of Schuessler's Test I and a variation of earlier procedures for computing "chance reproducibility" in that similar operations are performed to compute an expected number of errors which is converted to an expected reproducibility and compared with an observed reproducibility.[5] An example of the procedure is presented below using Schuessler's hypothetical responses of 100 individuals to three dichotomized items where 35 respondents gave a positive answer to item 1, 60 answered item 2 positively, and 80 answered item 3 positively. Table 16.1 presents the response patterns, the errors assigned to particular patterns, the distribution of proportions of the positive responses, the probability of each pattern, the expected and observed frequency of each pattern, and the expected and observed number of errors for each pattern.

In this example, a total of 26 cases with one error each would be expected, resulting in an expected reproducibility of 91.3, while the observed reproducibility (9 errors) is 97. This expected reproducibility has been called "chance reproducibility" and as such has been used as a rough check on the significance of observed reproducibilities. To eliminate this more or less subjective or intuitive decision it is suggested that the standard error of the expected proportion of non-error responses be used to decide when chance may be disregarded as an explanation of the pattern found.

Table 16.1. Observed and Expected Results for 100 Hypothetical Cases with Marginal Frequencies of 35, 60, 80

Response Pattern	No. of Errors	Distributed Marginal Probabilities			Probability of each Pattern	Frequency of Each Pattern		Errors	
						Exp.	Obs.	Exp.	Obs.
111	0	.35	.60	.80	.168	16.8	30	—	—
011	0	.65	.60	.80	.312	31.2	25	—	—
101	1	.35	.40	.80	.112	11.2	3	11.2	3
110	1	.35	.60	.20	.042	4.2	1	4.2	1
001	0	.65	.40	.80	.208	20.8	22	—	—
010	1	.65	.60	.20	.078	7.8	4	7.8	4
100	1	.35	.40	.20	.028	2.8	1	2.8	1
000	0	.65	.40	.20	.052	5.2	14	—	—

NOTE: The data presented here are adapted from Schuessler (Chapter 15).

5. See Matilda White Riley *et al.*, *Sociological Studies in Scale Analysis*, New Brunswick, New Jersey: Rutgers University Press, 1954.

Using Schuessler's hypothetical data again, we find that the standard error of the expected proportion of non-error responses (chance reproducibility) is 1.63. ($\sigma_p = \sqrt{\dfrac{PQ}{N}} = \sqrt{\dfrac{91.3 \times 8.7}{300}} = 1.63$); and the difference between the expected and observed reproducibilities divided by the standard error of the expected proportion of non-error responses is 3.47; $z = \dfrac{P - p}{\sigma_p} = \dfrac{91.3 - 97.0}{1.63} = -3.47$ (where P is the expected proportion of non-error responses [chance reproducibility], Q is the expected proportion of error responses, and N is the total number of responses [number of items $\times$ number of cases]).

This ratio indicates that a reproducibility this large could occur only rarely by chance.

Test of Assumption of Normality

It is important to note that this test is appropriate only if we assume that randomly determined reproducibilities for an infinite number of sets of data are normally distributed around chance reproducibility. To test this assumption, over sixty thousand data matrices were generated by computer, each set having specified numbers of items and cases and specified marginal frequencies.[6] Positive responses for each item in the matrix were randomly assigned until the specified number of positive responses had been distributed.[7] The number of such responses for each item was, of course, determined by the specified marginal frequencies. After each set of data was generated, errors were counted and the coefficient of reproducibility computed and distributed.[8]

From Sagi's work there was every reason to believe that these reproducibilities would distribute normally for three and four item sets with well balanced marginals. To confirm this belief, 30,000 sets of data with the characteristics used by Sagi were generated, 15,000 with marginals of 50, 50, 50, 50, and 15,000 with marginals of 20, 40, 60, 80. Tables 16.2 and 16.3 provide a comparison of Sagi's exact distributions with dis-

6. The computer work required for this analysis was performed at the Florida State University Computing Center on an IBM 709 computer which was financed in part by a National Science Foundation Grant (NSF–GP–671).

7. As a check on the results, two different randomizers were used. Both produced substantially the same results. However, for larger matrices, requiring the use of over 500,000 random numbers, a psuedo random number generator program provided quicker and more reliable results than a table of random digits which was used in preliminary runs.

8. Errors were counted according to the rule presented in footnote 4. The coefficient of reproducibility was computed using the formula $CR = 1 - E/NK$ where E is the number of errors, N the number of cases, and K the number of items.

tributions of coefficients of reproducibility of randomly generated sets of data. Plots of the cumulative proportions of these distributions on normal probability paper support the assumption of a normal distribution. In addition, when measures of skew and kurtosis were computed for each 1,000 sets of randomly generated data, the assumption of normality was supported. For each set the index of kurtosis was approximately three, while the index of skew and the mean approached zero.

When 100 case sets of data were generated with the same marginals the resulting distributions continued to pass all tests for normality and in this way supported the use of the standard error of the expected proportion of non-error responses as an appropriate statistical test. This was also true when the uneven marginals of Schuessler's hypothetical data were used. For each 1,000 cases generated, the mean was near zero, the standard deviation near one, the index of skew near zero, and the index of kurtosis around three. Sets of data were also generated for 6 item matrices and 9

Table 16.2. Comparison of Sagi's Exact Distribution of Error with the Distribution of Error for 15,000 Randomly Generated Sets of Data (N=10; Marginals 50, 50, 50, 50)

Number of Errors	Exact Distribution		Observed Distribution			Rep.
	Hypothetical Frequencies	Cumulative Percentage	Frequency	Per Cent	Cumulative Percentage	
0	1	.0000	...	...	...	1
1	50	.0003	...	...	...	.975
2	1,100	.0072	...	...	...	.950
3	13,825	.0936	...	...	...	.925
4	98,150	.7069	119 **	.79	.79	.900
5	417,252	3.3142	470	3.13	3.92	.875
6	1,544,200	12.9636	1,423	9.49	13.41	.850
7	3,012,150	31.7860	2,898	19.32	32.73	.825
8	4,516,175	60.0068	3,950	26.33	59.06	.800*
9	3,726,350	83.2921	3,519	23.46	82.52	.775
10	2,028,879	95.9702	1,875	12.50	95.02	.750
11	549,200	99.4021	639	4.26	99.28	.725
12	88,925	99.9578	102	.68	99.96	.700
13	6,550	99.9987	5 ***	.03	99.99	.675
14	200	100.0000	...	...	...	.650
15	1	100.0000	...	...	...	.625
Total	16,003,008		15,000			

* Chance reproducibility.
** Four or fewer errors.
***Thirteen or more errors.

Table 16.3. Comparison of Sagi's Exact Distribution of Error with the Distribution of Error for 15,000 Randomly Generated Sets of Data (N=10; Marginals 20, 40, 60, 80)

Number of Errors	Exact Distribution		Observed Distribution			
	Hypothetical Frequencies	Cumulative Percentage	Frequency	Per Cent	Cumulative Percentage	Rep.
0	2,520	.13	14	.09	.09	1
1	30,240	1.65	227	1.51	1.60	.975
2	158,340	9.63	1,245	8.30	9.90	.950
3	431,424	31.37	3,331	22.21	32.11	.925
4	638,470	63.54	4,746	31.64	63.75	.900*
5	502,432	88.86	3,737	24.91	88.66	.875
6	190,792	98.47	1,476	9.84	98.50	.850
7	29,232	99.95	211	1.41	99.91	.825
8	1,050	100.00	13	.09	100.00	.800
Total	1,984,500		15,000	°		

* Chance reproducibility.

item matrices and for larger numbers of cases and for extremely un-balanced marginals. In each case the assumption of normality was supported.[9]

Conclusion

This assumption of normality transforms chance reproducibility from a questionable index into a useful value and provides those working with scalogram analysis with a simple and direct way to decide if initial results are worth pursuing. In effect, it reduces the importance of arbitrary decisions in the scalogram procedure. If additional tests of this assumption reinforce the results reported above, statistical procedure should replace some of the traditional criteria used to decide when a set of data scale in a Guttman sense. For example, the requirement of a specified number of items may be eliminated when a statistical test is used, and the requirement that there be no substantial number of non-scale types may also be eliminated. The coefficient of reproducibility remains an important indication of the extent to which observed patterns approximate the theoretical cumulative pattern, but the range of the marginal frequencies would be unnecessary as a criterion for an acceptable scale if a statistical test were used.

9. Documentation will be provided by the author upon request.

Finally, it should be obvious that this procedure makes it possible for an investigator to rule out chance as an explanation of his results, but it will not rule out chance plus extensive rearrangement. If an initial test fails, response categories may be reclassified and items eliminated and the resulting data matrix may then pass a statistical test. This means that the statistical test presented in this paper, and any other test which accepts marginal frequencies and the dimensions of the data matrix as starting values of determination of the statistical significance of scalogram, is a limited test. It will help us answer the following kind of question: Given a specific set of marginals and specified numbers of items and cases, what is the probability that the observed results could be the product of chance factors? It will not help us answer this type of question: Given a set of uncombined responses, what is the probability that chance features of the data combined with repeated rearrangement and re-analysis will account for the observed pattern?

This limitation of the test does not make it useless. If it is applied, before purification, in a study where the response categories are logically determined by the nature of the problem, it may be extremely useful. In any situation where response categories and items to be used are determined beforehand, if the observed pattern of responses approximates a cumulative pattern and chance is ruled out as an explanation of the results, an investigator may then decide with confidence that scalogram purifying and scoring procedures are warranted. In the same situation, if chance is not ruled out, or if the cumulative pattern is not sufficiently approximated, scalogram analysis would be abandoned.

Chapter 17

A REVIEW AND COMPARISON OF SIMPLE STATISTICAL TESTS FOR SCALOGRAM ANALYSIS

ROLAND J. CHILTON
University of Massachusetts

Since its introduction after World War II, scalogram analysis has been repeatedly re-examined and several modifications of the original procedure have been suggested. Although a number of these attempts to improve the technique have produced some interesting results, perhaps the most important suggestions have been those which attempt to replace some of the subjective criteria for the acceptance of an observed pattern of responses as a cumulative scale with statistical procedures for achieving the same end (Green, 1956; Goodman, 1959; Sagi, 1959; Schuessler, 1961).

These statistical procedures have not been widely used and we can only speculate about the reasons for their neglect. It is possible that they have appeared to many social scientists too tedious or too difficult to compute. But it is also possible that many users of scalogram analysis have been unaware of the most important developments in this area. This is suggested by the fact that the most widely distributed computer program for scalogram analysis contains virtually no discussion of the possibility

Reprinted by permission from the American Sociological Association and the author from the *American Sociological Review*, Vol. 34, No. 2, pp. 238–45, 1969.

of capitalizing on chance factors in such analyses and provides no test of significance at all[1] (Dixon, 1965; Werner, 1966).

The review and comparison which follows is an attempt to evaluate the strengths and weaknesses of three statistical procedures for scalogram analysis by collating the work of several investigators and applying procedures suggested by them to two sets of "real" data and three sets of hypothetical data. In addition, two distributions of coefficients of reproducibility for 35,000 sets of computer generated data are used to illustrate the accuracy of one of the suggested procedures.

Suggested Statistical Procedures

A simple statistical procedure for scalogram analysis was suggested by Green in 1956 which required the computation of an expected proportion of non-error responses (chance reproducibility) and the computation of a standard error estimate for an observed coefficient of reproducibility.[2] However, Green (1956:81) indicated that such a value should be used cautiously because of the many approximations involved. In its place he recommended an index of consistency as a basis for deciding the scalability of a set of items.

Green also presented two different procedures for computing coefficients of reproducibility from summary information for which Goodman later developed approximate statistical methods. One of Green's coefficients, Rep_B, is selected for emphasis here because it is an extremely good approximation, producing accurate results for sets with as many as six items and very similar results for sets with as many as ten items.[3] More-

Revision of a paper read at the annual meeting of the American Sociological Association, August, 1968.

1. Although Werner's scalogram program (1966) includes several statistical tests, it is not widely distributed and contains some minor mistakes. Nevertheless, it is probably the most useful and promising attempt to include statistical procedures for scalogram analysis in a computer program.

2. The formula presented by Green and suggested earlier by Guttman (1950:77) for computing the standard error estimate for an observed coefficient of reproducibility is

$$\text{S.E. } Rep \approx \sqrt{\frac{(1 - Rep)\ (Rep)}{Nk}},$$

where N is the number of cases and k is the number of items.

3. Using a slightly modified form of Goodman's notations for uniformity within this paper, Green's formula is

$$Rep_B = 1 - (e/Nk) \text{ with } e = \sum_{i=1}^{k-1} f_{i,\ i+1} + \sum_{i=2}^{k-2} (f_{i,\ i+2} f_{i-1,\ i+1})\ /N$$

Where $f_{i,\ i+1}$ is the frequency of cases with a positive response to item i and a negative response to item $i+1$ and $f_{i,\ i+2}$ the frequency of cases with a positive response to item i and a negative response to item $i+2$ and N is the number of cases, k the number of items.

over, the expected value of Green's coefficient may be computed with a formula using the number of positive responses to each item, a procedure which is much less complicated than the standard procedure for computing chance reproducibility.[4]

Green's procedure for computing an approximation of the coefficient of reproducibility and his procedure for computing an expected coefficient of reproducibility are explicitly stated. But the procedure ascribed to him in this paper for estimating the importance of chance factors is not; it had to be inferred from his discussion of the possible use of the formula he presented.[5] Although he stated that it would be possible to calculate a more accurate variance figure for either of his coefficients of reproducibility, as opposed to the approximation presented above, he did not provide the necessary formula which he described as complicated and as one requiring long but straightforward computations.

In 1959, Goodman presented the formula, to which Green had alluded, for computing the variance of one of Green's coefficients of reproducibility (Rep$_B$), as well as the formula for computing the variances of Rep$_A$ and two additional measures of reproducibility not suggested by Green.[6] The formula for the variance of Green's Rep$_B$ is indeed complicated but, of course, consumes very little of the time of any contemporary computing machinery. For the reasons presented in the subsequent discussion, Goodman's formula for computing this value is probably the most important and useful of the statistical techniques now available. In comparison with other procedures which have been suggested, it appears to provide the most accurate estimate of the importance of chance factors in such analyses.

4. Using slightly modified form of Goodman's notation, this value is

$$Rep_I = 1 - (E/Nk) \text{ with } E = \frac{\sum_{i=1}^{k-1} f_i \bar{f}_{i+1}}{N} + \frac{\sum_{i=2}^{k-2} f_i \bar{f}_{i+2} f_{i-1} \bar{f}_{i+1}}{N^3},$$

Where f_i is the number of positive responses to item i, and $\bar{f}_i$ the number of negative responses to item i. For up to six items, this formula produces results identical to those produced by the conventional procedure for computing chance reproducibility. It produces very accurate results for as many as ten items. For details on the conventional procedure, see Riley (1954).

5. The procedure inferred from Green's discussion is

$$z = (Rep_B - Rep_I) / S.E. \; Rep_B$$

where Rep$_B$ and Rep$_I$ are computed using procedures recommended by Green. See notes 3 and 4 above.

6. Goodman's formula is:

$$S^2_n = \sum_{i=1}^{k-1} f_i \bar{f}_i f_{i+1} \bar{f}_{i+1}/N^3 + \left(\sum_{i=2}^{k-3} \left[S^2_{i, i+2} (E_{i-1, i+1} + E_{i+1, i+3})^2 \right] \right.$$

$$\left. + S^2_{1,3} E^2_{2,4} + S^2_{k-2,k} E^2_{k-3,k-1} \right) /N^2$$

where $E_{i,j} = f_i \bar{f}_j/N$ and $S^2_{i,j} = f_i \bar{f}_i f_j \bar{f}_j/N^3$

Working from a perspective slightly different from that of Green and Goodman, Schuessler later suggested two chi-square procedures for deciding the importance of chance factors in scalogram analyses. His first procedure requires the computation of an expected number of non-error response patterns (perfect scale types) and the corresponding expected number of response patterns, containing at least one error, which are then compared with a number of observed non-error response patterns and the corresponding number of observed error patterns to produce a chi-square value with one degree of freedom. The χ^2 obtained in this way is then converted to a z value by taking its positive square root.

Schuessler's second chi-square test requires a comparison of the observed and expected frequencies of scale types within specific score intervals. In practice, this utilizes the same expected frequencies computed for his first test, but, instead of comparing the total number of observed and the total number of expected frequencies of all perfect scale types, this procedures requires a number of additional comparisons.[7]

Table 17.1. Distribution of Marginal Probabilities, Assignment of Errors and Observed and Expected Frequencies for 100 Hypothetical Cases with Marginal Proportions of .20, .40, .60, .80

Row No.	Pattern	Distributed Marginal Probabilities				Probability of each Pattern	Frequency of each Pattern Exp.	Obs.	Errors Assigned ** G	S	L
0.*	0000	.80	.60	.40	.20	.0384	3.8	6	0	0	0
1.*	0001	.80	.60	.40	.80	.1536	15.4	15	0	0	0
2.	0010	.80	.60	.60	.20	.0576	5.8	2	1	1	1
3.*	0011	.80	.60	.60	.80	.2304	23.0	27	0	0	0
4.	0100	.80	.40	.40	.20	.0256	2.6	3	1	1	2
5.	0101	.80	.40	.40	.80	.1024	10.2	7	1	1	1
6.	0110	.80	.40	.60	.20	.0384	3.8	1	1	1	2
7.*	0111	.80	.40	.60	.80	.1536	15.4	19	0	0	0
8.	1000	.20	.60	.40	.20	.0096	1.0	4	1	1	3
9.	1001	.20	.60	.40	.80	.0384	3.8	2	1	1	2
10.	1010	.20	.60	.60	.20	.0144	1.4	1	2	2	3
11.	1011	.20	.60	.60	.80	.0576	5.8	3	1	1	1
12.	1100	.20	.40	.40	.20	.0064	.6	1	2	1	4
13.	1101	.20	.40	.40	.80	.0256	2.6	2	1	1	2
14.	1110	.20	.40	.60	.20	.0096	1.0	2	1	1	3
15.*	1111	.20	.40	.60	.80	.0384	3.8	5	0	0	0

* Indicates perfect scale type.
** G = Guttman System, S = Sagi System, L = Loevinger System.

7. A third procedure is also suggested by Schuessler but no computational details are provided; so no attempt has been made to compare it with the tests suggested above.

Error-Counting Procedures

Since all of these statistical procedures require explicit definition of error responses and error patterns, it is necessary to review some error-counting procedures. In the first such procedure, which we will refer to as the original Guttman procedure, the rule is clear and simple. Any response which would have to be changed in order that it conform to a scale pattern is an error response. Using this rule, the asterisked patterns in Table 17.1 are non-error response patterns and the response to item 3 in response pattern 2 is an error response. Response patterns 4 through 6 and 8 through 14 also contain error responses, with patterns 10 and 12 containing two each.

Variations on this error-counting procedure have been suggested for a variety of reasons, but it is not clear how much they have contributed to the original procedure. Although some of them may have been easier to compute at one time, they are clearly more difficult to interpret. Moreover, recent machine developments have made any computational advantages they might possess less important.

Examples of such coefficients are provided by Sagi (1959), one of which he attributes to Loevinger (1947). Of the two, only the one attributed to Loevinger (1947) produces a sharply different value in place of the original coefficient of reproducibility. Sagi's procedure requires that the items be ordered according to the number of positive responses, most popular items to the right, and that each adjacent pair of responses be examined for a step-down pattern ("10"); i.e. that an error be counted when a positive response ("1") to a less popular item is followed by a negative response ("0") to a more popular item (popularity here referring to the number of positive responses an item receives from the set of respondents). Table 17.1 illustrates this procedure for four items by presenting all possible response patterns. The reader will notice that, with one exception, the number of errors assigned to each pattern in this way is identical to the number assigned in the Guttman error-counting procedure. For larger sets of items the Sagi procedure fails to count larger numbers of errors.

Just as the Sagi procedure produces fewer error responses than the original Guttman procedure, the procedure attributed to Loevinger results in a much larger error count. This occurs because, after the items are ordered by the number of positive responses they contain, each response is compared with each of the more popular responses. Not only are adjacent step-down patterns ("10") counted as errors, but a step-down pattern for a non-adjacent pair of responses is also counted as an error. Thus the pattern "1100" is not assigned one error, as it would be by Sagi, or two errors, as it would be by Guttman, but rather four errors. Table 17.1 indicates the number of errors assigned to each possible response pattern for four dichotomized items by the Loevinger procedure.

With the increasing availability of computing equipment and computer programs, and in light of the contributions of Green and Goodman, there appears to be little advantage in these variations on the original error-counting procedure. Even without computing equipment, the simplest and most objective procedure requires only that the items be ordered according to the number of positive responses they contain and that the number of error responses be counted according to the original rule.

It is also appropriate to note at this point that the error-counting procedure which was developed as a paper and pencil approximation of the original Guttman procedure, the Cornell technique (Guttman, 1947), should clearly be abandoned now that simpler and more objective techniques are available. In addition, the practice of arraying both the positive and negative responses and counting as separate errors any out of place positive responses and their negative counterparts should be avoided (see Edwards, 1957). If this is done, the appropriate denominator is two times the number of items times the number of cases. Since this will produce exactly the same coefficient of reproducibility as the procedure, which utilizes only the positive response, there is little point in complicating the count.[8]

Comparison of Tests

When the tests suggested by Green, Goodman, and Schuessler are employed for the same sets of data, three different estimates of the probable importance or chance factors are produced. For example, the z value of a set of data with an observed coefficient of reproducibility of .970 for three items and 100 cases with marginals of .35, .60, and .80 is 5.75 when Green's standard error procedure is used; 5.57 when Goodman's procedure is employed; and 3.88 when Schuessler's Test I is computed. For a set of data with four items and 100 cases, marginals of .20, .40, .60, .80 and an observed coefficient of reproducibility of .925, a similar discrepancy appears with Green's procedure producing a z value of 2.02; Goodman's a value of 2.82; and Schuessler's a value of 2.17.

When this comparison is made for the dichotomized response of 99 people to six items, with an observed reproducibility of .90 and an expected reproducibility of .82, the corresponding z values are Green, 6.29; Goodman, 8.65; and Schuessler, 5.83. For seven items and 100 cases, the divergence is slightly larger with Green's procedure producing

---

8. The practice may have some practical advantage in the analysis of data using paper and pencil techniques, but there appears to be no advantage in adopting it for machine analysis.

a z value of 5.46; Goodman's a value of 8.15; and Schuessler's a value of 6.83. Finally, when eight items and 167 cases are tested, the z values are 7.47, 11.16, and 6.94.[9]

Goodman provides a comprehensive description of the mathematical bases for the use of the formula presented in Footnote 6. But as an illustration of the accuracy of his procedure, Tables 17.2 and 17.3 present the distribution of Guttman's coefficient of reproducibility for 35,000 sets of randomly generated data. These distributions not only illustrate the normal distribution of coefficients of reproducibility for randomly produced sets of data around "chance reproducibility," but also show the accuracy of the z values computed with Goodman's formula.[10]

The earlier procedure implied by Green, for example, results in a σ value of .0098 for an observed reproducibility of .970 for three items, with the marginals of those presented in Table 17.2. This value would suggest that approximately 18 errors, or a reproducibility of .940, would fall within 3 σ's of chance reproducibility, which is very close to the conclusion suggested by the distribution in Table 17.2. The standard error figure for the same data, produced when Goodman's procedure is followed, is .0102, which suggests that 17 errors or a reproducibility of .943 would fall within 3 σ's of chance reproducibility. For the distribution in Table 17.2 this is the more accurate figure.

A similar comparison of the standard error values produced by Green's and Goodman's systems for four items with marginals of .20, .40, .60, .80 is illustrated in Table 17.3. Following Goodman's procedure, we arrive at a standard error value of .0098 for an observed reproducibility of .925, which suggests that 37 errors, or a reproducibility of .907, would fall within 1 σ of chance reproducibility. However, when the Green approximation is used, the standard error for the same data is .0132, which would suggest that 68% of the randomly generated sets of data would have reproducibilities between .887 (fewer than 45 errors) and .913 (more than 35 errors). In this way, the distribution in Table 17.3 suggests that the Green approximation produces too large a standard error figure.

Schuessler's first procedure produces results which most frequently fall between those produced by the Goodman and Green approaches, and

9. For the six-item set of data, the marginals were .26, .33, .35, .40, .46, .68; for the seven-item set, .26, .34, .35, .40, .41, .47, .69; and for the eight-item set they were .28, .49, .49, .95, .101, .103, .123, .131. The observed and expected coefficients of reproducibility for the seven-item data were .873 and .804. For the eight-item data, they were .894 and .831.

10. For additional details on this procedure, see Chilton (1966). The article also presents a variation on Green's test, which is less accurate than the procedure attributed to Green in this paper and therefore ignored. Computer work for both papers was performed at the Florida State University Computing Center using machines financed in part by the National Science Foundation Grants (NSF–GP–671 and NSF 1981).

Table 17.2. Distribution of Error for 15,000 Randomly Generated
Sets of Data N=100, Marginals .35, .60, .80

No. of Errors	Frequency	Percent	Cumulative Percentage	Rep.
15, 16	13	.09	.09	. . . .
17	32	.21	.30	.943
18	71	.47	.77	.940
19	144	.96	1.73	.936
20	296	1.97	3.70	.933
21	482	3.21	6.91	.930
22	795	5.30	12.21	.926
23	1204	8.03	20.24	.923
24	1634	10.89	31.13	.920
25	1828	12.19	43.32	.916
26	1975	13.17	56.49	.913*
27	1931	12.87	69.36	.910
28	1534	10.23	79.59	.906
29	1207	8.05	87.64	.903
30	806	5.37	93.01	.900
31	539	3.59	96.60	.896
32	292	1.95	98.55	.893
33	135	.90	99.45	.890
34	49	.33	99.78	.886
35	22	.15	99.93	.883
36 plus	11	.07	100.00	. . . .
	15,000			

(Bracket annotations spanning the Percent column: 99.84, 96.82, 75.43)

* Chance Reproducibility.

this suggests that it, too, is less accurate than the Goodman procedure. It has an additional disadvantage in its requirement that the investigator count the number of perfect scale types and distribute the probabilities for 2^k response patterns, where k is the number of items. Its advantage is that the results are the same regardless of the error-counting procedure being used, because it employs only the observed and expected number of perfect scale types. However, since there is little reason for using the alternative error-counting procedures, this does not appear to be an important advantage.

Perhaps the best procedure in most situations, and one which should be included in any general scalogram program, is the computation of chance reproducibility according to the Green procedure and the computation of the standard error of the observed reproducibility according to Goodman's formula.[11] Neither of these procedures requires the

11. See notes 4 and 6 above.

Table 17.3. Distribution of Error for 20,000 Randomly Generated Sets of Data N=100, Marginals .20, .40, .60, .80

No. of Errors	Frequency	Percent	Cumulative Percentage	Rep.
26	1	.005	.005	.9350
27	8	.040	.045	.9325
28	20	.100	.145	.9300
29	35	.175	.320	.9275
30	75	.375	.695	.9250
31	142	.710	1.405	.9225
32	228	1.140	2.545	.9200
33	426	2.130	4.675	.9175
34	657	3.285	7.960	.9150
35	931	4.655	12.615	.9125
36	1309	6.545	19.160	.9100
37	1678	8.390	27.550	.9075
38	1942	9.710	37.260	.9050
39	2182	10.910	48.170	.9025
40	2154	10.770	58.940	.9000*
41	2083	10.415	69.355	.8975
42	1760	8.800	78.155	.8950
43	1432	7.160	85.315	.8925
44	1091	5.455	90.770	.8900
45	779	3.895	94.665	.8875
46	464	2.320	96.985	.8850
47	302	1.510	98.495	.8825
48	146	.730	99.225	.8800
49	96	.480	99.705	.8775
50	39	.195	99.900	.8750
51	12	.060	99.960	.8725
52	6	.030	99.990	.8700
53	2	.010	100.000	.8675
	20,000	100.000		

(Bracket annotations within the table: 99.815, 95.950, 66.155)

* Chance reproducibility.

distribution of the marginal probabilities into the k by 2^k matrix, which is required by the customary procedure for computing chance reproducibility. In addition, the results are so similar to those achieved by the earlier approach that the Green and Goodman procedures may be used interchangeably with great confidence.[12] Finally, since the simpler standard error formula suggested by Green usually results in a larger standard error value, the investigator working without the aid of a computer

12. Tests of this formula for sets of data, with as many as ten items, indicate that it is even more accurate than Green's formula for an observed reproducibility, Rep_B.

program may use it with the assurance that when it indicates that his observed reproducibility is significantly larger than the expected value, then the more accurate procedure would also support this conclusion. Only in borderline situations would it be necessary to use the more complicated formula. However, there appears to be no good reason why the more accurate procedure should not be included in any generally distributed scalogram program.

Conclusion

This re-examination and collation of the work of Goodman, Green, Sagi, and Schuessler suggest that scalogram analysis can be employed in an objective, replicable, and easily reported manner and that such reports can include reasonably accurate estimates of the importance of chance factors in the production of the observed results. If interpretations or conclusions are based upon a scalogram analysis, the reader should expect to find a clear definition of an "error response," the exact proportion of non-error responses contained in the data (the coefficient of reproducibility), the marginal frequencies or proportions, and the number of cases analyzed. With this information and knowledge of the number of original items discarded, the reader could intelligently assess the importance of the findings and conclusions presented.[13] In addition, users of generally available computer programs should expect scalogram programs to compute an expected reproducibility and an accurate standard error figure which will assist them in deciding the probable importance of chance factors.

13. The importance of reporting the number of discarded items is made very clear in Sagi's article in which he indicates the futility of applying statistical techniques after extensive purification of the original data.

References

CHILTON, ROLAND J. "Computer generated data and the statistical significance of scalogram." *Sociometry* 29 (June 1966) :175–181.

DIXON, WILFRED J. (Ed.) *Biomedical Computer Programs.* Los Angeles: Health Sciences Computing Facility of the University of California, 1965.

EDWARDS, ALLEN L. *Techniques of Attitude Scale Construction.* New York: Appleton-Century-Crofts, 1957.

GOODMAN, LEO A. "Simple statistical methods for scalogram analysis." *Psychometrika* 24 (March 1959) :29–43.

GREEN, BERT F. "A method of scalogram analysis using summary statistics." *Psychometrika* 21 (March 1956) :79–88.

GUTTMAN, LOUIS "The Cornell Technique for scale and intensity analysis." *Educational and Psychological Measurement* 7 (Summer 1947) :247–280.

————— "The basis for scalogram analysis," in Samuel A. Stouffer *et al., Measurement and Prediction,* Princeton: Princeton University Press, 1950.

LOEVINGER, JANE. "A systematic Approach to the Construction and Evaluation of Tests of Ability." *Psychological Monographs* 61. 1947.

RILEY, MATILDA W., JOHN W. RILEY, JR., JACKSON TOBY, and MARCIA L. TOBY. *Sociological Studies in Scale Analysis.* New Brunswick, New Jersey: Rutgers University Press, 1954.

SAGI, PHILIP C. "A statistical test for the significance of a coefficient of reproducibility." *Psychometrika* 24 (March 1957) :9–27.

SCHUESSLER, KARL F. "A note on statistical significance of scalogram." *Sociometry* 24 (September 1961) :312–318.

WERNER, ROLAND. "A Fortran Program for Guttman and Other Scalogram Analysis." *Computer Data Analysis Working Paper Number 1.* Syracuse, New York: Systems Research Committee, 1966.

Chapter 18

A NOTE OF EXTREME CAUTION ON THE USE OF GUTTMAN SCALES

CARMI SCHOOLER
National Institute of Mental Health

Guttman Scale (or Scalogram) analysis is one of the most widely used methods for assuring that all items in an index are drawn from the same universe of attributes. The present paper starts with a brief summary of the logic, apparently not universally known to the field, which demonstrates that Guttman Scales meeting generally accepted tests of unidimensionality can occur by chance alone. It then examines a method that has been proposed for statistically evaluating whether a given scale is merely the result of chance factors.[1] This method is then used to demonstrate that Scalogram analysis, as generally used at present, provides no definitive proof that all the items in a given scale are measures of the same dimension.

From its inception, proponents of Scalogram analysis have been concerned about the role chance might play in producing apparently acceptable Scales.[2] More recently, it has become evident through the work of Schuessler,[3] Sagi,[4] and Chilton[5] that apparently meaningful Guttman

Reprinted by permission from the University of Chicago Press and the author from the *American Journal of Sociology*, Vol. 74, No. 3, pp. 296–301, 1968.

1. Roland J. Chilton, "Computer Generated Data and the Statistical Significance of Scalogram," *Sociometry*, XXIX (June, 1966), 175–81.

Scales with coefficients of Reproducibility of 0.90 or more and coefficients of Scalability[6] of 0.60 or more can occur by chance alone. Such possibilities exist because the a priori likelihood of the occurrence of a particular scale type is equal to the joint probability of the occurrence of each of the constituent responses. Since the probability of the occurrence of a particular response to a question is its marginal frequency, the probability of the occurrence of all possible scale types can be calculated directly from knowledge of the marginal frequencies alone. This is done by multiplying the proportion of individuals making the appropriate response to each of the questions in the set. For example, the proportion of people who would fall by chance into the "yes, yes, no, no" scale type, defined by a response pattern involving saying "yes" to an item with which 80 percent of the sample agree, "yes" to an item with which 60 percent agree, "no" to an item with which 40 percent agree, and "no" to an item with which 20 percent agree, is 0.23 ($0.80 \times 0.60 \times 0.60 \times 0.80 = 0.23$). For any given set of marginal frequencies, the Reproducibility and Scalability attributable to chance can be obtained by calculating the chance probability of occurrence of each possible scale type, totaling the probabilities of the "correct" and "incorrect" patterns and entering the resultant totals into the appropriate formulas.

As can be seen from Table 18.1, given the theoretically conservative situation where 80 percent and 20 percent are the extreme cutting points, and the other cutting points are evenly distributed (e.g., 0.80, 0.65, 0.50, 0.35, and 0.20 for a five-item scale), chance Reproducibility would be well above 0.90 for three-item scales, would hover around 0.90 for four-item scales, and would drop only to 0.85 when six items are used. In fact, the use of ten items reduces chance Reproducibility only to 0.80. Furthermore, the conventionally accepted standard of 0.60 Scalability is met by chance alone by three-, four-, and five-item scales, and even the six-item scale has a chance Scalability of 0.59.[7]

Chilton has suggested a method "which transforms chance Reproducibility into a useful value and provides those working with Scalogram

2. See, for example, Louis Guttman's discussion in Samuel Stouffer, Louis Guttman, Edward A. Suchman, Paul F. Lazarsfeld, Shirley Star, and John A. Clausen (eds.), *Measurement and Prediction* (Princeton, N.J.: Princeton University Press, 1950), p. 82.

3. Karl F. Schuessler, "A Note on Statistical Significance of Scalogram," *Sociometry*, XXIV (September, 1961), 312–18.

4. Phillip O. Sagi, "A Statistical Test for a Coefficient of Reproducibility," *Psychometrica*, XXIV (March, 1959), 29–42.

5. Chilton, *op. cit.* (n. 1 above), p. 175.

6. Herbert Menzel, "A New Coefficient for Scalogram Analysis," *Public Opinion Quarterly* (Summer, 1953), pp. 268–80.

7. These figures would seem to support Guttman's early warning that "if the items are dichotomous ... it is probably desirable that at least 10 items be used" (Stouffer, *op. cit.* [n. 2 above], p. 71).

Table 18.1. Characteristics of Various Sized Guttman Scales, Calculated on a Probability Basis, Given Marginal Distributions with 80 Per Cent and 20 Per Cent as the Extremes of a Set of Even and Symmetrical Cutting Points

	Number of Items			
	3	*4*	*5*	*6*
Cutting points = (% cumulative "yes" response)	(80/50/20)	(80/60/ 40/20)	(80/65/ 50/35/ 20)	(80/68/ 56/44/ 33/20)
Probable no. errors per individual	0.200	0.4064	0.635	0.877
Chance Reproducibility	0.934	0.898	0.873	0.854
Chance Scalability:				
Item	0.778	0.661	0.604	0.612
Individual	0.762	0.691	0.649	0.594
% in largest error category	8.00	10.24	7.28	5.73
Highest proportion (category error) / (category frequency)	0.30	0.448	0.661	0.663
Highest % total errors per question	8.00	11.36	14.04	15.84

analysis with a simple and direct way to decide if initial results are worth pursuing."[8] This is done by dividing the difference between the expected and observed Reproducibilities by the standard error of the expected proportion of non-error responses, where the standard error is defined as

$$\sigma p = \sqrt{\frac{(\text{chance Reproducibility}) (1 - \text{chance Reproducibility})}{\text{number of cases}}}.$$

The method is based on the assumption "that randomly determined Reproducibilities for an infinite number of sets of data are normally distributed around chance Reproducibility."[9] Calculations based on 45,000 sets of machine-generated data suggest that the assumption is true. Chilton foresees that "the requirement of a specified number of items may be eliminated when a statistical test is used, and the requirement that there be no substantial number of non-scale types may also be eliminated."[10]

8. Chilton, *op. cit.* (n. 1 above), p. 180.
9. *Ibid.*, p. 178.
10. *Ibid.*, p. 180. Chilton's method assumes that only one set of cutting points is evaluated. The greater the number of potential cutting points, the greater the likelihood of capitalizing on chance so as to achieve acceptable levels of Reproducibility. Since high Reproducibilities can be obtained, irrespective of content, by the selection of items with appropriate cutting points, the number of potentially examinable cutting points directly affects the likelihood of an item's being spuriously related to other

A major difficulty, however, remains: Even if the Reproducibility of a given scale is significantly better than chance, this does not insure that all its component items are drawn from the same universe of attributes. The problem is best brought into focus by examining the probable chance error per individual in the situation where there is a symmetrical distribution and 80 percent and 20 percent are the extreme cutting points. Under such conditions, an individual will by chance make only 0.41 errors in a four-item scale and 1.95 errors in a ten-item scale. Thus, chance alone provides scales with relatively few errors. The explanatory effect of the probabilities, calculable on the basis of the marginal distributions alone, is so powerful that it leaves little unexplained variance. Not very much information is gained when we learn that a person has made significantly fewer than 1.95 errors in ten items. After the potential effects of the marginal distribution are taken into account, there may not be enough variance left to insure that *all* of the items are drawn from the same universe of attributes. Results significantly better than chance may come about merely because a small subset of the items is appropriately interrelated.

One way to explore this problem is through the use of artificial data. Sets of such data can be constructed which differ both in the numbers of pairs of items whose interrelationships meet the criteria of Scalogram analysis and in the number of items for which the responses are randomly supplied. If a Scalogram is to be effective there should be relatively few cases where an individual says "no" to one item and "yes" to another, less frequently endorsed, item. For a scale to be completely homogeneous, so that *all* of its items come from the same universe of attributes, this type of relationship has to exist between each possible pair of items. To the extent that a scale contains pairs of items between which such a relationship does not exist, it falls short of fulfilling the concept of unidimensionality which Scalogram analysis seeks. The effectiveness with which Scalogram analysis insures that such a relationship exists among *all* items within a scale can be assessed by evaluating those scales that would occur on a probability basis if some of the pairs of items were so

---

items. Thus, random responses to each of a set of four initially dichotomous items would result in Reproducibility of about 0.80, but if each of the items had had five potential cutting points from which a point of dichotomization could have been chosen, Reproducibilities in the 0.90's would have been reached. Given the extreme case, where the response to the items has been totally random, a kind of maximum efficiency is reached when the number of cutting points per item equals the potential number of correct scale types (i.e., 4 cutting points per item in a 3-item scale, 5 in a 4-item scale, etc.), because every item can then provide a cutting point which can be efficiently used. Thus, the final probability of a significant difference from chance is a function of both the particular marginal distribution used and the relative uniqueness of that distribution as a description of the data at hand.

related and others were not. Such scales can be constructed by hypothetically restricting the possible interrelationships between varying members of pairs of items so as to exclude that set of error patterns which occurs when an individual accepts the less frequently endorsed item in a given pair.

Table 18.2 shows the results of imposing such restrictions on varying numbers of items in scales of different sizes having the marginal distributions of those in Table 18.1. Scale types that could not occur under these restrictions were eliminated, the probable occurrence of the remaining types recalculated,[11] and the resultant level of Reproducibility compared with the chance Reproducibility that would occur if no such restrictions were imposed. Given 300 respondents and the marginal distributions of Table 18.1, relationships between *only three* items in four-, five-, and six-item scales can result in Scalabilities greater than 0.60 and Reproducibilities both greater than 0.90 and significantly greater than chance. Thus, it is possible that even when Reproducibility is greater than chance, one item in a four-item scale, two items in a five-, and three items in a six-item scale may bear no relationship to a putative common universe of attributes other than that their marginal frequencies happen or are constructed to fit convenient points in a cumulative distribution. Scalogram analysis, even when evaluated by the criteria of 0.60 Scalability, 0.90 Reproducibility, and significant Reproducibility according to Chilton's method, provides no proof that *all* of the constituent items are necessarily drawn from a homogeneous universe of attributes.

One set of criteria for acceptable Guttman Scales that does seem to reject the nonhomogeneous scales depicted in Table 18.2 are those presented by Ford.[12] His criteria are: (1) A specific error pattern may contain no more than 5 percent of the sample population. (2) The frequency of error resulting from a given response (i.e., rejection of a given item, or acceptance of a given item) should be less than half the frequency of that response. (3) Reproducibility must be 0.90 or more. (4) No question may provide over 15 percent of the total amount of error. Every scale in Table 18.2 violates at least one of these criteria.

These criteria, however, present several difficulties. First, they are not

11. I want to thank Isidore Chein for supplying the formula for recalculating the probability of occurrence of the remaining scale types. The formula is:
New probability of remaining response patterns =

$$\frac{\text{original probability}}{1 - \text{sum of the original probability of the removed response patterns}}.$$

12. Robert N. Ford, "A Rapid Scoring Procedure for Scaling Attitude Questions," in Matilda W. Riley, John W. Riley, Jr., Jackson Toby, in association with Marcia L. Toby, Richard Cohn, Harry C. Bredemeier, Mary Moore, and Paul Fine, *Sociological Studies in Scale Analysis* (New Brunswick, N.J.: Rutgers University Press, 1954).

Table 18.2 Characteristics of Various Sized Guttman Scales, Calculated on a Probability Basis, Given Marginal Distributions with 80 Per Cent and 20 Per Cent as the Extremes of a Set of Symmetrical Cutting Points; with Various Numbers of Error-Producing Relationships among Items Eliminated

	Number of Items							
Errors eliminated	Q1no-Q8yes Q4no-Q8yes	Q2no-Q4yes Q4no-Q8yes	Q2no-Q8yes Q4no-Q8yes	Q1no-Q 4yes Q4no-Q16yes	Q1no-Q2yes Q2no-Q4yes Q4no-Q8yes	Q2no-Q4yes Q4no-Q8yes	Q1no-Q 4yes Q4no-Q32yes	Q1no-Q2yes Q2no-Q4yes Q4no-Q8yes
	4	4	5	5	5	6	6	6
	Number of Items Related							
	3	3	3	3	4	3	3	4
Reproducibility	0.950	0.963	0.934	0.920	0.967	0.903	0.894	0.933
Significance of difference from chance (300 cases):								
$z=$	2.98	3.57	3.17	2.44	4.87	2.44	1.99	3.90
$p\leq$	.01	.001	.002	.02	.001	.02	.05	.001
Scalability:								
Item	0.843	0.878	0.794	0.700	0.896	0.698	0.667	0.792
Individual	0.837	0.884	0.787	0.719	0.891	0.718	0.694	0.747
% in largest error type	6.78	8.00	5.18	9.10	6.23	6.96	7.16	4.36
Highest proportion (category error)/(category frequency)	0.395	0.533	0.663	0.347	0.677	0.765	0.420	0.682
Highest % of total errors per question	11.3	10.7	13.3	15.41	16.64	17.95	15.12	18.31

stringent enough. The data presented in Table 18.2 represent the theoretically derived Reproducibilities and modal distributions of individuals into scale types, given various sets of marginal distributions and limited interrelationships among items. If repeated samples were taken with such distributions and interrelationships, both the Reproducibilities and numbers of individuals falling into the various scale types would vary around the theoretical central tendencies presented in Table 18.2. Since several of these scales barely miss satisfying all of Ford's criteria, it seems highly probable that the variation to be expected about the central tendencies depicted in Table 18.2 would result in some instances where scales with similar properties would meet all of Ford's criteria.

Second, Ford's criteria do not take account of the obvious differences among scales containing varying numbers of items. Given a random response pattern, the chance of an error scale type containing more than 5 percent of the sample or of a single question accounting for more than 15 percent of the total error is a direct function of the number of items in the scale.

Both of the above problems are functions of the most general difficulty with Ford's criteria—they are rules of thumb rather than deductions from the relationship between Scalogram analysis and probability theory. Consequently, there is no rationally based proof that all of the items in a scale that meet Ford's criteria are in fact drawn from the same universe of attributes. Even if more stringent versions of Ford's criteria are used, until such criteria are put on a theoretically meaningful and statistically assessable basis we have no real knowledge of the extent to which Scalogram analysis can be used to isolate a set of items, *all* of which come from a homogeneous domain of attributes.

Several approaches to statistically assessing the homogeneity of Guttman Scales can be suggested. One, proposed by John D. Campbell, would be to compare the observed interrelationship between each pair of items in a scale (e.g., Table 18.3, A), both with those that would occur by chance (Table 18.3, B) and with those would occur if a perfect scale existed (Table 18.3, C). With n number of items, there would be $n(n-1)/2$ pairs to examine. If the observed interrelationship between each pair was significantly better than chance and not significantly worse than that which would be expected if a perfect scale existed, then an argument could be made that the scale being tested is both better than chance and not significantly worse than a perfect scale.

A second approach would be to develop a computer program that would calculate the response patterns that would result from systematically varying, for scales containing differing numbers of items, the degree to which, among varying pairs of items, accepting a less frequently en-

Table 18.3. Campbell Assessment of Guttman Scale Homogeneity

	Item Accepted Less		
	Yes	No	Total
	A. "Observed" Relationship		
Item accepted more:			
Yes...................................	30	30	60
No....................................	10	30	40
Total............................	40	60	100
	B. Chance Relationship		
Item accepted more:			
Yes...................................	24	36	60
No....................................	16	24	40
Total............................	40	60	100
	C. Perfect Relationship		
Item accepted more:			
Yes...................................	40	20	60
No....................................	0	40	40
Total............................	40	60	100

dorsed item increases the probability of accepting a more frequently endorsed one. The resultant distributions of response patterns could then be examined to establish which criteria would best insure homogeneity.

A third possibility would be to consider Scalogram analysis as a test for heterogeneity rather than homogeneity. Although currently accepted procedures for Scalogram analysis do not insure that all of the items in an "acceptable" scale fulfil the underlying conception of unidimensionality, scales that do *not* meet current criteria are almost certainly heterogeneous.

PART IV

Summated Rating Methods

This section is concerned with summated rating scaling, or as it is frequently called, Likert scaling. This method is oriented more to attitude measurement than the others discussed in this book. Most other scaling methods can be employed in the measurement of attitudes and related phenomena and also in the scaling of other objects, events, items, and stimuli. To apply the Likert method to a scaling task one must be able to employ multiple response items. The items are used to order the respondents or subjects. The items are usually examined in regard to discriminate ability or sensitivity. If the items are ordered, it is in regard to these aspects and not in terms of the position relative to the dimension being scaled. Such items are usually used in the scaling of opinion, information, satisfaction, and values as well as attitudes.

Summated Rating Scaling is one of the most frequently used methods even though it has a somewhat limited applicability. The method has been used very consistently since it was introduced and an examination of the scaling literature as well as the substantive literature of the behavioral sciences would suggest that its use has not diminished. This is probably true because a sizable proportion of all scaling has been attitude scaling. Likert attitude scales have been developed for such areas as political attitudes, religious attitudes, familial attitudes, economic attitudes, etc. Likert scaling is also one of the simplest and easiest methods to use. This fact also may serve to explain, at least in part, its continuing popularity. Therefore, although this section of the present volume is not large, it is particularly important for attitude scaling.

In view of its popularity and simplicity, it is fortunate that Likert scales have been found to have reliability coefficients that compare very favorably with those of other types of scales. At least part of the material included in this section is concerned with comparative reliability. It should be noted that this comparison of reliabilities can be made only when the other methods are employed in attitude measurement. This fact is particularly relevant when one recognizes that if some other scaling method is used in the measurement of attitudes an additional set of assumptions and an additional task is required to secure individual attitude scores from stimuli or statements which have previously been scaled by the alternative methods.

Chapter 19, by Rensis Likert, presents the method for constructing a summated rating attitude scale. Likert presents criteria for the selection of statements or items, a description of the process of conducting an item analysis, and an internal consistency check.

Chapter 20, by Robinson, et al., is a more recent consideration of these and other matters relevant to someone who is constructing an attitude scale. This paper provides a discussion of content sampling, item wording, and item analysis. These are the matters of general concern to scale constructors. In addition, this selection presents a particularly important consideration of problems of response set, statistical standards, validation, reliability, and normative information. This selection provides an attitude researcher with an indication of what should be taken into account in making an attitude scale and what should be provided to the reader in the description of a scale.

Chapter 21 describes, compares, and criticizes three alternative techniques of scoring attitude scales using Likert and Thurstone procedures jointly. These techniques are Thurstone-weighted scoring and two scale product scoring procedures. These are attempts to score Likert items in ways that take into account the Thurstone position or score of an item, i.e., its strength.

The final selection in this section, chapter 22, compares the reliability and validity of Likert scale scoring, a scoring system developed with one of the scale product procedures described in the preceding chapter, regular Thurstone scoring, and an additional scoring method. The generally high reliability coefficients associated with regular summated rating scoring system suggests that the various alternatives have not yet found a sufficient justification for their general use. It is, however, of some importance to know that alternative scoring systems have been discussed and examined. The nature of these attempts should also be informative to persons being introduced to measurement and scaling.

Chapter 19

THE METHOD OF CONSTRUCTING AN ATTITUDE SCALE

RENSIS LIKERT
University of Michigan

I. The Selection of Statements

Each statement should be of such a nature that persons with different points of view, so far as the particular attitude is concerned, will respond to it differentially. Any statement to which persons with markedly different attitudes can respond in the same way is, of course, unsatisfactory.

The results obtained in constructing the present scales demonstrate the value of the following criteria. These criteria were kept in mind in collecting the statements for the original Survey of Opinions.

1. It is essential that all statements be expressions of *desired behavior* and not statements of *fact*. Two persons with decidedly different attitudes may, nevertheless, agree on questions of fact. Consequently, their reaction to a statement of fact is no indication of their attitudes. For example, a person strongly pro-Japanese and a person strongly pro-Chinese might both agree with the following statements:

"The League of Nations has failed in preventing Japan's military occupation of Manchuria."

Reprinted with permission of the author and the publisher from "A Technique for the Measurement of Attitudes," pp. 44–53. *Archives of Psychology,* #140, Columbia University 1932.

or

"Japan has been trying to create in Manchuria a monopoly of trade, equivalent to closing the 'open-door' to the trade of other countries."
To agree with them or believe them true is in no way a measure of attitude.

Rice (p. 184) has clearly stated the importance of recognizing this criterion when in discussing the Thurstone technique he says:

> What is the possibility that the acceptance or rejection by a subject of a statement upon the completed scale may represent a rational judgment concerning the truth or falsity of the statement made? It would seem to exist. If so, the validity of the statement as an index of attitude is destroyed or impaired.

In dealing with expressions of desired behavior rather than expressions of fact the statement measures the present attitude of the subject and not some past attitude. The importance of dealing with present rather than past attitudes has been emphasized by Thurstone and Murphy (p. 615). A very convenient way of stating a proposition so that it does involve desired behavior is by using the term *should*.

2. The second criterion is the necessity of stating each proposition in *clear, concise, straight-forward statements*. Each statement should be in the simplest possible vocabulary. No statement should involve double negatives or other wording which will make it involved and confusing. Double-barreled statements are most confusing and should always be broken in two. Often an individual wishes to react favorably to one part and unfavorably to the other and when the parts are together he is at a loss to know how to react. Thus in the following illustration a person might well approve one part and disapprove another part: "In order to preserve peace, the United States should abolish tariffs, enter the League of Nations, and maintain the largest army and navy in the world." To ask for a single response to this statement makes it meaningless to the subject. This statement should be divided into at least three separate statements.

The simplicity of the vocabulary will, of course, vary with the group upon whom the scale is intended to be used, but it is a desirable precaution to state each proposition in such a way that persons of less understanding than any member of the group for which the test is being constructed will understand and be able to respond to the statements. Above all, regardless of the simplicity or complexity of vocabularly or the naiveté or sophistication of the group, each statement *must avoid every kind of ambiguity*.

3. In general it would seem desirable to have each statement so worded

that the modal reaction to it is approximately in the middle of the possible responses.

4. To avoid any space error or any tendency to a stereotyped response it seems desirable to have the different statements so worded that about one-half of them have one end of the attitude continuum corresponding to the *left or upper* part of the reaction alternatives and the other half have the same end of the attitude continuum corresponding to the *right* or *lower* part of the reaction alternatives. For example, about one-half the statements in the Internationalism scale have the international extreme corresponding with "Strongly approve" while the other half has it corresponding with "Strongly disapprove." These two kinds of statements ought to be distributed throughout the attitude test in a chance of haphazard manner.

5. If multiple choice statements are used, the different alternatives should involve *only a single attitude variable* and not several.

II. Constructing the Scale

It is usually desirable to prepare and select more statements than are likely to be finally used, because after trying the statements upon a group, some may be found to be quite unsatisfactory for the intended purpose. For this reason after selecting a good number of statements they should be given to the group or a part of the group whose attitudes we wish to measure. The sample used should be sufficiently large for statistical purposes.

For purposes of tabulation and scoring, a numerical value must be assigned to each of the possible alternatives. If five alternatives have been used, it is necessary to assign values of from one to five with the three assigned to the undecided position on each statement. The "one" end is assigned to one extreme of the attitude continuum and the "five" to the other; this should be done consistently for each of the statements which it is expected will be included in the scale. Thus if we arbitrarily consider the "favorable to the Negro" extreme "five" and the "unfavorable to the Negro" extreme "one," the alternative responses to the following statements would be assigned the values shown in the material on page 236.

Some may object to the designation made, saying that the terms "favorable" and "unfavorable" are ambiguous or that the favorable attitude is just opposite to that here considered favorable.

Numerical
Value

"How far in our educational system (aside from trade education) should the most intelligent negroes be allowed to go?

1   (a) Grade school.
2   (b) Junior high school.
3   (c) High school.
4   (d) College.
5   (e) Graduate and professional schools."

Numerical
Value

"In a community where the negroes outnumber the whites, a negro who is insolent to a white man should be:

5   (a) excused or ignored.
4   (b) reprimanded.
3   (c) fined and jailed.
2   (d) not only fined and jailed, but also given corporal punishment (whipping, etc.).
1   (e) lynched."

"All negroes belong in one class and should be treated in about the same way."

	STRONGLY APPROVE	APPROVE	UNDECIDED	DISAPPROVE	STRONGLY DISAPPROVE
Value	(1)	(2)	(3)	(4)	(5)

"Where there is segregation, the negro section should have the same equipment in paving, water, and electric light facilities as are found in the white districts."

	STRONGLY APPROVE	APPROVE	UNDECIDED	DISAPPROVE	STRONGLY DISAPPROVE
Value	(5)	(4)	(3)	(2)	(1)

So far as the measurement of the attitude is concerned, it is quite immaterial what the extremes of the attitude continuum are called; the important fact is that persons do differ quantitatively in their attitudes, some being more toward one extreme, some more toward the other. Thus, as Thurstone has pointed out in the use of his scales, it makes no difference whether the zero extreme is assigned to "appreciation of" the church or "depreciation of" the church, the attitude can be measured in either case and the person's reaction to the church expressed.

The split-half reliability should be found by correlating the sum of the odd statements for each individual against the sum of the even statements. Since each statement is answered by each individual, calculations can be reduced by using the sum rather than the average.

An objective check ought then to be applied to see (1) if the numerical values are properly assigned and (2) whether the statements are "differentiating." One possible check is item analysis which calls for calculating the correlation coefficient of each statement with the battery. If a negative correlation coefficient is obtained, it indicates that the numerical values are not properly assigned and that the one and five ends should be reversed. If a zero or very low correlation coefficient is obtained it indicates that the statement fails to measure that which the rest of the statements measure. Such statements will be called undifferentiating. Thurstone refers to them as irrelevant or ambiguous. By "undifffferentiating" we merely mean that the statement does not measure what the battery measures and hence to include it contributes nothing to the scale. A statement which is undifferentiating for a scale measuring one attitude continuum may be quite satisfactory for a scale measuring another attitude continuum. The following are some of the reasons why a statement may prove undifferentiating:

1. The statement may involve a different issue from the one involved in the rest of the statements, that is, it involves a different attitude continuum.

2. The statement may be responded to in the same way by practically the entire group. For example, the response to the following statement was practically the same upon the part of all students—some two thousand—to whom it was given: "Should the United States repeal the Japanese Exclusion Act?"

3. The statement may be so expressed that it is misunderstood by members of the group. This may be due to its being poorly stated, phrased in unfamiliar words, or worded in the form of a double-barreled statement.

4. It may be a statement concerning fact which individuals who fall

at different points on the attitude continuum will be equally liable to accept or reject.

It is, of course, desirable in constructing an attitude scale that the experimenter exercise every precaution in the selecting of statements so as to avoid those that are undifferentiating. However, item analysis can be used as an objective check to determine whether the members of a group react differentially to the statement in the same manner that they react differentially to the battery; that is, item analysis indicates whether those persons who fall toward one end of the attitude continuum on the battery do so on the particular statement and vice versa. Thus item analysis reveals the satisfactoriness of any statement so far as its inclusion in a given attitude scale is concerned.

No matter for what *a priori* reasons the experimenter may consider a statement to belong in a scale, if the statement, when tried on a group, does not measure what the rest of the statements measure, there is no justification for keeping that statement in the battery. After all, we are interested in measuring the attitudes of the members of the group, not those of the experimenter.

There is no reason to expect that the logical analysis of the person who selects the statements will necessarily be supported by the group. Quite often, because of a lack of understanding of the cultural background of the group, the experimenter may find that the statements do not form the clusters or hierarchies that he expected. It is as important psychologically to know what these clusters are as it is to be able to measure them.

The degree of inclusion, i.e. the size of the correlation coefficient between the item and the battery, required for a particular statement will no doubt be a function of the purpose for which the attitudes are being measured. If a general survey type of study is being undertaken the degree of inclusion required will be less than when a more specialized aspect of attitudes is being studied. A similar relationship is to be noted in the measurement of intelligence.

The only difficulty in using item analysis is that the calculation of the necessary coefficients of correlation is quite laborious. The criterion of internal consistency was tried and the results obtained were found to be comparable with the results from item analysis. Table 19.1 shows a comparison of the results obtained from item analysis and the criterion of internal consistency. It will be noted that the relation between the order of excellence for the different statements as determined by item analysis and the criterion of internal consistency as expressed by rho is +.91. Since the criterion of internal consistency is much easier to use than item analysis and yet yields essentially the same results, its use is suggested.

In using the criterion of internal consistency the reactions of the group

Table 19.1. Comparison of the Results Obtained from the Application of the Criterion of Internal Consistency and Item Analysis to the Negro Scale for Groups "A" and "B" Combined— (N=62)

Column 1	Column 2	Column 3	Column 4	Column 5
1	.69	1.7	2	5
2	.64	1.5	6	6
3	.51	1.7	10	11
4	.18	0.4	14	14
5	.62	1.3	7	8
6	.40	0.7	11	13
7	.12	0.1	15	15
8	.39	1.1	12	10
9	.26	0.9	13	12
10	.65	2.7	5	1
11	.60	1.2	8	9
12	.54	1.4	9	7
13	.67	2.3	4	2
14	.74	2.0	1	3
15	.68	1.6	3	4

rho (Column 4 vs. Column 5) $= +.91$

Column 1—Statement numbers.

Column 2—Coefficient of correlation between the score on the individual statement and the average score on all fifteen statements.

Column 3—Difference between the average score of the highest 9 individuals and the lowest 9 individuals.

Column 4—Order of excellence as determined by item analysis based upon the coeficients of correlation shown in Column 2.

Column 5—Order of excellence as determined by the criterion of internal consistency based upon the differences shown in Column 3.

that constitute one extreme in the particular attitude being measured are compared with the reactions of the group that constitute the other extreme. In practice approximately ten percent from each extreme was used. Table 19.2 shows the criterion of internal consistency applied to the Internationalism scale for Group D. This criterion acts as an objective check upon the correct assigning of numerical values in that if the numerical values are reversed on a particular statement the extreme high group will score low on that statement and the extreme low group will score high, i.e. we will obtain a negative difference between the two extreme groups on that question. Furthermore, if a statement is undifferentiating it will not differentiate or discriminate the two extreme groups, i.e. the high group will not score appreciably higher than the low group upon that statement.

Table 19.2. Criterion of Internal Consistency Applied to the Internationalism Scale for Group "D"—(N=100)

STATEMENT NUMBERS

HIGH GROUP

Indiv. No.	Score	Three-Point Statements												Five-Point Statements											
		1	2	3	4	5	6	7	8	9	10	11	12	13	14	15	16	17	18	19	20	21	22	23	24
85	108	4	4	4	4	4	4	4	4	4	4	4	4	5	5	5	5	5	5	5	5	5	5	5	5
65	104	4	4	3	4	4	4	4	4	4	4	4	4	5	5	5	5	5	5	5	5	5	3	5	4
13	102	4	4	4	4	4	4	4	4	4	4	4	4	5	5	5	5	5	5	5	5	5	3	3	5
10	101	4	4	4	4	4	4	4	4	4	4	4	4	4	3	3	4	5	5	5	5	5	3	3	5
71	101	2	4	4	4	4	2	2	4	4	4	4	4	5	3	4	5	5	5	5	4	5	4	5	5
98	100	4	4	4	4	4	4	4	4	4	4	4	4	5	3	5	4	5	5	3	5	5	5	3	4
27	98	4	2	4	4	4	4	4	4	4	4	4	4	4	5	3	5	5	4	4	4	5	2	5	5
60	98	4	4	4	4	4	4	2	4	2	4	4	4	5	5	3	4	5	5	3	5	5	4	5	5
64	98	4	4	4	4	4	4	4	4	2	4	4	4	4	3	4	3	5	5	5	5	5	4	5	4
Sum of 9-high		34	34	35	36	36	36	32	36	32	36	36	36	42	37	35	40	45	45	44	40	45	35	41	42
Sum of 9-low		18	20	20	28	24	29	21	20	22	21	34	23	21	24	22	15	31	22	15	22	24	17	14	22
D		16	14	15	8	12	7	11	16	10	15	2	13	21	13	13	25	14	23	25	22	21	18	27	20
D/9		1.8	1.6	1.7	.9	1.3	.8	1.2	1.8	1.1	1.7	.22	1.4	2.3	1.4	1.4	2.8	1.6	2.6	2.8	2.4	2.3	2.0	3.0	2.2
Order		1.5	5	3.5	10	7	11	8	1.5	9	3.5	12	6	6.5	11.5	11.5	2.5	10	4	2.5	5	6.5	9	1	8

Table 19.2. continued
(3-point statements and 5-point statements treated separately)

LOW GROUP

17	49	2	2	2	2	2	2	2	4	2	2	2	2	1	2	2	1	2	2	4	2	2	1	2
77	54	2	2	2	2	2	4	2	2	4	3	4	2	1	3	2	1	2	2	2	2	2	1	2
22	60	2	2	2	2	4	4	2	2	3	3	4	2	3	3	2	2	2	2	4	2	2	1	2
35	61	2	2	2	4	3	3	4	2	3	2	4	3	1	2	2	3	1	3	3	2	1	2	1
53	62	2	2	2	4	4	2	2	2	2	3	4	2	3	3	4	2	2	4	1	2	1	2	4
69	62	2	2	2	2	2	4	2	2	3	3	4	4	3	3	2	2	2	5	4	3	2	1	2
94	63	2	2	2	4	3	4	4	2	3	4	4	2	3	3	4	2	2	2	3	3	2	2	2
21	64	2	2	2	4	2	4	2	2	2	2	4	4	2	3	4	2	2	2	2	3	2	2	4
88	64	2	4	2	4	2	2	2	2	4	3	4	4	4	3	4	2	1	2	1	2	2	3	3
Sum of 9-low		18	20	20	28	24	29	21	20	22	21	34	23	21	24	22	15	31	22	15	22	24	17	14	22

Finally, on the basis of the results obtained from item analysis or the criterion of internal consistency and having due regard for all the factors concerned, one should select the most differentiating statements for the final form or forms of the attitude test. If, through this selection of the more differentiating statements, statements concerning a particular aspect of the attitude being measured are eliminated, then, obviously, the final scale can only be said to measure the attitude continuum represented by the remaining statements. For example, if it is found by the use of these objective checks that statements concerning the economic status of the Negro involve an attitude continuum other than that of statements having to do with the social equality of the Negro, the former should not be mixed with the latter. On the contrary, two attitude scales should be constructed. If, on the other hand, these two groups of statements are found to involve the same attitude continuum, they can be combined into a single scale. As previously stated, the degree of inclusion required or desired will generally be a function of the purpose for which the attitude scales are being used.

A sufficient number of statements should be used in each form to obtain the desired reliability. In preparing the final form or forms, it would be desirable to apply the fourth criterion stated under "The Selection of Statements."

Because a series of statements form a unit or cluster when used with one group of subjects which justifies combining the reactions to the differents statements into a single score, it does not follow that they will constitute a unit on all other groups of persons with the same or different cultural backgrounds. For example, an examination of the statements in the Imperialism scale will reveal that it contains statements having to do with imperialism both in China and Latin America, and while it is true that these statements form a sufficient cluster to justify their being treated as a unit with the groups used, still with other groups of persons with markedly different attitudes toward China or Latin America it is probable that this single scale would have to be divided into two or more scales.

The ease and simplicity with which attitude scales can be checked for split-half reliability and internal consistency would seem to make it desirable to determine the reliability and examine the internal consistency of each attitude scale for each group upon which it is used. It is certainly reasonable to suppose that just as an intelligence test which has been standardized upon one cultural group is not applicable to another so an attitude scale which has been constructed for one cultural group will hardly be applicable in its existing form to other cultural groups.

References

MURPHY, G. and MURPHY, L. B. *Experimental social psychology.* New York. Harper. 1931.

RICE, S. A. *Report,* Inst. of methods of rural sociol. research. U. S. Dept. of Agriculture. 1930. 11–20.

THURSTONE, L. L. and CHAVE, E. J. *The measurement of attitude.* Chicago. Univ. of Chicago Press. 1929.

Chapter 20

CRITERIA FOR AN ATTITUDE SCALE

JOHN P. ROBINSON, JERROLD G. RUSH, AND KENDRA B. HEAD
University of Michigan

I.

The first step for the scale builder, and the first criterion on which his work can be evaluated, is writing or finding items to include in the scale. It is usually assumed that the scale builder knows enough about the field to construct an instrument that will cover an important theoretical construct well enough to be useful to other researchers in the field. If it covers a construct for which instruments are already available, the author should demonstrate sound improvements over previous measures. There are three further preliminary considerations that represent the minimum that an adequately constructed scale ought to possess.

PROPER SAMPLING OF CONTENT

Proper sampling is not easy to achieve, nor can exact rules be specified for ensuring that it is done properly (as critics of Guttman's phrase "universe of content" have been quick to point out). Nevertheless, there is little doubt of the critical nature of the sampling procedure in scale construction. Future research may better reveal the population of behaviors,

Excerpted with permission of the author and the publisher from *Measures of Political Attitudes* by John P. Robinson, Chapter 1, pp. 9–21. Ann Arbor, Michigan: Survey Research Center, 1968.

objects, and feelings that ought to be covered in any attitude area, but some examples may suggest ways in which the interested researcher can provide better coverage in designing scales. Investigators of the "authoritarian personality" lifted key sentiments expressed in small group conversations, personal interviews, and written remarks and transformed them into scale items; some of these items in fact consisted of direct verbatim quotations from such materials. In the job statisfaction area, we gave detailed consideration to the analysis from representative samples of responses to open-ended questions that ask the respondent, "What things do you like best (or don't you like) about your job?" We feel that these responses offer invaluable guidelines to the researcher both as to the universe of factors he should be covering and the probable weight that should be given to each factor. Other instruments in the job satisfaction area were built either on the basis of previous factor analytic work, or on responses to questions about critically satisfying or dissatisfying situations at work, or on both of these.

Difficult decisions remain to be made about the number of questions needed to cover each factor (probably a minimum of two in any lengthy instrument) but the important first step is to make sure that the waterfront has been covered.

SIMPLICITY OF ITEM WORDING

One of the great advantages of securing verbatim comments from group discussions or open-ended questions (as people in advertising have apparently discovered) is that such attitudes are couched in language easily comprehended and recognized by respondents. One of the most obvious advantages of more recently constructed scales is that item wording has become far less stuffy, lofty, or idealistic. Even today, however, survey researchers still must adapt items developed from college samples for use on heterogeneous populations.[1]

There are other item wording practices that are, thankfully, going out of style as well: double-barrelled items that contain so many attitudes that it is hard to tell why the person agrees or disagrees with it (e.g., "The government should provide low-cost medical care because too many people are in poor health and doctors charge so much money"); items that are so vague they mean all things to all people ("Everybody should receive adequate medical care"); or items that depend on knowledge of little-known facts ("The government should provide for no more medical care than that implied in the Constitution"). Other considerations about writing items, such as negative *vs.* positive wording, will be covered under our discussion of response set.

1. The process is often referred to as "farmerization," i.e., making items intelligible to the less sophisticated.

ITEM ANALYSIS

While item wording is something the investigator can manipulate to ensure coverage of attitudinal areas, there is no guarantee that respondents will reply to the items in the manner intended by the investigator. Item analysis is one of the most efficient methods whereby the investigator can check whether people are responding to the items in the manner intended. We have encountered too many instances in the literature where authors inadvertently assume that their *a priori* division of scale items corresponds to the way their respondents perceive these items.

There have been many methods of item analysis proposed, and, in fact, complex multidimensional analyses (described below under homogeneity, in our detailing of statistical procedures) can be seen as the ultimate item analytic procedure. The researcher need not go so far as factory analyzing his data to select items to be included or discarded, but an item intercorrelation matrix (on perhaps a small subsample or pretest sample) is certainly the most convenient basis of doing item analysis. If it is hypothesized that five items in a large battery of items (say those numbered 1, 2, 6, 12, and 17) comprise a scale of authoritarianism, then the majority of the ten inter-item correlations between these five items should be substantial. At the minimum they should be significant at the .05 level. While this minimum may seem liberal, it is in keeping with the degree to which items in the most reputable scales intercorrelate for heterogeneous populations. If items 1, 2, 12, and 17 intercorrelate substantially with each other but 6 does not correlate well with any of them, then item 6 should be discarded or rewritten.

Measuring the degree to which each of the five items correlates with some external criterion is a further valuable device for the selection of items. This is usually referred to as the item-validity method.

We learned one valuable lesson about writing items from a certain item analysis we performed. A previous study had uncovered four dimensions of value—authoritarianism, expression, individualism, and egalitarianism—and we wished to incorporate measures of these factors into a study of political attitudes. One individualism item—"It is the man who starts off bravely on his own who excites our admiration"—seemed in particular need of farmerization. Accordingly, the item was reworded, "We should all admire a man who starts out bravely on his own." Item analysis revealed this revised statement to be more closely associated with *authoritarian* items than with the other individualism items. It became clear that a seemingly logical wording change can unexpectedly alter the entire implication of an item.

Often a researcher does not have the benefit of pretest groups in order to eliminate or revise unsatisfactorily items. In such a case, the item-analysis phase of scale construction should be incorporated into the determinination of the dimensionality, scalability, or homogeneity of the test items. This will ensure that there is empirical as well as theoretical rationale for combining the information contained in various items.

II.

The second large area of evaluation is the concern that the scale builder has given to the avoidance of "response set" in the items. Response set refers to a tendency on the part of individuals to respond to attitude statements for reasons other than the content of the statements. Thus, a person who might want to appear agreeable and thus fail to disagree with any attitude statement is said to show an "ageement response set." Only through experience and by constant revision can the researcher insulate his scale from this potentially dangerous side effect. As a basic guard against response set, the researcher should try to make the scale as interesting and pleasant for the respondent as possible. If the respondent finds the instrument to be dull or unpleasant, there is a greater chance that he will try to speed through it as quickly as possible. It is in such a setting that the scale is most liable to response set contamination, such as indiscriminate agreement or checking off in a certain column.

There are two main sources of response set that are most difficult to control.

ACQUIESCENCE

Most of us have been seen (or perhaps been) people whose attitudes change in accord with the situation. Such people are said to "acquiesce" in the presence of opposition from others. In the same way, some people are "yea-sayers," willing to go along with anything that sounds good, while others (perhaps optimists) are unwilling to look at the bad side of anything. These dispositions are thus reflected in people's responses to attitude questions. How then is it possible to separate their "real" attitudes from their personality dispositions?[2]

There are various levels of attack, all of which involve forsaking simple affirmative item format. The first involves at least an occasional switching of response alternatives between positive and negative. For simple "yes-no" alternatives, a few "no-yes" options should be inserted. Similarly, for the "strongly agree-agree-uncertain-disagree-strongly disagree" or Likert format, the five alternatives occasionally should be listed in the opposite order. This practice will offer some possibility of locating people who

2. Rorer (1965) points out many relevant objections to attempting separation of the acquiescent response set from item content.

choose alternatives on the sole basis of the order in which they appear. It may also alert an overly casual respondent to think more about his answers.

It is more difficult to vary the item wording from positive to negative, as those who tried to reverse authoritarianism items have found. A logician can argue that the obverse of "Obedience is an important thing for children to learn" is not "Disobedience is an important thing for children to learn," and the investigator is on shaky ground in assuming that a respondent who agrees with both the first statement and the second is completely confused. Along the same line, the practice of inserting a single word in order to reverse an item can produce some pretty silly-sounding items, while changing one word in an unusual context has produced items in which the ordinary respondent will not notice a change. In sum, writing item reversals requires sensitivity. The interested researcher would be well advised to check previous competent work on the subject (Christie et al., 1958) before undertaking such a task. However, the literature is still ambiguous as to the real value of item reversals (e.g., Wrightsman, 1966).

A third and more difficult, yet probably more effective, approach concerns the construction of "forced-choice" items. Here two (or more) replies to a question are listed and the respondent is told to choose only one: "The most important thing for children to learn is (obedience) (independence)." Equating the popularity or "social desirability" of each alternative requires even more intensive effort for both the scale constructor and the respondent. Since the factor of social desirability is an important response set variable in its own right, we give it individual attention next.

SOCIAL DESIRABILITY

In contrast to the theory that the acquiescent person reveals a certain desire for subservience in his willingness to go along with anything, Edwards (1957) has proposed more positively that these people are just trying to make a good impression. As yet research has been unable to determine clearly whether the overly high incidence of positive correlation among questionnaire items is ultimately due more to bias from acquiescence or to social desirability (Christie and Lindauer, 1963). The methods of lessening social desirability bias, in any event, usually involve the use of forced-choice items in which the alternatives have been equated on the basis of social desirability ratings. In more refined instruments, the items are pretested on social desirability, and alternative-pairings (or item-pairings) which do not prove to be equated are dropped or revised.

One further method consists of using the respondent's score on the

Crowne-Marlowe social desirability scale as a correction factor. Smith (1967) gives an explicit example of the mechanics of this approach.

III.

We have mentioned the major sources of response set contamination but there are others of which the investigator should remain aware. One of the more prevalent sources of contamination is the faking of responses according to some preconceived image that the respondent wants to convey. On a job satisfaction scale, for example, the respondent may try to avoid saying anything that might put his supervisor in a bad light or might involve a change in work procedures. College students may be aware of a professor's hypothesized relationship between two variables and try to answer so as to make this prediction work out or fail. Other undesirable variations of spurious response patterns that the investigator might want to minimize can result from the respondent's wanting (a) to appear too consistent, (b) to use few or many categories in his replies, or (c) to choose extreme alternatives.

The third and final area of evaluation for each instrument is the various statistical and psychometric procedures incorporated into its construction. While each of these statistical considerations—sampling, norms, reliability, homogeneity, and validation—is important, an inadequate performance on any one of them does not render the scale worthless. Nevertheless, inadequate concern with most of them certainly does indicate that the scale should be used with reservation. Fortunately, scale constructors in the past few years appear to have paid more heed to these considerations than did the vast majority of their predecessors. Still, even today few scales rate optimally on all these factors. It is very seldom indeed that one runs across scales which overcome (or even attempt to overcome) the distortion due to restricted samples or incomplete validation procedures.

We have chosen seven statistical standards we hope cover the basic requirements involved in the construction of competent scaling instruments.

REPRESENTATIVE SAMPLE

In this day and age, it is hoped, researchers are aware of the fallacy of generalizing results from samples of college students[3] onto an older and much less well-educated general population. Significant differences are even likely to be found between freshmen and seniors, engineering and

3. Some statisticians contend that a sample of single class should be treated as having a sample size of one, not the number of students in the class.

psychology students, and college A and college B so that one must be careful in expecting results from one class to hold for all college students. In the political attitude area, we shall see that there are great dangers in expecting findings from political elites to hold for typical citizens (or even in using scales developed on elites with such typical samples).

This is not meant to discourage researchers from improving the representativeness of whatever populations they do have available for study, but rather to caution them against implying that their findings hold for people not represented by their samples. Nor is it meant to imply that samples of college students are a useless basis on which to construct scales. In some areas (attitudes toward foreign affairs, for example), one might well argue that college exposure is probably the best single criterion of whether a person can truly appreciate the intricacies of the issues involved.

But an instrument constructed from replies of a random cross-section of all students in a university has much more to offer than the same instrument developed on students in a single class in psychology (even if there are many more students in the class than in the university sample). The prime consideration is the applicability of the scale and scale norms to respondents who are likely to use them in the future.

NORMATIVE INFORMATION

The adequacy of norms (e.g., mean scale scores, percent agreements, etc.) is obviously dependent on the adequacy of the sample. The absolute minimum of normative information, which should be available for the researcher to be aware of any differences between his sample and the sample on which the scale was developed, is the mean scale score and standard deviation for the sample on which the scale was constructed. There are further pieces of statistical data that are extremely useful: item means (or percent agreements) and standard deviations, median scores (if the scales scores are skewed), or more obscure statistics like the interquartile range.

Most helpful are means and standard deviations for certain well-defined groups (men or women, Catholics or Baptists) who have high or low scale scores. When such differences have been predicted, the results bear on the *validity* of the scale, which is discussed below. Validity, reliability, and homogeneity also constitute needed normative information, of course, and they are covered below in the detail required by their complexity.

RELIABILITY (TEST-RETEST)

Unfortunately, one of the most ambiguous terms in psychometrics is "reliability." There are at least three major entities to which the term can refer: (1) the correlation between the same person's score on the same items at two separate points in time; (2) the correlation between two different sets of items at the same time (called "parallel-forms" if the items are presented in separate format, and "split-half" if the items are all presented together); and (3) the correlation between the scale items for all people who answer the items. The latter two indices refer to the internal structure or homogeneity of the scale items (the next criterion), while the former indicates stability of a person's item responses over time. It is unfortunate that test-retest measures, which require more effort and sophistication on the part of the scale developer and show lower reliability figures for his efforts, are available for so few instruments in the literature. While the test-retest reliability level may be approximately estimated from indices of homogeneity, there is no substitute for the actual test-retest data.

HOMOGENEITY

In addition to split-half, parallel forms, and inter-item indices of the internal homogeneity of the test items, there exist other measures of this desirable property. Some of these item-test and internal consistency measures, as Scott (1960) has shown, bear known statistical relationships with one another. Included in this collection are certain indices of scalability for Guttman items, although not the most often-used Coefficient of Reproducibility. Even between such "radically" different procedures as the traditional psychometric and Guttman cumulative, however, there likely exist reasonably stable relationships between indices based on inter-item, item-test, and total test characteristics; as yet, however, these have not been charted. For now, the major difference between the indices seems to lie in the researcher's preference for large or small numbers. Inter-item correlations and homogeneity indices based on Loevinger's concepts seldom exceed .40; if one prefers larger numbers, a Reproducibility Coefficient or split-half reliability coefficient computed on the same data could easily exceed .90. Thus, since it seems at present to be the only way of relating the various indices, one is apparently forced to rely on the imperfect criterion of statistical significance in order to evaluate instruments for which different indices have been employed. To make the job even more difficult, statistical distributions of these various indices are not always available so that significance can be ascertained.

Of all the indices that have been proposed, however, probably none combines simplicity with amount of information contained as well as the inter-item correlation matrix. Computing Pearson r correlation coefficients for more than five items is certainly a time-consuming operation on a hand calculator. However, for the researcher who does not have access to a computer that prints out such a matrix, there are some simple rank-order correlation formulas that can be calculated by hand in a few minutes, so that even a ten-item scale inter-item correlation matrix can be put together in a few hours. The job is too lengthy if there are too many alternatives or over 100 subjects, but in the case of dichotomous items, the coefficients Y or γ (defined in a statistical appendix to this chapter) can be easily calculated to determined inter-item significance. These, however, constitute only rule-of-thumb procedures for deciding whether a group of items deserves to be added together to form a scale or index. Similarly, the criterion of significance level is proposed only because it is a standard which remains fairly constant across the myriad of measures that are now, or have been, in vogue. Probably it is only the minimum to be expected before one can talk about a scale which can be reasonably called "homogeneous." Hopefully, more satisfactory norms may be proposed in the future.

When the number of items goes beyond ten, however, the inter-item matrix is indeed quite cumbersome to compute by hand calculator for any coefficient, and the researcher is well advised to look for a computer specialist and a correlation matrix program. Computers have the ability to generate 50 to 100 item intercorrelations in less than ten minutes, given a reasonably-sized sample. This does not work out to burdensome cost if the researcher has put much effort into his data collection. At this level of analysis (i.e., more than ten items), the researcher might as well proceed to invest in a factor analysis or cluster analysis of his data. This type of analysis will help him locate the groups of items that go together much faster than could be done by inspecting the correlation matrix.[4] There are many kinds of factor analysis programs and options; under most circumstances, however, the differences between them usually do not result in radical changes in the structure that is uncovered.

To say that factor analytic programs do not usually vary greatly in their output is not to imply that structures uncovered by factor analysis may not lead to serious ambiguities in the interpretation of data. There is one common type of attitudinal data arrangement in particular for which the factor structure seems indeterminant. This is the case where almost all the items are correlated from say .15 to .45. Sometimes only a single factor will emerge from such a matrix and sometimes a solution will be generated which more clearly reflects item differentiation on a series of factors. We have encountered one instance where an instrument—supposedly constructed carefully to reflect a single dimension of inner- vs. other-directedness, according to a forced-choice response format—

4. However, the researcher should not be deceived by what appear to be high factor loadings. Items having factor loadings which reach levels of .50 or .70 are equivalent to correlation coefficients of .25 and .49.

was found when analyzed in Likert format to contain eight factors. Thus, one can offer no guarantee that inter-item significance will always yield unidimensional scales. Nor does it seem possible to offer any better advice or to recommend any competent practical literature on the inconsistencies into which factor analysis can lead one. On balance, however, one is further ahead performing such analyses than not doing so.

The length of this discussion clearly shows that we feel the determination of homogeneity to be a crucial step in scale construction. Only by these procedures can the analyst properly separate the apples, oranges, and coconuts from the salad of items he has put together. In a future volume we hope to be able to go into the detailed rationale for the conclusions and recommendations rather cursorily made in this section.

One final word of caution is in order: It is possible to devise a scale with very high internal consistency merely by writing the same item in a number of different ways. Sampling of item content then can be a crucial component in internal consistency.

DISCRIMINATION OF KNOWN GROUPS

This is where the value of a scale is truly tested—the aspect of validity. Nevertheless, group discrimination is not necessarily the most challenging hurdle to demonstrate validity. It is pretty hard to construct a liberalism-conservatism scale that will *not* show significant differences between John Birchers and Students for Democratic Society, or a religious attitude scale that will not separate Mormons from Jews or ministerial students from engineers. The more demanding hurdle is the ability of the scale scores to reliably single out those liberals or conservatives, agnostics or believers, in heterogeneous groups—or to predict which of them will demonstrate behavior congruent with their hypothesized attitudinal state. A still more definitive test is cross-validation, a test to which very few attitudes scales have been subjected.

CROSS-VALIDATION

A test of cross-validation requires two different samples and measures of some criterion variable on each sample. The question to be answered by the test is whether the combination of items for sample A that best correlates with the criterion variable in sample A will also work for sample B's criterion, and whether the best set of sample B items works on the sample A criterion. Note that the crux of the procedure involves picking (and, if necessary, weighting) the *items* from the sample A experience which work best on sample B.

An even more refined method, and probably the utimate standard now available, is the multi-trait multi-method matrix as proposed by Campbell and Fiske (1959). The method requires more than one index of each of the several constructs (say x, y, and z) we want to measure by our instrument. It is best to include as many measures or indices of each construct as possible, as well as to measure for control purposes such variables as intelligence or agreement response set which could be at the root of any apparent relationship. In the resulting correlation matrix, the various indices of the single construct (say x) should correlate higher among themselves than any index of x correlates with any indices of y, z, or the control variables.

Needless to say, this comprises a gross oversimplification of the Campbell-Fiske method. The reader should peruse the authors' article thoroughly before attempting comparable analyses. It is worth noting that the authors find only a couple of personality scales which meet their conditions. To our knowledge, no attitude scales have as yet advanced the claim.

OTHER PROCEDURES

Since there are many methods used in constructing scales beyond our recommended procedures, we should also note alternative methods that may be employed. Such alternatives may include special precautions taken to ensure better items, better testing conditions, or adequate validation—although at times the precautions have had the opposite effects from that intended.

One interesting procedure to which researchers have become increasingly attracted involves the use of positive and negative items. Sometimes, as we have noted, items intended as negative are responded to as negative correlates of positive items; in other instances, this does not work. A procedure that may provide valuable insights into the response patterns of the sample is the separation of the high and low scores on both the positive and negative scales. There are four groups to be examined; *yea-sayers* (who score high on both the negative and positive items), *nay-sayers* (who score low on both), *assenters* who score high on the positive items and low on the negatives, and the *dissenters*, who follow the opposite pattern. This division can be seen more clearly in the following diagram:

		Positive Items	
		Low	High
Negative Items	Low	Nay-sayers	Assenters
	High	Dissenters	Yea-sayers

A parallel analysis for Likert scales (or procedures which demand more than a simple dichotomous item response) is the separation of the group at the mean into those who are ambivalent (combining extreme positive responses with extreme negative responses) from those who fall in the middle by taking an extreme position on very few items.

IV.

In certain chapters of one of our other volumes, where a sufficient number of instruments to warrant comparison was present, we actually tried to rate each scale on the above twelve considerations. If these ratings prove helpful enough, it might be worthwhile compiling them for all the attitude scales in the future.

It is very important that the reader realize that even this extensive list of proposed criteria is not exhaustive. The actual choice of an instrucment, where possible, should be dictated by decision-theoretic considerations. Thus, the increasing of homogeneity by adding questionnaire items needs to be balanced against corresponding increases in administrative analysis and cost (and against respondent fatigue and noncooperation) before reaching a decision about how many attitude items to use. For assessing general levels of some attitude state (e.g., merely to separate believers from atheists), well-worded single items may do the job just as well as longer scales no matter how competently the scales are devised. For an excellent theoretical exposition of the decision-theoretic approach for psychometric problems Cronbach and Gleser (1965) is recommended. In this extended version of their earlier volume, the authors provide a number of relevant examples.

Appendices to this series will deal with (a) measures of occupational attitudes and characteristics and (b) authoritarianism, alienation, and other social-psychological values and attitudes. Resources permitting, a general sourcebook on attitude methodology will be produced. The general methodology report should be most valuable in making clear the rationale on which our instrument evaluation is based and in explaining why we feel the above factors to be the most crucial considerations out of vast numbers that have been proposed. For now, we highly recommend the American Psychological Association's publication (1966) as an invaluable guidebook for scale construction and evaluation.

Statistical Appendix

COMPUTATION OF Y TO DETERMINE INTER-ITEM CORRELATION

$$Y = \frac{\sqrt{ad} - \sqrt{bc}}{\sqrt{ad} + \sqrt{bc}} \quad \text{and}$$

		Item 1		
		Yes	No	
Item 2	Yes	a	b	a + b
	No	c	d	c + d
		a+c	b+d	N

The significance of Y can be computed by calculating its standard error for the case where Y is hypothesized to be 0. Thus when Y exceeds

$$Y = \left(\frac{N\sqrt{N}}{4}\right) \left(\frac{1}{(a+b) \quad (b+d) \quad (a+b) \quad (c+d)}\right)^{\frac{1}{2}}$$

by a factor of 2, the items are significantly related at the .05 level, and when it exceeds Y itself by a factor of 2.5 the items are related at the .01 level (assuming the number of respondents is greater than 30).

Goodman and Kruskal's (1959) gamma, γ, is a measure that can be called into use when the number of item alternatives is greater than 2. Approximate sampling distributions of the statistic have recently become available (Rosenthal, 1966). The reader may be interested to know that for the dichotomous case, gamma reduces to the formula for Y with the square root signs removed; hence, gamma tends to take on larger values than Y for the same data.

References

AMERICAN PSYCHOLOGICAL ASSOCIATION. *Technical Recommendations for Psychological Tests and Diagnostic Methods, Psychological Bulletin Supplement,* 1966, 51, #2. (Available from the APA at 1200 Seventeenth Street, N.W., Washington, D. C. 20036.)

BONJEAN, C., HILL, R., and McLEMORE, S. *Sociological Measurement.* San Francisco: Chandler Publishing Co., 1968.

CAMPBELL, D., and FISKE, D. "Convergent and Discriminant Validation by the Multi-trait Multi-method Matrix," *Psychological Bulletin,* 56, 1959, 81–105.

CHRISTIE, R. and LINDAUER, F. "Personality Structure," *Annual Review of Psychology,* 1963, 14, 201–30.

CHRISTIE, R. et al. "Is the F Scale Irreversible?" *Journal of Abnormal and Social Psychology,* 1958, 56, 143–59.

CRONBACH, L. and GLESER, GOLDINE. *Psychological Tests and Personnel Decisions,* Second Edition. Urbana: University of Illinois Press, 1965.

EDWARDS, A. *The Social Desirability Variable in Personality Assessment and Research.* New York: Dryden Press, 1957.

MILLER, D. *Handbook of Research Design and Social Measurement.* New York: David McKay Co., 1964.

RORER, L. "The Great Response Style Myth," *Psychological Bulletin,* 63, 1965, 129–56.

ROSENTHAL, IRENE. "Distribution of the Sample Version of the Measure of Association, Gamma." *Journal of the American Statistical Association,* 1966, 440–53.

SCOTT, W. "Measures of Test Homogeneity," *Educational and Psychological Measurement,* 20, 1960, 751–57.

SHAW, M. and WRIGHT, J. *Scales for the Measurement of Attitudes.* New York: McGraw-Hill, 1967.

SMITH, D. "Correcting for Social Desirability Response Sets in Opinion-Attitude Survey Research," *Public Opinion Quarterly,* 31, 1967, 87–94.

SNIDER, J. and OSGOOD, C. *The Semantic Differential: A Sourcebook.* Chicago: Aldine, 1968.

WRIGHTSMAN, L. *Characteristics of Positively-Scored and Negatively-Scored Items from Attitude Scales.* Peabody Teachers' College (Nashville, Tennessee), 1966.

Chapter 21

A NOTE ON THE SCALE PRODUCT AND RELATED METHODS OF SCORING ATTITUDE SCALES

H. J. BUTCHER
University of Sussex, England

I. Thurstone's and Likert's Methods of Scoring

It has long been recognized that both the Thurstone and Likert methods of scoring attitude scales suffer from certain weaknesses, and several attempts have been made to develop a new method which would combine the advantages and avoid the defects of both. This note will not be concerned with alternative methods of *constructing* attitude scales; a good case appears to have been made for the superiority of the Thurstone technique of equal-appearing intervals (Ferguson, 1939), and all the scales to be discussed were constructed by that technique.

The Thurstone method of *scoring* seems more open to question. The respondent is asked to mark those items on a list of attitude statements with which he agrees, and the median scale value of these items then becomes his score. This dichotomization of the items into those "agreed with" and "disagreed with" is somewhat crude and fails to extract the

Reprinted with permission of the author and the publisher from the *British Journal of Psychology*, Vol. 47, 1956, pp. 133–39.

I should like to thank Prof. R. A. C. Oliver and Prof. P. E. Vernon for their generous encouragement and helpful criticism during the writing of this note.

maximum information. Not only is the respondent likely to agree to quite different degrees with the various selected items on which his score is directly based; in the case also of the items which he does not mark he would presumably express *degrees* of disagreement if the form of questionnaire permitted.

One theoretical weakness of the Likert system of scoring is equally obvious. Of two items one may have a Thurstone scale value of 9, and another, much less extreme, a scale value of 6; yet under the Likert system approval (or disapproval) of these two items will earn the same mark.

The respective advantages and disadvantages of the two methods seem, therefore, to be as follows. The Thurstone method accurately weights each item, although it does not allow for shades of agreement or disagreement; the Likert method permits a number of possible responses to each item, ranging, for example, from "strongly agree" to "strongly disagree," but gives equal weight to items which should ideally receive differential weighting. On balance, Likert's technique appears slightly preferable, both because scores are easier to compute and because it has been shown by Likert, Roslow & Murphy (1934) to produce rather higher coefficients of reliability. It is obviously tempting, however, to try to produce a type of scoring which will take into account both the scale values of items and the graded responses of individuals.

Three such attempts are those of Likert, Roslow & Murphy (1934), Eysenck & Crown (1949) and Castle (1953). All these newer techniques combine features of Thurstone's and Likert's methods, but they differ slightly in details of procedure and in the results claimed for them.

II. Thurstone-weighted Scoring

GENERAL CHARACTERISTICS

Likert, Roslow & Murphy describe what they call "Thurstone-weighted" scoring as follows:

> Another method which is in a way a combination of the Thurstone and simpler (i.e. Likert) methods was also tried. . . . The same statements were used in this scoring as were used in the simpler method, but instead of assigning 1–2–3–4–5 to the respective alternatives, the values assigned were based upon the scale values of the Thurstone method. The "strongly agree" alternative of each question was given the value shown by the Thurstone key for that statement.
>
> The "strongly disagree" alternative of each statement was given a value equal to 11.0 minus the value of the "strongly agree" alternative for that particular statement. The question mark was always given the value of 5.5 and the "agree" alternative for each statement was given a value midway between 5.5

and the value of the "strongly agree" alternative for the same statement. The "disagree" position was similarly scored.

Certain features of the Thurstone-weighted technique, as described by Likert, Roslow & Murphy, merit further attention. First, the scoring of the "strongly agree" and "strongly disagree" responses to each item is weighted according to the Thurstone scale values of the items, and the same principle applies to a lesser degree to the "agree" and "disagree" responses, but the "uncertain" responses all receive the same score, regardless of the scale value of the item concerned. This is probably a slight theoretical weakness, since a subject who is uncertain about a statement at the positive end of a Thurstone-Constructed scale may well be manifesting a very different attitude from another subject who is uncertain about a statement at the negative end, yet with the Thurstone-weighted technique as with the Likert, uncertainty on either of the two items will contribute the same mark to the subject's total score.

Another characteristic of this kind of scoring is worth noting for purposes of comparison. The method of computation ensures automatically that, as with Likert scoring, the *sum* of the five weights is the same for each item.

RELIABILITY

Likert, Roslow & Murphy scored several Thurstone-constructed attitude-scales both by the ordinary Likert and by the Thurstone-weighted techniques, and found high agreement between the two sets of scores, with correlations between +0.95 and +0.98. (The correlations of Thurstone-weighted scores with ordinary Thurstone scores were markedly lower, ranging from +0.83 to +0.87.) The respective reliability coefficients resulting from the Likert and the Thurstone-weighted scoring techniques were not specifically compared by the authors, who were satisfied by the high correlations between the two that little or nothing was to be gained by using the considerably more laborious Thurstone-weighted method. It is possible to extract and compare some of the reliabilities from data provided in separate tables in the article. This comparison (and tests of significance) confirm the conclusions of Likert, Roslow & Murphy. None of the differences in reliability, as between the two methods of scoring, is statistically significant (see Table 21.1).

III. Scale Product Scoring

EYSENCK AND CROWN'S DESCRIPTION

The above comparison of reliabilities is of particular interest in view of the claim made for the rather similar Scale Product technique devised by Eysenck & Crown (1949), who describe their method as follows:

Table 21.1

N	Attitude scale no.	Likert scoring. Split-half reliability	Thurstone-weighted scoring. Split-half reliability
61	23	+0.91	+0.86
65	30 (groups D and E)	+0.89	+0.88
50	30 (group F)	+0.89	+0.90

In this method, the weight given to the various responses to each item by the Likert method is multiplied by the scale position of the item as found in the Thurstone method. The rationale of this procedure is obvious. As an example, let us take two items, both of which are answered "strongly agree," and would therefore score 5 points on the Likert system. One of these items, say, is mildly anti-Semitic, the other extremely so. (The relative Thurstone scale positions might be 6 and 10, where 5 denotes neutrality.) Clearly the endorsement of the second item is a much stronger indication of anti-Semitism than the endorsement of the first, and should consequently be weighted much more heavily. This weighting is accomplished automatically by multiplying the Likert weight (s) by the scale position (6 and 10 respectively), so that the respective weights would be, according to the method of Scale Products, 30 and 50.

This method gives rather unwieldy numbers, and can be improved by dividing all products by a constant, and rounding off. This brings the scale products down to manageable proportions. The actual values finally selected for the anti-Semitism scale are given in Table 2 [21.2] under the heading "Original Weights."

Eysenck & Crown's Table 2 is reproduced as Table 21.2

RELIABILITY

Eysenck & Crown then report that this Scale Product method of scoring, when applied to their scale of anti-Semitism, produced a reliability of +0.94, which was significantly higher at the $P=0.01$ level than either of the corresponding values produced by the Thurstone and Likert methods. Eysenck (1954) also claims that: "This new method has been shown to be more reliable than the other two, and it might be made more reliable still by using an empirical weighting system rather than the fairly mechanical one shown in the text." This claim is important but appears to rest on the scoring of only one scale. The present writer had hoped to test it further, but encountered certain difficulties in interpreting Eysenck & Crown's description of the technique.

SOME DIFFICULTIES OF INTERPRETATION

First, it obviously will not do simply to multiply the Thurstone scale value of each item by the Likert weights 1–2–3–4–5, which is what Eysenck

Table 21.2. Original Weights

Item no.	Scale position	Strongly agree	Agree	Uncertain	Disagree	Strongly disagree
1	3.7	5	5	4	3	3
2	7.8	1	3	4	5	7
3	4.9	5	4	4	4	3
4	8.5	1	2	4	6	7
5	1.3	7	6	4	2	1
6	2.0	7	5	4	3	1
7	4.6	5	4	4	4	3
8	9.2	0	2	4	6	8
9	2.9	6	5	4	3	2
10	6.2	3	3	4	5	5
11	0.5	8	6	4	2	0
12	6.9	2	3	4	5	6
13	7.3	2	3	4	5	6
14	1.0	8	6	4	2	0
15	8.1	1	3	4	5	7
16	1.7	7	6	4	2	1
17	6.5	3	3	4	5	5
18	8.9	1	2	4	6	7
19	2.5	7	5	4	3	1
20	5.3	5	4	4	4	5
21	5.7	3	4	4	4	5
22	3.4	6	5	4	3	2
23	9.7	0	2	4	6	8
24	4.1	5	5	4	3	3

& Crown might seem to suggest. Suppose one has two items whose scale values are 3 and 10. The "strongly agree" response of the first item would then be scored 15, and the "strongly disagree" response 3. Similarly, the second item (with scale value 10) would receive a score of 10 for the "strongly agree" response and a score of 50 for the "strongly disagree" response. But in this case the latter item (and all the other items with high scale values) would exercise a disproportionate influence on each respondent's total score, although the direction of scoring is of course arbitrary for an attitude scale, and the scale values might equally well run from 10 to 0 as from 0 to 10. Presumably the correct procedure is to express each scale value as a deviation from the theoretical neutral point of the scale before multiplying it by the Likert weights. This, as we shall see later, is what Castle does.

Furthermore, on turning to the table of weights, it is very difficult to see how Eysenck & Crown's figures can be obtained by even a modified version of the procedure they recommend. In Table 21.2 all the "uncertain" replies receive the same score, regardless of the scale value of the item. If, however, the different scale values of the items (or the different deviations from neutrality) are multiplied by Likert weights and are then all divided by a constant, which is what Eysenck & Crown appear to suggest, the "uncertain" responses ought surely to receive differing Scale Product weights in the same way as all the other responses. Secondly, for each item the five weights add up to 20 (except in one case, which is clearly a misprint). If the scale values or deviations had been multiplied by Likert weights and then divided by a constant, it is hard to see how this could have occurred.

At first sight, an alternative interpretation of Eysenck & Crown's description of the procedure might be that *for each item* the Scale Product weights are divided by a constant, or in other words that one constant is not used for the whole table but a number of different constants, such that for every item the "uncertain" response is scaled down to a value of 4. But it turns out that this procedure could not produce the set of figures in the table. All it would produce would be a set of slightly magnified Likert weights (1–3–4–5–7) identical for each item.

The more one examines the table in Eysenck & Crown's article, the more difficult it becomes to reconcile it with the account in the text, for the table seems to have been constructed by some method rather more analogous to that used by Likert, Roslow & Murphy (1934). At least it has been shown to possess two features which characterize their method (all "uncertain" responses scored alike and equal sums of weights for each item), but which are hard to reconcile with Eysenck & Crown's own description.

This analysis of Eysenck & Crown's method is necessarily detailed and critical. It should be added that their account of the Scale Product method is highly condensed and forms only a small part of a long and important paper.

IV. Castle's Scale Product Method

GENERAL CHARACTERISTICS

The third paper is that of Castle (1953), who uses the method *advocated* by Eysenck & Crown with one or two unimportant variations. Although he makes no claim to originality of technique, his is the only work known to the writer which describes the actual employment of this method. It will therefore be convenient to refer to it as "Castle's method" to distinguish it from the method exemplified in Eysenck & Crown's table.

Castle's scale products are obtained by first computing the deviation of each item's scale value from the neutral point of the scale, and then multiplying this deviation by the Likert weights; the "uncertain" responses thus receive different Scale Product weights according to item in the same way as the other responses, and the sum of five weights assigned also varies according to the scale value of the item. One slight peculiarity of his method is that he takes 6.5 as the neutral point of the scale, although his items were originally sorted into 11 piles.

A more serious point of criticism arises from his use of minus weights. The three methods previously discussed which allow for graded responses (i.e. the Likert, Thurstone-weighted, and Eysenck & Crown methods) all use reversed scoring for half of the items in the scale. With the Likert method, for example, it is usual for the items falling at one end of the continuum to be scored 1–2–3–4–5 and those at the other end 5–4–3–2–1. The same principle applies to the Thurstone-weighted and Eysenck & Crown methods, which possess the property that if the scale values of two items are equidistant from and on different sides of the zero point, these items will receive the same weights but in reverse. Thus in all these methods the *range* of weights for the different items is the same for the "strongly agree" responses as for the "strongly disagree." As a consequence of his use of minus weights, this is not true of Castle's scoring, and the defect is aggravated by the large numbers he uses. The result is that the weights he assigns to "strongly agree" responses range from $+23$ to -21 (for the 5 items tabulated in his paper), and those of the "strongly disagree" only from $+5$ to -4. This sort of asymmetry seems likely to produce some curious results.

RELIABILITY

Castle, following Eysenck & Crown, suggested that his scale was probably more reliable than if it had been scored by the Likert method. But, as we have seen, Castle's method differs in principle from that used by Eysenck & Crown, as inferred from their table of weights. It therefore seemed of interest to determine empirically whether a type of Scale Product scoring approximating to Castle's would produce a higher split-half coefficient of reliability than the ordinary Likert method.

The material used consisted of Part I of a questionnaire on attitudes to education constructed by Oliver (1953). This scale contained 30 items, of which the following are examples: "Mathematics is valuable for the training it gives in abstract reasoning," and "Naturalness is more important than good manners in children." On a sample of 57 Manchester University evening students it had been found to have a split-half reliability (corrected) of $+0.92$ when scored by the Likert method.

Scale Product weights were calculated on the same principle as Castle's, and the same 57 answer sheets re-scored. The scoring technique differed from Castle's in the following respects: (a) the theoretical mid-point of the scale was taken to be not 6.5 but the actual mean scale value of the 30 items, (*b*) reversed scoring was used for half of the items in order to preserve the symmetry of the scale, (*c*) the resulting Scale Product weights were scaled down, as recommended by Eysenck & Crown, both in order to avoid unwieldly numbers and to make them more comparable with Likert weights. The split-half reliability (corrected) was now found to be +0.89. This is lower than the previous value, but a test of significance (Fisher's z test) showed the difference not to be significant at the $P=0.05$ level.

V. Comparing Reliabilities: Some Further Considerations

To apply a test of significance of this kind to the difference between two reliability coefficients is possibly of doubtful legitimacy when these have been obtained from the same data. The assumption underlying such a test is that the data are independent and uncorrelated, so that both the above comparison and the earlier one between the reliability of the Likert and the Thurstone-weighted methods should be treated with caution. The same is true of Eysenck & Crown's claim for the superior reliability of their Scale Product method. On the other hand, it may well be that the result of applying such a test of significance to coefficients obtained from the same data is to *underestimate* the significance of the difference. In this case it would be only the negative results of the tests of significance reported in the present paper that should be considered suspect. The main point of both comparisons, that neither the Thurstone-weighted nor the Castle technique is more reliable than the Likert, remains in any case unaffected.

Another consideration relevant to the above comparisons of reliability has been mentioned by Vernon (1953), who suggests that the use of both Thurstone-type scaling and item analysis in scale construction tends towards higher reliability. Item analysis does not appear to have been used in the construction either of Eysenck and Crown's or of Castle's scales, but was used in addition to Thurstone scaling in the construction of Oliver's. Thus it is possible that the Scale Product method, in the version either of Eysenck & Crown or Castle, is more reliable than Likert scoring when applied to an ordinary Thurstone scale, but not when applied to one that has been modified by item analysis as used in the Likert method of construction.

References

CASTLE, P. F. C. (1953). A note on the Scale Product method of constructing Attitude Scales. *Occup. Psychol.* 27, 104–8.

EYSENCK, H. J. (1954). *The Psychology of Politics*. London: Routledge and Kegan Paul.

EYSENCK, H. J. & CROWN, S. (1949). An experimental study in opinion-attitude methodology. *Int. J. Opinion Attitude Res.* 3, 47–86.

FERGUSON, L. W. (1939). The requirements of an adequate Attitude Scale. *Psychol. Bull.* 26, 665–73.

LIKERT, R., ROSLOW, S. & MURPHY, G. (1934). A simple and reliable method of scoring the Thurstone Attitude Scales. *J. Soc. Psychol,* 5, 228–38.

OLIVER, R. A. C. (1953). Attitudes to Education. *Brit. J. Educ. Studies,* 2, 31–41.

VERNON, P. E. (1953). *Personality Tests and Assessments.* London: Methuen.

Chapter 22

A COMPARISON OF FOUR METHODS OF SCORING AN ATTITUDE SCALE IN RELATION TO ITS RELIABILITY AND VALIDITY

PAMELA K. POPPLETON AND G. W. PILKINGTON
University of Sheffield, England

A. Introduction

In the course of a study previously reported (Poppleton & Pilkington, 1963), two parallel forms of a scale for the measurement of religious attitudes were constructed using the Thurstone method for the compilation of statements. They were used in a preliminary survey among a group of 120 subjects before conducting an item analysis and drawing up the final version of the scale. Each form consisted of 22 items, subjects being asked to respond to each item by endorsing one of five categories: "Strongly Agree," "Agree," "Uncertain," "Disagree," and "Strongly Disagree." Half the group had Form A first, and half Form B, the other form being given three weeks later. They were scored by four different methods:

1. Ordinary Thurstone scoring on items which were endorsed.
2. Likert scoring. Response categories were weighted 5–4–3–2–1, and the weights reversed at the mid-point of the scale.
3. The Scale–Product method. A modified version used by Castle (1953)

Reprinted with permission of the author and the publisher from the *British Journal of Social and Clinical Psychology*, Vol. 2, 1963, pp. 36–39.

was adopted, in which the Likert weight is multiplied by the deviation of the Thurstone scale value of the item from the mid-point of the scale (5.5).

4. A method suggested by Guilford (1954) which uses empirically derived weights for the response categories, and which will be referred to here as "the method of weighted proportions."

B. Results and Discussions of Results

Data concerning reliability and validity are shown in Tables 22.1 and 22.2.

RELIABILITY

A Thurstone score based on the median position of each respondent is essentially a limen score indicating the subject's central response tendency towards the attitude. But variability is introduced because respondents may endorse as few as three or as many as 14 items, so that the size of the correlation between limen scores tends to be reduced. The Likert method of scoring by summing the responses in each category has commonly been found to yield higher reliability coefficients. For example, Mosier (1941) investigated the relationship between the limen score and the summation score of a mental test, and found the reliability of the first to be 0.88 compared with 0.94 for the second. The coefficients yielded here by our attitude scale are closely comparable.

Table 22.1. Statistics on Four Methods of Scoring Parallel Forms

	Form A	Form B	Product moment correlation N = 120
Thurstone Scoring			
Mean	4.9	4.6	
S.D.	1.7	2.0	} 0.85
Range	7.24	7.33	
Likert Scoring			
Mean	63.0	66.4	
S.D.	14.3	13.4	} 0.95
Range	60	59	
Scale Product Method			
Mean	14.1	23.4	
S.D.	47.8	45.4	} 0.88
Range	206	199	
Mehtod of Weighted Proportions			
Mean	87.1	90	
S.D.	20.5	19.6	} 0.95
Range	74	74	

Table 22.2. Validity. r_{bis} *between Scale Scores and Reported Religious Activities (Form B)*

	Standard error r_{bi}, shown in brackets Method of scoring (N = 120)				
Criterion	Thurstone	Likert	Scale— Product	Weighted proportions	
(a) Active church membership	0.84 (0.05)	0.86 (0.05)	0.84 (0.05)	0.88 (0.04)	49
(b) Church attendance (twice or more) in a month	0.74 (0.06)	0.79 (0.06)	0.81 (0.06)	0.86 (0.05)	44
(c) Daily private prayers	0.72 (0.07)	0.73 (0.07)	0.73 (0.07)	0.72 (0.07)	28
(d) Membership of a student religious group	0.60 (0.09)	0.55 (0.09)	0.57 (0.09)	0.60 (0.09)	32
(e) Expressing some form of religious belief	0.93 (0.03)	0.96 (0.02)	0.94 (0.03)	0.97 (0.02)	63

The Scale–Product method represents the major attempt to combine the Thurstone and Likert methods. The results reported, however, do not support Butcher's (1956) hypothesis that this method gives a higher reliability coefficient than Likert scoring before item analysis has been applied. Our reliability coefficient of 0.88 may be compared with Butcher's (0.89) and Castles' (0.81), although Eysenck and Crown (1949) quote one of 0.94 for their Anti-Semitism scale. But Butcher presents a detailed criticism of their use of the Scale–Product method, suggesting that the weights actually used are difficult to reconcile with their description of the method. It appears that their weights approximate closely to those achieved by the method of weighted proportions, so that the resulting scores cannot really be called "scale products." The lower reliability of true Scale–Product scoring can probably be explained by the fact that the Thurstone scale values of items in parallel forms are close but not identical, and multiplying them by response weights exaggerates the differences. Table 22.3 shows four pairs of items with approximately the same Thurstone scale values. The deviation of each of these from the mid-point of the scale is multiplied by five to arrive at the weight for the "Strongly Agree" category.

The two sets of items may be regarded as parallel to all intents and purposes, but they nevertheless receive different weights on the two forms. Adding these differences means that subjects who obtain identical scores on the Likert system can be separated by as much as ten or eleven points using scale–product scores. It would seem, therefore, that this method can

Table 22.3. Thurstone Scale Values of Parallel Items and Their Derived Weights

	Form A			Form B	
Thurstone Scale value of item	Deviation from mid-point of scale (5.5)	Weight for "Strongly Agree" category	Thurstone Scale value of item	Deviation from mid-point of scale (5.5)	Weight for "Strongly Agree" category
3.15	2.35	12	2.90	2.6	13
3.64	1.86	9	3.32	2.18	11
3.97	1.53	8	3.87	1.63	8
4.41	1.09	5	4.22	1.28	6

never give a high coefficient of equivalence because of its tendency to exaggerate small differences in the original scale values. It is as if one were taking measurements with a piece of elastic fastened at the middle.

An alternative method of retaining information about extremeness and intensity is to use weights derived from the proportion of respondents who endorse an item in each of the five categories. Guilford (1954) gives an account of such a method. The summation scores are divided at the mean, and the proportion of high and low scores for each item in each category is calculated. Weights are then assigned by entering the proportions in an abac given by Guilford (p. 446). This was the procedure we followed. Eysenck (1954) suggests that this method may prove more reliable than the Scale–Product method, but this has not until now been tested. Our results show a very satisfactory reliability coefficient, although it is still no higher than that yielded by Likert scoring.

VALIDITY

Table 22.2 shows that the correlation coefficients given by each criterion for the four different methods of scoring are quite closely comparable. There is no one "best order," but on practically all indices the method of weighted proportions yields the optimum weights for high validity. Weighting has been shown to be important in relation to validity when the number of items is not large (Guilford, 1954; Gulliksen, 1958). But when, as with our scale, there are over 20 items, Guilford suggests that weights can depart from their optimal values without seriously affecting validity. Moreover, the different weighting systems could not, in principle, make much difference, because all used positive values in the same directions, and so might be expected to yield comparable results.

C. Conclusions

This study upholds the general finding that summation scores give higher reliability coefficients than limen scores. The Likert method is also found to provide as high a reliability as the method of weighted proportions. The Scale–Product method, however, when used as intended, does not give as high a reliability as the latter two methods. It is argued that this type of scoring is unsatisfactory. The method of weighted proportions also provides a good indication of validity. But when the number of items is more than 20 the system of weights used does not affect validity seriously. In general, the method of weighted proportions has much to recommend it, but the simpler Likert system does nearly as well.

References

BUTCHER, H. J. (1956). A note on the scale product and related methods of scoring attitude scales. *Brit. F. Psychol.* 47, 133–139.

CASTLE, P. F. C. (1953). A note on the scale-product technique of attitude scale construction. *Occup. Psychol.* 27, 104–109.

EYSENCK, H. J. (1954). *The Psychology of Politics.* London: Routledge & Kegan Paul.

EYSENCK, H. J. & CROWN, S. (1949). Primary social attitudes: II. An experimental study in opinion-attitude methodology. *Internat. F. Opin. Attit. Res.* 3, 47–86.

GUILFORD, J. P. (1954). *Psychometric Methods,* 2nd edn. London: McGraw-Hill.

GULLIKSEN, H. (1958). *Theory of Mental Tests.* New York: Wiley.

MOSIER, C. (1941). Psychophysics and mental test theory. *Psychol. Rev.* 48, 235–249.

POPPLETON, P. K. & PILKINGTON, G. W. (1963). The measurement of religious attitudes in a university population. *Brit. F. soc. clin. Psychol.* 2, 20–36.

PART V

Unfolding Theory

This section describes a method of scaling that has been developed at the University of Michigan by Clyde H. Coombs. The process is identified as unfolding. A brief examination of the theory and procedures, as they are described in the following articles and selections, will demonstrate and explain why this particular name has been employed. A reading of the selections also will reveal that this method of scaling is relatively unlike the others described in this volume. The uniqueness of the method in no way diminishes its elegance but may well have contributed to the method's relatively infrequent use in research.

Research oriented behavioral scientists, like human beings in general, are more willing to use relatively old, more familiar, less elegant, and sometimes even less appropriate methods to secure scale scores than they are to use a unique but perhaps more powerful method. Often they are more interested in securing a set of scores that may be analyzed than they are in using measurement techniques that are more closely related to a metric that is more isomorphic to their observations. Many researchers have been less concerned with measurement and scaling than they have been with matters of design, analysis, and hypothesis testing. This frequently leads the researcher to employ only single questions and not scales to measure variables, in spite of the accompanying loss of discrimination and reliability that this involves. The tendency to neglect the new and less familiar measurement techniques may be reinforced by the greater availability of computer programs for the older more established scaling methods. One frequently has the feeling that many researchers allow the availability of various computer programs to determine the direction and

mode of measurement and analysis they employ. If the method of unfolding is to be more widely used, it is necessary for its procedures to be
more generally known and its utility and value more fully appreciated.
It is hoped that Part V will produce such a result.

The process of unfolding allows us to discern the metric relations "inherent" in data. The technique does not provide more information than
is appropriate. Researchers who "know what they want" will continue to
use and even prefer scaling procedures that provide more metric, even if
the metric is supplied by assumptions instead of observed relations. Others
who wish to be more attentive to scaling and who "want to know" what
relations are resident in their data will prefer a method like unfolding.

The initial selection in this section, chapter 23, extends Stevens' discussion of scales of measurement included in Part I of this volume. This
extension grows from an identification of two aspects of scales: the order of
the elements and the order of distances between elements. The types of
scales that result from systematically examining the logical combinations
of these two characteristics constitute a more elaborate and extensive
system than that proposed by Stevens. The resulting scales include those
that are more rigorous than ordinal scales but less rigorous than interval
scales. Included among these scales are the types that can be developed
through the process of unfolding.

The next selection in this part, Chapter 24, describes the basic model
and the procedure of unfolding. Chapter 25 presents an application of
this technique. The selection also contains a description of the process of
converting an ordered metric scale into an interval scale—by making additional assumptions. The final selection in this part, by David Goldberg and
C. H. Coombs, is the application of unfolding to a specific research problem. This illustration is provided since unfolding is a much less familiar
technique than most of the others described and an additional example
is warranted.

Chapter 23

SOME ASPECTS OF THE METATHEORY OF MEASUREMENT

C. H. COOMBS
University of Michigan

There are two major aspects to every theory of measurement; on the one hand, there is the formal, logical aspects, and on the other hand, the experimental or operational side. By the formal or logical aspect is meant a set of axioms which specify operations and relationships among a set of elements. By the experimental or operational side is meant a set of operations on the objects themselves by means of which they can be observed, in respect to some attribute, to satisfy certain axioms. These two aspects correspond respectively to what Woodger [1] calls structure and meaning, or alternatively syntax and semantics.

As a consequence of the fact that operations on objects do not always reveal the same axioms to be satisfied, there has gradually developed a series of what are called scales, each scale corresponding in principle, to a set of axioms. If it is possible to perform operations on objects such that the objects are observed to satisfy the axioms of one of the scales, then the

Reprinted with permission of the author and the publisher from *A Theory of Psychological Scaling* by Clyde H. Coombs, 1952. Engineering Research Bulletin, No. 34, University of Michigan Press, pp. 1–6.

1. Woodger, J. H. "The Technique of Theory Construction," *International Encyclopedia of Unified Science*, vol. 2, no. 5, p. 6 ff. Chicago: University of Chicago Press, 1939.

entire mathematical development based on those axioms may be validly substituted for operations on the objects.

The development of physics as "the queen of the sciences" may be regarded as a consequence, in part, of two fundamental conditions: (1) the development of a powerful mathematics built on the axioms of arithmetic, and (2) the fact that these axioms appear to be realized in vast areas of physical phenomena. Consequently, scientists in other areas of knowledge, and in particular social scientists, have not only respected physics but have tried to build their own science in its image. Unfortunately, the second condition above is, in general, not satisfied by social psychological phenomena. It behooves the social scientist to try to formulate axiom systems which satisfy the behavior he is interested in and to encourage the development of the appropriate mathematics. In its broadest sense, then, this is simply the recommendation that the social scientist attend to his measurement theory.

To return to the subject of the different scales of measurement, S. S. Stevens [2] recognizes four scales of measurement, the ratio scale, the interval scale, the ordinal scale, and the nominal scale. There is no one place in the literature in which the axioms for these scales are given, certainly not in a systematic fashion. Axioms for the ratio and ordinal scales are most clearly stated by Nagel,[3] who gives twelve axioms of quantity. The first six generate an ordinal scale and the entire twelve generate a ratio scale. Alternative postulational bases for an interval scale are contained in the first chapter of von Neumann and Morgenstern's *Theory of Games and Economic Behavior* [4] and in J. Marschak's recent paper in *Econometrica*.[5] It would be desirable to formulate a single set of axioms such that each scale could be characterized by a subset of these axioms, then the logical distinctions between the scales would be clear. In the summary of one of the sessions of a seminar on the application of mathematics to the social sciences, there is contained such a set of axioms prepared by Dr. Howard Raiffa. One of the natural consequences of the abstract approach is that it generates other types of scales not previously recognized experimentally or theoretically.

Interesting differences between these scales occur with respect to the assignment of numbers in the process of measurement. If the axioms for

2. Stevens, S. S. "On the Theory of Scales of Measurement." *Science*, 1946, vol. 103, p. 677–680.

3. Nagel, Ernest. *On the Logic of Measurement*. Ph. D. thesis. Columbia University, 1930.

4. von Neumann, J., and O. Morgenstern. *Theory of Games and Economic Behavior*. 2nd ed., Princeton: Princeton University Press, 1947. p. 641.

5. Marschak, Jacob. "Rational Behavior, Uncertain Prospects, and Measurable Utility," *Econometrica*, 1950, vol. 18, p. 111–141.

a ratio scale are satisfied by an attribute, the assignment of any one number to any one object is sufficient for fixing a unique number for each of all the other objects. This is equivalent to the selection of a unit of measurement. If the axioms for an interval scale are satisfied by an attribute, the assignment of any two different numbers to any two different objects is sufficient to fix a unique number for each of all the other objects with respect to this attribute. This is equivalent to the selection of an origin and a unit of measurement. It is evident that for an interval scale the differences between the numbers assigned to objects form a ratio scale. If the axioms for an ordinal scale are satisfied, the assignment of any number to any one object fixes an upper bound or a lower bound for the number to be assigned to any other object, and no other knowledge about the number for any other object is contained in the measurement. In the case of a nominal scale, the assignment of a number to an object or class of equal objects merely uses up that number and puts no other constraint on the assignment of other numbers to other classes of objects.

There are a number of useful psychological scales which may be added to Stevens' list. Among these is the type of scale called an ordered metric,[6] in which the order of objects on a continuum is known, and also the distances between objects may themselves be at least partially ordered. It is evident that this type of scale is weaker than an interval scale but more powerful than an ordinal scale. With this type of scale, the assignment of any two different numbers to a *particular pair* of objects (the two adjacent ones that are farthest apart) fixes a pair of numbers for every other object, one an upper bound and the other a lower bound.

Another type of scale which can be added to the system is the partially ordered scale. For immediate purposes it suffices to say that it is useful in the treatment of intransitive paired comparisons or, more generally, what will be called conjunctive and disjunctive behavior as distinct from summative behavior.[7] It is evident that the partially ordered scale falls between an ordinal and a nominal scale. The assignment of any number to any one object in a partially ordered system provides either an upper or a lower bound for the number to be assigned to *some* other objects.

From a study of these scales it is evident that they may be ordered on the basis of the axioms they satisfy. The ratio scale satisfies the most axioms and may be regarded as the most powerful scale. The interval scale comes next, then the ordinal scale, and, finally, the nominal scale is the weakest. The axioms satisfied by one of these scales are included in the

6. Coombs, C. H. "Psychological Scaling Without a Unit of Measurement," *Psychol. Rev.*, 1950, vol. 57, p. 145–158.

7. Coombs, C. H. "Mathematical Models in Psychological Scaling," *J. A. S. A.*, 1951, vol. 46, p. 480–489.

axioms satisfied by a more powerful scale and include those satisfied by a weaker scale. Thus, an interval scale is also an ordinal and a nominal scale, and a ratio scale includes them all.

A more complete listing of scales, however, reveals that they are partially ordered. This is made evident by regarding scales as being composite scales. A scale may be regarded as constituted of two kinds of elements, the objects themselves and the distances between objects. The objects may be scaled on a nominal, partially ordered, or ordered scale, for example. The distances between objects, regarded as elements may also be scaled on one of these scales. Listing the scale that holds for the objects themselves, first, and the scale that holds for the distances between objects, second, the scale called here a nominal scale may be regarded as a nominal-nominal scale. The partially ordered scale becomes a partially ordered-nominal scale. The ordinal scale is seen to be an ordinal-nominal scale. The ordered metric includes the ordinal-partially ordered, and the ordinal-ordinal.

On this basis a lattice of scales would consist of the following:

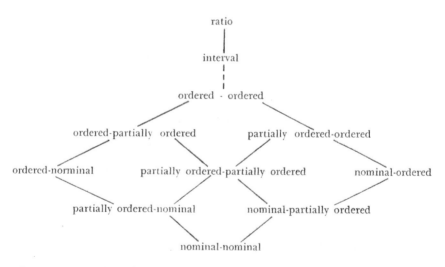

where a connecting line between two scales indicates that the axioms of the lower scale are included in the axioms of the higher scale. Referring to the distances or intervals between objects as first-order differences, the above lattice is constructed on zero and first-order differences between objects. This lattice could be extended by including second-order differences between objects, which could be elements of an abstract system and on which there could also be relations such as "greater than." Thus, an example of a second-order difference relation would be the statement that the difference between the distances between stimuli A and B and stimuli B and C is greater than the differences between the distances between stimuli C and D and stimuli D and E.

The relation of the interval scale to the ordered-ordered is complex. Mr. Donald Mela has pointed out[8] that if the number of stimuli is infinite, the ordered-ordered scale is identical with the interval scale. But if the number of stimuli is finite, it would require an infinite succession of differences between differences to make them identical.

Elsewhere,[9] an arbitrary distinction has been drawn, for purposes of exposition, between psychological measurement and psychological scaling on the basis of the axioms that are satisfied. In brief, if the operations of arithmetic are permissible on the scale elements, i.e., numbers, assigned to objects or their differences, these scales are classified as measurement; otherwise they are classified as scaling. On this basis, ratio and interval scales are both classified as measurement, and the remaining scales, including the ordered metric, ordinal scale, partially ordered scale, and nominal scale are classified as scaling. The distinction is arbitrary but convenient. In the broadest sense, however, all scales represent degrees of measurement, a ratio scale representing the highest degree, a nominal scale the lowest, and the other scales partially ordered between these two. This partial ordering of the scales themselves on degree of measurement represents an ordering of the power of the scales. The more axioms of quantity that are used in the basis of a scale the more elaborate the mathematical development of the abstract system. There are more operations that can be performed on the numbers and more relations to be deduced. It is in this sense that one scale is spoken of as being more powerful than another. There is a great difference in this respect between the interval and the ordinal scales, and it is natural that psychologists have frequently been willing to make the necessary assumptions to achieve an interval scale in preference to an ordinal scale.

There is a relationship, perhaps, between the power of a scale, on the one hand, and the generality of its applicability, on the other. The more powerful the scale the more axioms it is based on. For a scale to be validly applicable, an object system must also satisfy the axioms of the scale. In general, one would expect that the more axioms the fewer the object systems that could satisfy it. Thus, the power of a system and its generality tend to be inversely related.

A characteristic of much of the development of measurement theory in psychology and the social sciences has been an effort to achieve more

8. Personal communication.
9. Coombs, "Mathematical models."

powerful scales without sacrificing generality. One consequence of this is that the meaning associated with the numbers has moved toward more gross statistical interpretations and away from individual interpretations. This is evident in Gulliksen's paper on paired comparisons and the logic of measurement.[10] Judgments of individuals incompatible with the general solution are regarded as discriminal error, but the prediction of the distribution of a number of judgments is frequently highly accurate.

Guttman's work in scaling theory[11] may be regarded as an effort to avoid the more gross statistical interpretations of scale values to permit interpretation of the individual case. The result is that he drops from an interval scale to an ordinal scale, from a more powerful system to a more general system.

A characteristic of the philosophy which underlies most psychological measurement theory is that it is directed toward trying to *construct* scales, to *make* scales. In the approach presented here the point of view is that the existence of a scale has psychological significance[12] and that the conclusion that a scale exists cannot be drawn if the measurement technique and theory guarantees the construction of a unidimensional scale.

10. Gulliksen, Harold. "Paired Comparisons and the Logic of Measurement," *Phychol. Rev.*, 1946, vol. 53, p. 199–213.

11. Guttman, Louis. Chapters 2, 3, 6, 8, 9 in: Samuel A. Stouffer, *et al.*, *Measurement and Prediction*. Princeton: Princeton University Press, 1950.

12. Coombs, C. H. "Some Hypotheses for the Analysis of Qualitative Variables," *Psychol. Rev.*, 1948, vol. 55, p. 167–174.

Chapter 24

PSYCHOLOGICAL SCALING WITHOUT A UNIT OF MEASUREMENT

C. H. COOMBS
University of Michigan

I. Introduction

The concept of measurement has generally meant the assignment of numbers to objects with the condition that these numbers obey the rules of arithmetic (1). This concept of measurement requires a ratio scale— one with a non-arbitrary origin of zero and a constant unit of measurement (3). The scales which are most widely made use of in psychology are regarded as interval scales in that the origin is recognized to be arbitrary and the unit of measurement is assumed to be constant. But

Reprinted with permission of the author and the publisher from the Psychological Review, Vol. 57, 1950, pp. 145–58. Copyright 1950 by the American Psychological Association.

This paper is a condensation of some of the ideas contained in a chapter of a general theory of psychological scaling developed in 1948–49 under the auspices of the Rand Corporation and while in residence in the Department and the Laboratory of Social Relations, Harvard University. While the author carries the responsibility for the ideas contained herein, their development would not have been possible without the criticism and stimulation of Samuel A. Stouffer, C. Frederick Mosteller, Paul Lazarsfeld, and Benjamin W. White in a joint seminar during that year. Development of the theory before and after the sojourn at Harvard was made possible by the support of the Bureau of Psychological Services, Institute for Human Adjustment, Horace H. Rackham School of Graduate Studies, University of Michigan.

this type of scale should be used only if it can be experimentally demonstrated by manipulation of the objects that the numbers assigned to the objects obey the laws of addition. The unit of measurement in psychology, however, is obtained by a combination of definitions and assumptions, which, if regarded as a first approximation and associated with a statistical theory of error, serves many practical purposes. But because we may sometimes question the meaning of the definitions and the validity of the assumptions which lead to a unit of measurement, it is our intent in this paper to develop a new type of scale not involving a unit of measurement. This type of scale is an addition to the types set up by S. S. Stevens (3). Stevens recognized ratio, interval, ordinal, and nominal scales. The type which we shall develop falls logically between an interval scale and an ordinal scale. We shall make no assumption of equality of intervals, or any other assumption which leads to a unit of measurement. We shall find, however, that on the basis of tolerable assumptions and with appropriate technique we are able to *order* the magnitude of the intervals between objects. We have called such a scale an "ordered metric." We shall develop the concepts first in an abstract manner with a hypothetical experiment and then illustrate the ideas with an actual experiment. Under the limitations of a single paper we shall not present the psychological theory underlying some of the concepts and we shall place certain very limiting conditions on our hypothetical data in order to simplify the presentation.

II. The Problem

When we set up an attitude scale by any of a variety of methods, for example the method of paired comparisons and the law of comparative judgment, we order statements of opinion on the attitude continuum and assign a number to each statement. We recognize in this instance that the origin for the numbers is arbitrary. We then follow one of several possible procedures (determining which statements an individual will indorse, for example) to locate the positions of individuals on this same continuum. Because both individuals and stimuli have positions on this continuum we shall call it a joint distribution, joint continuum, or J scale. In general, with a psychological continuum, we might expect that for one individual the statements of opinion, or stimuli, have different scale position than for another individual. Thurstone (4) has provided the concept of stimulus dispersion to describe this variability of the scale positions on a psychological continuum. We have recently (2) discussed an equivalent concept for the variability of scale positions which an individual may assume in responding to a group of stimuli. These two concepts have been basic to the development of a general theory of scaling to which this paper is an introduction.

For didactic purposes we shall achieve brevity and simplicity for the presentation of the basic ideas underlying an ordered metric scale if we impose certain extreme limiting conditions on the variability of the positions of stimuli and individuals on the continuum. These conditions are that the dispersions of both stimuli and individuals be zero. In other words these conditions are that each stimulus has one and only one scale position for all individuals and that each individual has one and only one scale position for all stimuli. For purposes of future generalization we shall classify these conditions as Class 1 conditions. Stimuli will be designated by the subscript j and the position of a stimulus will be designated the Q_j value of the stimulus on the continuum. The position of an individual on the continuum will be designated the C_i value of an individual i.

If we conceive of the attribute as being an attitude continuum, the C_i value of an individual is the Q_j value of that statement of opinion which perfectly represents the attitude of that individual. In this case the C_i value of an individual is his ideal or norm. We shall assume that the degree to which a stimulus represents an individual's ideal value is dependent upon the nearness of the Q_j value of the stimulus to the C_i value of the individual.

We shall then make the further assumption that if we ask an individual which of two statements he prefers to indorse he will indorse that statement the position of which is nearer to his own position on the continuum.

Thus if asked to choose between two stimuli j and k, the individual will make the response

$$j > k \qquad\qquad [1]$$

if

$$|Q_j - C_i| < |Q_k - C_i|$$

where $j > k$ signifies the judgment "stimulus j preferred to stimulus k."

Under the extreme limiting conditions we have imposed on the C_i and Q_j values the method of paired comparisons would yield an internally consistent (transitive) set of judgments from each individual, though not necessarily the same for each, and each different set of such judgments could be represented by a unique rank order for the stimuli for that individual. We shall call the rank order of the stimuli for a particular individual a qualitative I scale, or, in general, an I scale.

Thus if an individual placed four stimuli in the rank order A B C D as representing the descending order in which he would indorse them, then, this would be equivalent to the consistent set of judgments $A > B$, $A > C$,

A $>$D, B $>$C, B $>$D, C $>$ D; where the symbol "$>$" signifies "prefer to indorse," as before. The order, A B C D, is the qualitative I scale of this individual. Hence for Class 1 conditions it is sufficient to collect the data by the method of rank order; the greater power of the method of paired comparisons would be unnecessary and wasted.

Let us assume now that we have asked each of a group of individuals to place a set of stimuli in rank order with respect to the relative degree to which he would prefer to indorse them. Our understanding of the results that would follow will be clearer if we build a mechanical model which has the appropriate properties. This is very simply done by imagining a hinge located on the J scale at the C_i value of the individual and folding the left side of the J scale over and merging it with the right side. The stimuli on the two sides of the individual will mesh in such a way that the quantity $|C_i - Q_j|$ will be in progressively ascending magnitude from left to right. The order of the stimuli on the folded J scale is the I scale for the individual whose C_i value coincides with the hinge.

It is immediately apparent that there will be classes of individuals whose I scales will be qualitatively identical as to *order* of the stimuli and that these classes will be bounded by the midpoints between pairs of stimuli on the J scale. For example, suppose that there are four stimuli, A B C D, whose Q_j values or positions on a joint continuum are as shown in Figure 24.1 and that there is a distribution function of the positions of individuals on this same continuum as indicated.

If we take the individual whose position is at X in Figure 24.1, the I scale for that individual is obtained by folding the J scale at that point and we have the scale shown in Figure 24.2.

The qualitative I scale for the individual at X is A B C D.

If we take the individual in position Y as shown in Figure 24.1 and construct his I scale, we have the scale shown in Figure 24.3.

The qualitative I scale for the individual at Y is C D B A.

Consider all individuals to the left of position X on the J scale in Figure 24.1. The I scales of all such individuals will be quantitatively different for different positions to the left of X. For every one of them, however, the *order* of the stimuli on the I scale will be the same, A B C D. We shall

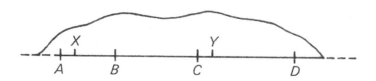

Figure 24.1. A joint distribution of stimuli and individuals.

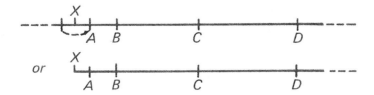

Figure 24.2. The I scale of an individual located at X in the joint distribution.

regard these I scales as being qualitatively the same. As a matter of fact, the I scales of individuals to the right of X continue to be qualitatively the same until we reach the midpoint between stimuli A and B. For an individual immediately to the right of this midpoint the qualitative I scale will be B A C D. I scales immediately to the right of the midpoint AB will continue to be qualitatively the same, B A C D, until we reach the midpoint between stimuli A and C. Immediately past this midpoint the qualitative I scale is B C A D. Continuing beyond this point a complicating factor enters in which we shall discuss in a later section under metric effects.

The distinction which has been made here between quantitative and qualitative I scales is of fundamental importance to the theory of psychological scaling. In almost all existing experimental methods in psychological scaling we do not measure the magnitudes $|C_i - Q_j|$, but only observe

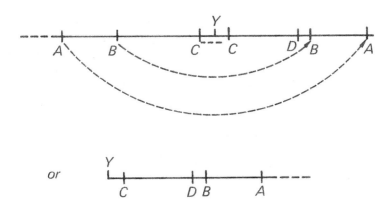

Figure 24.3. The I scale of an individual located at Y in the joint distribution.

the ordinal relations for a fixed C_i.[2] The kind of information which is obtained by the experimenter is essentially qualitative in nature.

As we shall see, data in the form of I scales may tell us certain things:

1. whether there is a latent attribute underlying the preferences or judgments,
2. the order of the stimuli on the joint continuum,
3. something about the relative magnitudes of the distances between stimuli,
4. the intervals which individuals are placed in and the order of the intervals on the continuum, and
5. something about the relatives magnitudes of these intervals.

III. A Hypothetical Example

Let us now conduct a hypothetical experiment designed merely to illustrate the technique. Of course this experiment, if actually conducted, would not turn out as we shall construct it, because we shall assume the extreme limiting conditions on Q and C values that were previously imposed.

Let us imagine that we have a number of members of a political party and that we have four individuals who are potential presidential candidates. Let us ask each member of the party to place the four candidiates, designated A, B, C, and D, in the rank order in which he would prefer them as President. With four stimuli the potential number of qualitatively different rank orders is 24—the number of permutations of four things taken four at a time. If there were no systematic forces at work among the party members we would get a distribution of occurrences of these 24 I scales which could be fitted by a Poisson or Binomial distribution, everything could be attributed to chance, and the experiment would stop there. Instead, let us imagine for illustrative purposes a different result equally extreme in the opposite direction. Let us imagine that from the N individuals doing the judging only seven qualitatively different I scales were obtained and these were the following:

I_1	A	B	C	D
I_2	B	A	C	D
I_3	B	C	A	D
I_4	C	B	A	D
I_5	C	B	D	A
I_6	C	D	B	A
I_7	D	C	B	A

2. The ordinal relations of $| C_i - C_j |$ may also be obtained experimentally for a fixed Q_j, over a set of C_i. We call such scales S scales by analogy with I scales. For sake of simplicity they are not treated here but actually the treatment for I scales and S scales is identical if the roles of stimuli and individuals are merely interchanged.

The significance of the deviation of such results as these from pure chance would be self-evident. Consequently we would look at these seven I scales to see if there was some systematic latent attribute represented by a joint continuum such that individual differences and stimulus differences on such a continuum could account for these manifest data. Studying the set of seven I scales we observe that two of them are identical except in reverse order, A B C D and D C B A. Furthermore we see that in going from one to the next, two adjacent stimuli in the one have changed positions in the next. These are the characteristics of a set of I scales which have been generated from a single J scale. Seven I scales is the maximum number that one can obtain from a single quantitative J scale of four stimuli under the conditions of Class 1. The systematic latent attribute underlying this set of I scales is represented by the J scale which generates them. Our objective, then, is to recover this J scale and discover its properties or characteristics.

To recover the J scale we proceed as follows. Every complete set of I scales has two and only two scales which are identical except in reverse order. These are the I scales which arise from the first and last intervals of the J scale. Consequently these two I scales immediately define the ordinal relations of the stimuli on the J scale, in this case A B C D (the reverse order, of course, is equally acceptable). From the seven I scales we can order on the J scale the six midpoints between all possible pairs of stimuli. In going down the ordered list of I scales as previously determined, the pair of adjacent stimuli in one I scale which have changed places in the next I scale specify the midpoint on the J scale which has been passed.

Thus in the first interval (Figure 24.4), we have all the individuals to the left of the midpoint between stimuli A and B. The second I scale is B A C D, and as stimuli A and B have changed places in going from the I_1 scale to the I_2 scale we have passed the midpoint AB. In going from I_2 to I_3 stimuli A and C exchange orders on the two I scales and hence the midpoint between A and C is the boundary between I_2 and I_3. If we con-

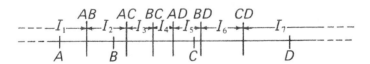

Figure 24.4. An example of how the midpoints of four stimuli may section the joint distribution into seven intervals, each characterized by an I scale.

tinue this process we see that the order of the six midpoints is as follows: AB, AC, BC, AD, BD, CD. These six boundaries section the joint distribution into seven intervals which are ordered as also are the stimuli. From the order of the six midpoints in the case of four stimuli we have one and only one piece of information about metric relations on the joint continuum. Because midpoint BC precedes AD we know that the distance between stimuli C and D is greater than the distance between stimuli A and B. We shall discuss these points and characteristics in more detail in the section on metric effects. There are then an infinite variety of quantitatively different J scales which would yield this same set of seven I scales—but there is only one qualitative J scale. The J scale in Figure 24.4 meets the conditions necessary to yield the manifest data.

It must be emphasized that all *metric* magnitudes in Figure 24.4 are arbitrary except that the distance from stimulus A to B is less than the distance from stimulus C to D.

With the qualitative information obtained in this experiment about the latent attribute underlying the preferences for presidential candidates the next task is the identification of this attribute. Here all the experimenter can do is to ask himself what it is that these stimuli have and the individuals have to these different degrees as indicated by their ordinal and metric relations. One might find in this hypothetical case that it appears to be a continuum of liberalism, for example, or of isolationism. One would then have to conduct an independent experiment with other criteria to validate one's interpretation.

IV. Metric Effects

While the data with which we deal in the vast majority of scaling experiments are qualitative and non-numerical there are certain relations between the manifest data and the metric relations of the continuum. These relations have not all been worked out and general expressions have yet to be developed. The complexity of the relations very rapidly increases with the number of stimuli; therefore to illustrate the effect of metric we shall take the simplest case in which its effect is made apparent, the case of four stimuli.

With four stimuli, A, B, C, and D, there are 24 permutations possible. Thus it is possible to find 24 qualitatively different I scales. Also, obviously, each of these 24 orders could occur as a J scale which could give rise to a set of I scales. Half of the J scales may be regarded as merely mirror images of the other half. Thus if we have a J scale with the stimuli ordered B A D C and identify the continuum as liberalism-conservativism, then, in principle, we also have the J scale with the stimuli ordered C D A B and would identify it as conservativism-liberalism. Hence there are only twelve J scales which may be regarded as qualitatively distinguishable on the basis of the order of the four stimuli. I scales which are mirror images of each other, though, are definitely not to be confused. They may well represent entirely different psychological meanings. The direction of an I scale is defined experimentally—the direction of a J scale is a matter of choice.

Each J scale of four stimuli gives rise to a set of seven qualitative I scales. We are interested in knowing, of course, whether the J scale deducted from a set of I scales obtained in an experiment is qualitatively unique. The answer to this appears to be yes and immediately obvious when it is recognized that a set of I scales generated from a J scale has two and only two I scales which are mirror images of each other; that these two I scales must have been generated from the intervals on the opposite ends of the J scale, and that the order of the I scales within a set is unique. These statements are still to be developed as formal mathematical proofs and hence must be regarded as tentative conclusions.

However, a given qualitative J scale does not give a unique *set* of I scales. For example, with four stimuli, we may have the qualitative J scale A B C D. This order of four stimuli on a J scale you can yield *two different sets* of I scales as follows:

Set 1					Set 2		
A	B	C	D		A	B	C	D
B	A	C	D		B	A	C	D
B	C	A	D		B	C	A	D
B	C	D	A	———————	C	B	A	D
C	B	D	A		C	B	D	A
C	D	B	A		C	D	B	A
D	C	B	A		D	C	B	A

It will be noticed that these two sets of seven I scales from the same qualitative J scale are identical except for the I scale from the middle interval. This arises from the following fact. There are six midpoints for the four stimuli on the J scale. These are as follows: AB, AC, AD, BC, BD, CD. The order and identity of the first two and the last two are immutable; they must be, in order: AB, AC,　,　, BD, CD. But the order of the remaining two midpoints is not defined by the qualitative J scale but by its quantitative characteristics. If the interval between stimuli A and B is greater than the interval between C and D, then the midpoint AD comes before the midpoint BC and the set of seven I scales will be set 1 listed above. If the quantitative relations on the J scale are the reverse and the midpoint BC comes before AD, then the set of seven I scales which will result are those listed in set 2 above.

Thus we see that in the case of four stimuli, a set of I scales will uniquely determine a qualitative J scale and will provide one piece of information about the metric relations. For five or more stimuli the number of pieces of information about metric relations exceeds the minimum number that are needed for ordering the successive intervals. *However, the particular pieces of information that are obtained might not be the appropriate ones for doing this.* It is interesting to note here that this is a new type of scale not discussed by Stevens. This is a type of scale that falls between what he calls ordinal scales and interval scales. In ordinal scales nothing is known about the intervals. In interval scales the intervals are equal. In this scale, which we call an ordered metric, the intervals are not equal but they may be ordered in magnitude.

As the number of stimuli increases, the variety of different *sets* of I scales from a single qualitative J scale increases rapidly. This means that a great deal of information is being given about metric relations. For example a J scale of five stimuli yields a set of eleven I scales (in general n stimuli will provide $\binom{n}{2} + 1$ different I scales from one J scale). Depending on the relative magnitudes of the four intervals between the five stimuli on the J scale, the same qualitative J scale may yield *twelve different sets* of I scales. This means that for a given order of five stimuli on a J scale there are twelve experimentally differentiable quantitative J scales. Previously, in the case of four stimuli, we found only two differentiable quantitative J scales for a given qualitative J scale.

The particular set of I scales obtained from five stimuli *may* provide up to five of the independent relations between pairs or intervals. For example, suppose we have the qualitative J scale A B C D E. Among the twelve possible sets of I scales which could arise are the following two, chosen at random:

	Set 1						Set 2		
A	B	C	D	E		A	B	C	D	E
B	A	C	D	E		B	A	C	D	E
B	C	A	D	E		B	C	A	D	E
B	C	D	A	E		B	C	D	A	E
C	B	D	A	E	———	B	C	D	E	A
C	D	B	A	E	———	C	B	D	E	A
D	C	B	A	E	———	C	D	B	E	A
D	C	B	E	A		D	C	B	E	A
D	C	E	B	A		D	C	E	B	A
D	E	C	B	A		D	E	C	B	A
E	D	C	B	A		E	D	C	B	A

Let us see what information is given by each of these sets about the relative magnitudes of the intervals between the stimuli on the J scales.

Consider set 1 first. The order of the ten midpoints of the five stimuli according to set 1 is as follows: AB, AC, AD, BC, BD, CD, AE, BE, CE, DE. We know immediately, from the fact that the midpoint BC comes after AD, that the interval between stimuli A and B ($\overline{AB}$) is greater than the interval between the stimuli C and D ($\overline{CD}$). We have summarized this in the first row of the table below. The other rows contain the other metric relations which can be deduced from set 1.

Set 1

Order of midpoints		Relative magnitude of intervals on J scale
AD,	BC	$\overline{CD} < \overline{AB}$
CD,	BE	$\overline{BC} < \overline{DE}$
BD,	AE	$\overline{AB} < \overline{DE}$
CD,	AE	$\overline{AC} < \overline{DE}$

Or, in brief form, the I scales contained in set 1 indicate that the following relations must hold between stimuli on the J scale.

$$\overline{CD} < \overline{AB} < \overline{DE}$$
$$\overline{AB} + \overline{BC} < \overline{DE}$$

In the same manner we may study the implications of set 2 for the metric relations between stimuli on the J scale. The midpoints for this set are in the following order: AB, AC, AD, AE, BC, BD, CD, BE, CE, DE.

Set 2

Order of midpoints		Relative magnitude of intervals on J scale
AD,	BC	$\overline{CD} < \overline{AB}$
CD,	BE	$\overline{BC} < \overline{DE}$
AE,	BD	$\overline{DE} < \overline{AB}$
AE,	BC	$\overline{CE} < \overline{AB}$

Or, in brief, the relative magnitudes of the intervals between stimuli on the J scale are known to the following extent.

$$\overline{BC} < \overline{DE} < \overline{AB}$$
$$\overline{DE} + \overline{CD} < \overline{AB}$$

The different implications of these two sets of I scales for the metric relations on the J scale may be illustrated by sketching two quantitative J scales which have the appropriate metric relations (Figure 24.5).

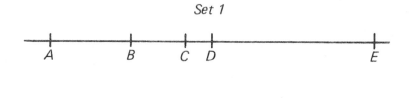

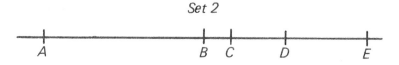

Figure 24.5. An example of two joint distributions with the same order of stimuli but different metric relations obtained from different sets of I scales.

The two sets of I scales which are illustrated here were only two of twelve possible different sets which could be generated from a single *qualitative* J scale of five stimuli. Each of the twelve sets of I scales would imply a different set of quantitative relations among the distances between stimuli on the J scale. The two sets of I scales used here happened to differ from each other in three of their particular members. If we take the twelve potential sets and make a frequency distribution of the number of pairs of sets which have 1, 2, 3, 4, or 5 particular I scales different or 10, 9, 8, 7, or 6 I scales in common, we get the following distribution.

Number of identical ordinal positions in a pair of sets with the same qualitative I scale	Number of such pairs of sets
10	18
9	24
8	17
7	6
6	1
	$66 = \begin{pmatrix} 12 \\ 2 \end{pmatrix}$

The surface has not even been scratched on the generalizations which can be developed. Enough has been presented here to provide a general idea of the type of information which can be derived.

V. An Experiment

In order to study the feasibility of this unfolding technique and to compare several different psychological scaling techniques an experiment was conducted in several classes in the Department of Social Relations at Harvard University. A questionnaire was administered pertaining to grade expectations in a course. Data were collected suitable for four different kinds of analyses on the same content area from the same individuals. The four types of analyses for which data were collected were (1) the generation of the joint continuum by the unfolding of I scales obtained by the method of rank order, (2) the generation of the joint continuum by the unfolding of I scales obtained by the method of paired comparisons, (3) the generation of the joint continuum by what we shall call the Guttman triangular analysis, and (4) the generation of the joint continuum by what we shall call the parallelogram analysis.

We shall present here the analysis of only the rank order data. The experiment was first conducted in a graduate course in statistics and then, with a slight change in the wording of some questions, in an undergraduate course in sociology. Despite these differences in subjects and questions the general results were practically identical in the two groups and because our primary interest is in the technique and not the content of this experiment we have lumped the data of the two groups.

The questionnaire was arranged as a small booklet with each question on a different page. The questions appropriate to the different techniques were deliberately mixed. The nature of the instructions, the separate paging for the questions, and the mixture of the questions were part of a deliberate effort to induce inconsistency, or at least to minimize a deliberate and artificial consistency. We felt this was accomplished but the high degree of consistency was surprising.

The content of the questionnaire follows.

page 1

Instructions: This is an experiment to test certain theoretical aspects of psychological scaling techniques. It is entirely voluntary and you need not answer the questionnaire if you so choose. However, you, as an individual, will not be identified; complete anonymity is preserved. We are interested *only* in certain internal relations in the data. This will become obvious to you because it will appear that we are getting the same information repeatedly in different ways.

You are free, of course, to mark these items entirely at random. It is our hope, however, that enough of you will take a serious attitude toward the experiment and make an effort to respond to each item on the basis of considered judgment. The questions pertain to your grade expectations in this course. Of course, everyone wants an A or B, but we would like to ask you to give serious consideration to what you really can expect to get. We want you to be neither modest nor self-protective. If you think you will get an A or flunk the course, make your judgments accordingly. Remember: there is complete anonymity.

There is one item or question on each of the following pages. Consider each question or item independently of the others. Answer each one without looking back at previous answers. Treat each item as an item in its own right and do not concern yourself with trying to be logical or consistent. Work quickly.

page 2

In the following list of grades circle the two grades which best represent what you expect to get in this course.

A B C D E

page 3

I expect to get a grade at least as good as a B.

yes   no

page 4

Of these two grades, which is nearer the grade you expect to get?

A   D

page 5

Of these two grades, which is nearer the grade you expect to get?

B   C

page 6

Of these two grades, which is nearer the grade you expect to get?

A   B

page 7

I expect to get a grade at least as good as a D.

yes   no

page 8

Of these two grades, which is nearer the grade you expect to get?

D   E

page 9

Of these two grades, which is nearer the grade you expect to get?

C   A

page 10

I expect to get a grade at least as good as an A.

yes   no

page 11

Of these two grades, which is nearer the grade you expect to get?

B   E

page 12

Of these two grades, which is nearer the grade you expect to get?

E   C

page 13

I expect to get a good grade.

yes   no

page 14
Of these two grades, which is nearer the grade you expect to get?
D   B
page 15
I expect to get a grade at least as good as a C.
yes   no
page 16
Of these two grades, which is nearer the grade you expect to get?
E   A
page 17
Of these two grades, which is nearer the grade you expect to get?
C   D
page 18
Place the five grades in rank order below such that the one on the left is the grade you most expect to get, then in the next space is the grade you next most expect to get, and so on, until finally at the right is the grade you least expect to get.

1	2	3	4	5
the grade I *most* expect to get				the grade I *least* expect to get

Page 18 of the questionnaire contained the rank order data. The total number of subjects for whom usable rank data were obtained was 121 (statistics class 40, sociology class 81). The individuals not included in the data which follow were people who introduced new grades (F), plus or minus grades, or left blanks. All individuals who wrote down the five letters A B C D E in some order are contained in the analysis. The I scales obtained by the method of rank order and the number of people in each of the classes who so responded are given in Table 24.1.

Let us first consider the two I scales in the bottom two rows of Table 24.1. These two scales, B A C E D and C A B D E, were each given by one individual. There is no way that either of these two scales could have arisen from a J scale on which the stimuli are in the order A B C D E regardless of the metric relations on the J scale. All the evidence, both *a priori* and from the other response patterns, indicates that the order of stimuli on the J continuum is A B C D E. So we must regard these two I scales as errors on the part of respondents or assume some esoteric psychologics to explain them, the latter completely unjustified. Hence we shall drop these two individuals from further consideration.

Let us now look at the first eight scales listed in Table 24.1. From five stimuli we can have eleven different I scales to correspond to the eleven intervals into which the J scale is sectioned by the ten midpoints between stimuli. Consequently these eight I scales constitute a *partial* set. But because one of the missing intervals (interval 8) has two alternative I

Table 24.1. Number of People in Two Classes Giving Each of the Rank Order I Scales

I scale	Statistics class	Sociology class
A B C D E	14	6
B A C D E	10	22
B C A D E	6	21
C B A D E	1	4
C B D A E	1	11
C B D E A	2	3
C D B E A	1	0
D E C B A	1	0
B C D A E	3	7
B C D E A	0	6
B A C E D	0	1
C A B D E	1	0
Total	40	81

scales, there are two possibilities for the complete set of which these eight are a partial set, as follows:

I scale		Total N
A B C D E		20
B A C D E		32
B C A D E		27
C B A D E		5
C B D A E		12
C B D E A		5
C D B E A		1
D C B E A — C D E B A		0
D C E B A		0
D E C B A		1
E D C B A		0

There were three intervals toward the low end of the J scale which were not occupied by any students. Apparently not very many students expected to get a low grade. The fact that one of the intervals on the J scale (interval 8) is blank and could be represented by two alternative I scales means that one of the metric characteristics of the intervals between stimuli on the J scale is not experimentally given. But from the remaining I scales the order on the J scale and some of the metric effects are determined.

The indications are that the order of stimuli on the joint continuum is

A B C D E. We know this, of course, from *a priori* grounds, but the point is that this fact need not be known beforehand. Secondly, the individuals are placed in intervals and the intervals ordered on the continuum. The number of ordered intervals is eleven but individuals occupy only eight of the eleven intervals. Thirdly, we know certain metric relations among the intervals between stimuli. These are obtained as follows. From the order of the I scales within the set, the successive midpoints between stimuli are: AB, AC, BC, AD, AE, BD, $\left(\dfrac{DC}{BE} \right)$, $\left(\dfrac{BE}{DC} \right)$, CE DE. Because interval 8 was unoccupied and there are two alternative I scales which satisfy it, it is not known whether the midpoint DC or BE comes first. Hence one piece of metric information is lacking. In the table below are the metric relations which may be deduced and the basis for the deduction.

Order of midpoints	Metric relations
BC, AD	$AB < \overline{CD}$
AE, BD	$\overline{DE} < A\overline{B}$

Or, in brief $\overline{DE} < \overline{AB} < \overline{CD}$. The psychological distance between the grades D and E is the least and the distance between the grades C and D the largest, with the distance from A to B in between. No information is given of the relative magnitude of the distance between the grades B and C.

But now for another portion of the students there is a different interpretation. There are two I scales in Table 24.1, B C D A E and B C D E A, which have not yet been considered. They are also members of a set that is as valid as the first set. If we remove I scales 4 and 5 from the partial set of 8 and substitute these two we have the following:

I scale	Total N
A B C D E	20
B A C D E	32
B C A D E	27
B C D A E	10
B C D E A	6
C B D E A	5
C D B E A	1
D C B E A — C D E B A	0
D C E B A	0
D E C B A	1
E D C B A	0

This set differs from the preceding only in I scales 4 and 5. If we analyze the significance of the set to the metric relations we have the following:

Order of midpoints	Metric relations
AE, BC	$\overline{AB} > \overline{CE}$

It appears from this that for the 16 individuals who gave the two I scales B C A D E and B C D E A, if they are treated as members of the same set, the psychological distance between the grades A and B is greater than either the distance from C to D or from D to E or, in fact, is greater than the sum of these two distances. This is in contrast to the 17 individuals in the first set who gave the I scales C B A D E and C B D A E in positions 4 and 5 for whom the relative distances between grades was in the order $\overline{DE}$, $\overline{AB}$, $\overline{CD}$.

The reader must be aware of the fact that it is only these 33 individuals for whom these metric relations are deduced. There is no way of knowing, for example, that the 20 individuals who gave the I scale A B C D E did so from the first interval on one of these two possible quantitative J scales or from any one of the many more differing in metric relations. An individual yielding the I scale A B C D E is known to be to the left of the midpoint AB and the relative distances $\overline{BC}$, CD, and DE do not affect his I scale. It is just the critical I scales which provide information about metric and this information is valid only for the individuals who yield these I scales. By putting them together we can construct a total picture, but our only evidence that they go together is one of internal consistency.

The general picture of metric relations among grades given by these two sets of I scales is the following:

1. For 17 individuals the metric relations are $\overline{DE} < \overline{AB} < \overline{CD}$
2. For 16 individuals the metric relations are $\overline{AB} > \overline{CE}$

Thus we find from the data that some of these individuals are simply on a different continuum from other individuals. To somehow compute scale values for the stimuli which will be assumed to hold for all these individuals is to do violence to the experimental evidence. From the data we have learned where an individual is on the continuum in relation to the stimuli, and in addition something about how the whole continuum looks to him.

VI. Summary

We have presented a new type of scale called an ordered metric and have presented the experimental procedures required under certain limiting conditions to secure such a scale.

We have pointed out that the information which could be obtained under these conditions is as follows:

1. the discovery of a latent attribute underlying preferences,
2. the order of the stimuli on the attribute continuum,
3. something about the relative magnitudes of the distances between pairs of stimuli,
4. the sectioning of the continuum into intervals, the placing of people in these intervals, and the ordering of these intervals on this attribute continuum,
5. something about the relative magnitudes of these intervals.

These were illustrated with a hypothetical example and an experiment.

References

1. CAMPBELL, N. R. Symposium: Measurement and its importance for philosophy, in *Action, perception, and measurement*, The Aristotelian Society, Harrison and Sons, Ltd., 1938.
2. COOMBS, C. H. Some hypotheses for the analysis of qualitative variables. *Psychol. Rev.*, 1948, 55, 167–74.
3. STEVENS, S. S. On the theory of scales of measurement. *Science*, 1946, 103, 677–80.
4. THURSTONE, L. L. The law of comparative judgment. *Psychol. Rev.*, 1927, 34, 273–86.

Chapter 25

A REAL EXAMPLE OF UNFOLDING

C. H. Coombs

University of Michigan

This example is from the grade expectations of the students in an introductory course in mathematical psychology given in the fall of 1959, the summer of 1960, and the fall of 1961. This is an elementary course required of all the graduate students in psychology at The University of Michigan who are *not* planning to do any further work in mathematical psychology. Students interested in mathematical psychology skip this course and begin with a more technical one.

Early in the course the students were given a dittoed form containing all pair comparisons of the letter grades A +, A, A—, B+, B, B—, C+, in a random sequence. For each pair comparison the students were instructed to indicate which grade was more nearly the grade they expected to get in the course. Each subject then tabulated his own pair comparisons and, if they were transitive, converted them into a rank order *I* scale. At this point in the course no instruction had been given pertaining to scaling theory in general or to the unfolding technique in particular.

The *I* scales obtained are presented in Table 25.1 in which the letter grades from A+ to C+ are represented respectively by the symbols *A* to *G* in order. The total number of subjects is 62, of whom 58 gave transi-

Reprinted with permission of the author and the publisher from *A Theory of Data* by Clyde H. Coombs, pp. 92–102. New York: John Wiley & Sons, Inc., 1964.

tive *I* scales and 4 did not. These latter are indicated in the table with the intransitive stimuli contained in parentheses. These stimuli in parentheses are to be read clockwise, that is, the *I* scale $(F^EG)D\ C\ B\ A$ indicates that this subject "preferred" *F* to *E, E* to *G,* and *G* to *F.*

When the *I* scale number for each *I* scale is determined we find, among others, that *I* scale numbers 5, 9, and 10 are represented by more than one *I* scale. This reflects the fact that the order of midpoints and hence the metric relations are not the same for the subjects with different *I* scales with the same *I* scale number. It is clear that the same qualitative *J* scale satisfies all subjects which is, of course, not surprising, but what would be of psychological interest is the metric relations.

Although no single set of metric relations satisfies all the *I* scales, we might be interested in that particular set of metric relations which satisfies the most data. This is not too difficult to determine by trial and error, ·and the *I* scales listed in the first column are the ones which satisfy this dominant quantitative *J* scale. Because this set of *I* scales is incomplete, some metric information is unobtainable, but such information as can be obtained is given in Table 25.2. The metric information is presented in

Table 25.1. Data on Grade Expectations

I Scale Number	I Scales and Their Respective Frequencies of Occurrence			
1	A B C D E F G 6			
2	B A C D E F G 4			
3	B C A D E F G 5			
4	C B A D E F G 1			
5	C B D A E F G 8	B C D E A F G 2		
6	C B D E A F G 2	C D B A E F G 1		
7	C D B E A F G 2	C B D E F A G 1	D C B A E F G 1	
8	C D B E F A G 0			
9	D C B E F A G 2	C D B E F G A 1	D C E B A F G 1	
10	D C E B F A G 5	D C B E F G A 1	D E C B A F G 2	C D (B^EF)G A 1
11	D C E F B A G 1			
12	D E C F B A G 1	D E C B F G A 1		
13				
14	D E F C B G A 1			
15		E D C F G B A 1		
16	E D F C G B A 2	E D F (B^CG) A 1	F E D C B A G 1	
17	E D F G C B A 3			
18	E F D G C B A 0			
19		F (E^DG) C B A 1		
20	F E G D C B A 2			
21	F G E D C B A 0	(F^EG) D C B A 1		
22	G F E D C B A 0			

Table 25.2. Midpoint Order and Metric
Implications for Data in Table 1

Midpoint Order*	Metric Implications
AB	
AC	
BC	BC, AD $\Rightarrow \overline{CD} > \overline{AB}$
AD	
AE	AE, BD $\Rightarrow \overline{AB} > \overline{DE}$
BD	
AF	BD, AF $\Rightarrow \overline{DF} > \overline{AB}$
CD	CD, BE $\Rightarrow \overline{DE} > \overline{BC}$
BE	CD, AG $\Rightarrow \overline{DG} > \overline{AC}$
BF	BF, CE $\Rightarrow \overline{BC} > \overline{EF}$
CE	BF, AG $\Rightarrow \overline{FG} > \overline{AB}$
CF, AG	CF, DE $\Rightarrow \overline{CD} > \overline{EF}$
DE, BG	CF, BG $\Rightarrow \overline{FG} > \overline{BC}$
CG	AG, DE $\Rightarrow \overline{AD} > \overline{EG}$
DF, EF, DG	DE, CG $\Rightarrow \overline{EG} > \overline{CD}$
BG	CG, DF $\Rightarrow \overline{CD} > \overline{FG}$
FG	

* The weak ordering is due to the incompleteness of the set of
I scales.

the form of a partial order in Figure 25.1 with the symbols converted back into the letter grades.

Three psychological inferences may be drawn from the metric relations pictured in Figure 25.1. (1) The spread of the A's on this subjective scale is greater than the spread of the B's, that is, $A + A - > B + B -$. (2) Changing the letter grade has more psychological effect than adding a plus or minus, i.e., the largest single interval is from $A -$ to $B +$ and the second largest is from $B -$ to $C +$. (3) Adding a plus to a grade adds a larger psychological increment than a minus subtracts, that is, $A + A > AA -$ and $B + B > BB -$.

The metric relations and hence these inferences drawn from them are a composite from the relations between the different *I* scales. There is no evidence from an analysis of preferential choice data that all these metric relations hold for all individuals on the *J* scale. Furthermore, we do not know the extent to which generalizations may be drawn. It would not be surprising if these same individuals had a different metric for their grade expectations in another course or even a different metric for this same course after a midterm examination. In fact such research might contribute to the literature on motivation and level of aspiration.

The results of this analysis are quite typical, in my experience, of the findings in an unfolding analysis of the data generated by what is presumed to be a unidimensional *J* scale on a priori grounds—a dominant *J* scale may usually be found satisfying a majority of the cases, 72% in this instance; there will be some *I* scales which require a different metric and others which even require a different qualitative *J* scale; and there will be some intransitive *I* scales (if the data are collected by unreplicated pair comparisons).

Whether the cases that do not fit the dominant *J* scale merely represent unreliability in the data or are significant departures can best be determined by using more powerful methods of collecting the data in the first place, for example, methods with redundancy which permit control of inconsistency. An instance of this is contained in the Coombs-Pruitt study of variance and skewness preferences.

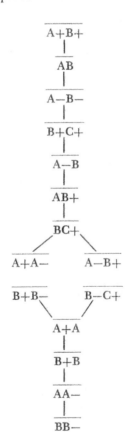

Figure 25.1. Metric relations for grade expectations.

Conversion of an Ordered Metric Scale to an Interval Scale

It is very useful to have some reasonable method by which numbers can be assigned to the stimuli on a J scale so as to satisfy the given metric relations. There are, of course, an infinite number of such representations, so we have the further problem of selecting a particular one. Frank Goode* of The University of Michigan has developed and explored a method for assigning a reasonable set of numerical scale values to stimuli on an ordered metric scale. His method has potentialities beyond the use made of it here.

We begin with a partial order on the distances between stimuli. The general idea is to set the smallest interval equal to some positive but unknown quantity Δ_1, and successively larger distances are obtained by introducing additional unknown positive Δ's. A matrix equation may be set up which expresses the scale values of the stimuli as a linear function of these Δ's. Values assigned to the Δ's lead to interval scale values for the ordered metric scale.

We illustrate the procedure with an example worked out in detail. We take the partial order on the metric for grade expectations, as given in Figure 25.1 and construct an interval scale. For the purpose we restore the letters from A to G in alphabetical order to represent the grades from A + to C +.

The procedure is best carried out using a worksheet, as illustrated in Table 25.3 for this problem. On the left-hand side of the table is the partial order as recoded from Figure 25.1. Each element in the partial order of distances is made up of one or more contiguous single intervals, and this relation is indicated in the second column of the worksheet. The rest of the worksheet indicates the sequence of steps in working out a solution for this particular case, the little arrows indicating footnotes to the table.

Beginning at the bottom of the partial order we see that $\overline{EF}$ is the smallest interval, and it is, of course, a single interval. We set it equal to some positive but unknown quantity, Δ_1. The distance $\overline{BC}$ is the next largest, so it is as great as Δ_1 plus some positive unknown quantity Δ_2. By similar reasoning, $\overline{DE}$ is equal to $\Delta_1 + \Delta_2 + \Delta_3$ and $\overline{AB}$ is equal to $\Delta_1 + \Delta_2 + \Delta_3 + \Delta_4$.

At this point the partial order is seen to divide. Both $\overline{DF}$ and $\overline{FG}$ are larger than $\overline{AB}$ but are themselves not comparable. We may proceed by going up either branch of the partial order, and it makes no difference in the ultimate outcome, although one usually turns out to be easier than the other. We note that one side of the partial order involves $\overline{DF}$, a sum of two single intervals, and the other involves $\overline{FG}$, a single interval alone,

---

* Personal communication, 1957. His work is still in progress.

Table 25.3. Worksheet for Expressing Intervals between Stimuli as Linear Functions of a Set of Positive Quantities

Interval	Decomposition	Intermediate	Final
$\overline{AD}$	$\overline{AB}+\overline{BC}+\overline{CD}$	$=2\Delta_2+\Delta_3+3\Delta_4+2\Delta_5+\Delta_6$ (6)	$=4\Delta_2+2\Delta_3+2\Delta_4+\Delta_6+4\Delta_7+3\Delta_8$
$\overline{BE}$	$\overline{BC}+\overline{CD}+\overline{DE}$	$=2\Delta_2+\Delta_3+2\Delta_4+\Delta_5+\Delta_6+\Delta_7$	$=4\Delta_2+2\Delta_3+\Delta_4+\Delta_6+4\Delta_7+3\Delta_8$
$\overline{CF}$	$\overline{CD}+\overline{DE}+\overline{EF}$	$=2\Delta_2+\Delta_3+2\Delta_4+\Delta_5+\Delta_6+\Delta_7$	$=3\Delta_2+2\Delta_3+\Delta_4+\Delta_6+4\Delta_7+3\Delta_8$
$\overline{DG}$	$\overline{DE}+\overline{EF}+\overline{FG}$	$=2\Delta_2+\Delta_3+2\Delta_4+\Delta_5+\Delta_6+\Delta_7$	$=3\Delta_2+2\Delta_3+\Delta_4+\Delta_6+3\Delta_7+3\Delta_8$
$\overline{CE}$	$\overline{CD}+\overline{DE}$	$=2\Delta_2+\Delta_3+\Delta_4+\Delta_5+\Delta_6$	$=3\Delta_2+2\Delta_3+\Delta_4+\Delta_6+3\Delta_7+2\Delta_8$
$\overline{BD}$	$\overline{BC}+\overline{CD}$	$=2\Delta_2+\Delta_3+\Delta_4+\Delta_5+\Delta_6$	$=3\Delta_2+2\Delta_3+\Delta_4+\Delta_6+3\Delta_7+2\Delta_8$
$\overline{EG}$	$\overline{EF}+\overline{FG}$	$=2\Delta_2+\Delta_3+\Delta_4+\Delta_5+\Delta_6$	$=3\Delta_2+\Delta_3+\Delta_4+\Delta_6+3\Delta_7+2\Delta_8$
$\overline{AC}$	$\overline{AB}+\overline{BC}$	$=2\Delta_2+\Delta_3+3\Delta_4+2\Delta_5$ (2)	$=2\Delta_2+\Delta_3+\Delta_4+\Delta_6+2\Delta_7+2\Delta_8$
$\overline{DF}$	$\overline{DE}+\overline{EF}$	$=\Delta_2+\Delta_3+2\Delta_4+2\Delta_5$ (5)	$=2\Delta_2+\Delta_3+\Delta_4+\Delta_6+2\Delta_7+\Delta_8$
$\overline{CD}$	$\overline{CD}$	$=\Delta_2+\Delta_3+2\Delta_4+\Delta_5$	$=2\Delta_2+\Delta_3+\Delta_4+\Delta_6+2\Delta_7+\Delta_8$
$\overline{FG}$	$\overline{FG}$	$=\Delta_2+\Delta_3+\Delta_4+\Delta_5$	$=2\Delta_2+\Delta_3+\Delta_4+\Delta_6+\Delta_7+\Delta_8$
$\overline{AB}$	$\overline{AB}$	$=\Delta_2+\Delta_4+\Delta_6$	$=2\Delta_2+\Delta_3+\Delta_4+2\Delta_7+2\Delta_8$
$\overline{DE}$	$\overline{DE}$	$=\Delta_2$	$=\Delta_2+\Delta_3+\Delta_4+2\Delta_7+2\Delta_8$
$\overline{BC}$	$\overline{BC}$	$=\Delta_2$	$=\Delta_2+\Delta_3+\Delta_4+2\Delta_7+\Delta_8$
$\overline{EF}$	$\overline{EF}$	$=\Delta_4+\Delta_5$	$=\Delta_2+\Delta_7+\Delta_8$

Intermediate forms (column with Δ_1, steps (1)–(4)):

$$(1)\quad = 2\Delta_1+\Delta_2+\Delta_3$$
$$= \Delta_1+\Delta_2+\Delta_3+\Delta_4$$
$$(3)\quad = \Delta_1+\Delta_2+\Delta_3$$
$$= \Delta_1+\Delta_2$$
$$(2)\quad = \Delta_1$$

(1) $\overline{DF} > \overline{AB} \Rightarrow \Delta_1 > \Delta_4$, so (2) let $\Delta_1 = \Delta_4$, (3) $\overline{EG} > \overline{AC} \rightleftharpoons \overline{EG} = \overline{AC} + \Delta_6$. (4) $\overline{FG} = \overline{EG} - \overline{EF}$.

(5) $\overline{EG} > \overline{CD} \Rightarrow \Delta_4 + \Delta_5 > \Delta_7$, so (6) let $\Delta_4 + \Delta_5 = \Delta_7$, so $\Delta_4 + \Delta_5 = \Delta_7 + \Delta_5$.

and we suspect that it is generally easier in the long run to take first the side of the partial order that involves sums of single intervals. This is what we do here.

The distance $\overline{DF} = \overline{DE} + \overline{EF}$ and hence is equal to $2\Delta_1 + \Delta_2 + \Delta_3$. However, $\overline{DF} > \overline{AB}$, so $2\Delta_1 + \Delta_2 + \Delta_3 > \Delta_1 + \Delta_2 + \Delta_3 + \Delta_4$, and hence

$$\Delta_1 > \Delta_4$$

So we let

$$\Delta_1 = \Delta_4 + \Delta_5$$

In the next column of the worksheet, then, we rewrite $\overline{DF}$ as $\Delta_2 + \Delta_3 + 2\Delta_4 + 2\Delta_5$, and for all the elements of the partial order below $\overline{DF}$ we substitute $\Delta_4 + \Delta_5$ for Δ_1. We then continue up the partial order:

$$\begin{aligned}
\overline{AC} &= \overline{AB} + \overline{BC} \\
&= (\Delta_2 + \Delta_3 + 2\Delta_4 + \Delta_5) + (\Delta_2 + \Delta_4 + \Delta_5) \\
&= 2\Delta_2 + \Delta_3 + 3\Delta_4 + 2\Delta_5
\end{aligned}$$

and we must check to see that $\overline{AC}$ is at least as great as $\overline{DF}$ for every component and greater in at least one, that is, $\overline{AC}$, as a vector, must dominate $\overline{DF}$. To illustrate the procedure in detail, the coefficients of the components for $\overline{AC}$ are, in order, $(2, 1, 3, 2)$ and for $\overline{DF}$, $(1, 1, 2, 2)$. We see then that $\overline{AC}$ will necessarily be greater than $\overline{DF}$, as required by the partial order.

We come next to $\overline{EG}$, which must be larger than $\overline{AC}$, so we add a positive but unknown quantity Δ_6 to $\overline{AC}$ and we obtain $\overline{EG} = 2\Delta_2 + \Delta_3 + 3\Delta_4 + 2\Delta_5 + \Delta_6$.

At this point we go back and check the other branch of the partial order that leads up to $\overline{EG}$, that is, $\overline{FG} < \overline{CD} < \overline{EG}$. Because $\overline{EG} = \overline{EF} + \overline{FG}$, we may solve for $\overline{FG}$ directly by subtracting $\overline{EF} = \Delta_4 + \Delta_5$ from $\overline{EG}$; and we obtain $\overline{FG} = 2\Delta_2 + \Delta_3 + 2\Delta_4 + \Delta_5 + \Delta_6$. We must check, then, to make sure that $\overline{FG}$ dominates $\overline{AB}$. The $\overline{FG}$ vector is $(2, 1, 2, 1, 1)$ and the $\overline{AB}$ vector is $(1, 1, 2, 1, 0)$, so dominance is assured.

The partial order shows $\overline{CD}$ to be greater than $\overline{FG}$, so we introduce a new positive quantity Δ_7 and set $\overline{CD} = \overline{FG} + \Delta_7$. But because

$$\overline{EG} > \overline{CD}$$
$$2\Delta_2 + \Delta_3 + 3\Delta_4 + 2\Delta_5 + \Delta_6 > 2\Delta_2 + \Delta_3 + 2\Delta_4 + \Delta_5 + \Delta_6 + \Delta_7$$

and hence

$$\Delta_4 + \Delta_5 > \Delta_7$$

So we let

$$\Delta_4 + \Delta_5 = \Delta_7 + \Delta_8$$

and now Δ_5 may be eliminated from all expressions.

We may then continue up the partial order, which is now a simple procedure because all the remaining elements of the partial order are compounded from elements we have already determined. For example, the next element in the partial order is $BD = BC + CD$. This process is completed to insure that no further changes are necessary in the vector components of the single intervals.

Table 25.4. Worksheet for Scale Values of Stimuli

Successive Single Intervals	In Vector Notation	Scale Values of Stimuli
$\overline{AB} = \Delta_2 + \Delta_3 + \Delta_4$	$+ \Delta_7 + \Delta_8$ $(1, 1, 1, 0, 1, 1)$	$B = (1, 1, 1, 0, 1, 1)$
$\overline{BC} = \Delta_2$	$+ \Delta_7 + \Delta_8$ $(1, 0, 0, 0, 1, 1)$	$C = (2, 1, 1, 0, 2, 2)$
$\overline{CD} = 2\Delta_2 + \Delta_3 + \Delta_4 + \Delta_6$	$+ 2\Delta_7 + \Delta_8$ $(2, 1, 1, 1, 2, 1)$	$D = (4, 2, 2, 1, 4, 3)$
$\overline{DE} = \Delta_2 + \Delta_3$	$+ \Delta_7 + \Delta_8$ $(1, 1, 0, 0, 1, 1)$	$E = (5, 3, 2, 1, 5, 4)$
$\overline{EF} =$	$+ \Delta_7 + \Delta_8$ $(0, 0, 0, 0, 1, 1)$	$F = (5, 3, 2, 1, 6, 5)$
$\overline{FG} = 2\Delta_2 + \Delta_3 + \Delta_4 + \Delta_6$	$+ \Delta_7 + \Delta_8$ $(2, 1, 1, 1, 1, 1)$	$G = (7, 4, 3, 2, 7, 6)$

The next step is to obtain the scale values of the stimuli in terms of the Δ's. We arbitrarily assign the scale value of zero to stimulus A, then the scale value of stimulus $B = \overline{AB}$, $C = \overline{AB} + \overline{BC}$, and so on. A worksheet for this is shown in Table 25.4. On the left-hand column of the worksheet are the successive single intervals expressed in terms of the final Δ components, $(\Delta_2, \Delta_3, \Delta_4, \Delta_6, \Delta_7, \Delta_8)$, as determined in Table 25.3. The middle column of the worksheet expresses the single intervals in vector notation. The column on the right of the worksheet contains the scale values of the stimuli in vector notation, each succeeding stimulus vector being a component-wise sum of the single interval vectors.

The scale values of the stimuli in the last column of Table 25.4 may be written as the successive rows of a matrix, V, which postmultiplied by a column vector Δ generates a new column vector S, which is a feasible solution to the scale values of the stimuli, with stimulus A at zero.

The column vector Δ may be any set of numbers greater than zero. One simple solution is the "equal Δ" solution for which the column vector Δ is a unit vector. This solution is

$$
\begin{pmatrix}
1 & 1 & 1 & 0 & 1 & 1 \\
2 & 1 & 1 & 0 & 2 & 2 \\
4 & 2 & 2 & 1 & 4 & 3 \\
5 & 3 & 2 & 1 & 5 & 4 \\
5 & 3 & 2 & 1 & 6 & 5 \\
7 & 4 & 3 & 2 & 7 & 6
\end{pmatrix}
\begin{pmatrix}
1 \\ 1 \\ 1 \\ 1 \\ 1 \\ 1
\end{pmatrix}
=
\begin{pmatrix}
5 \\ 8 \\ 16 \\ 20 \\ 22 \\ 29
\end{pmatrix}
$$

$$V \qquad \Delta \ = \ S \qquad \qquad [1]$$

An infinite number of solutions satisfy the same metric relations, of course, and this "equal Δ" solution is only one. We face the problem, then, of selecting a particular scale from among these many. The following comments are offered to assist in resolving this problem.

Table 25.5. General Expressions and Scale Values for the Stimuli

	Stimuli						
	A	B	C	D	E	F	G
General Expression	0	$\Delta_1 + 3$	$2\Delta_1 + 4$	$3\Delta_1 + 10$	$4\Delta_1 + 12$	$5\Delta_1 + 12$	$6\Delta_1 + 17$
$\Delta_1 = 2$	0	5	8	16	20	22	29
$\Delta_1 = 10$	0	13	24	40	52	62	77

It will be recalled that $\Delta_1 = \Delta_7 + \Delta_8$ and is the smallest single interval. We may keep all the other Δ's constant and equal to *one* while we set $\Delta_8 > 1$ (that is, $\Delta_1 > 2$). This is equivalent to keeping certain *differences* between distances constant while we otherwise stretch or contract the scale, and has the corresponding effect of depressing or dramatizing the differences in distances between stimuli.

An easy way to see what is going on is to write a new expression for each scale value in terms of Δ_1 as the only unknown. Thus, the scale value of stimulus B, where $\Delta_1 = \Delta_7 + \Delta_8$ and all other Δ's are equal to *one*, is

$$B = \Delta_2 + \Delta_3 + \Delta_4 + \Delta_7 + \Delta_8 = \Delta_1 + \Delta_2 + \Delta_3 + \Delta_4 = \Delta_1 + 3$$

and the scale value of C is

$$C = 2\Delta_2 + \Delta_3 + \Delta_4 + 2\Delta_7 + 2\Delta_8$$
$$= 2\Delta_1 + 2\Delta_2 + \Delta_3 + \Delta_4 = 2\Delta_1 + 4$$

A table of these values for all the stimuli, with Δ_1 fixed at the value 2 and at the value 10, is contained in Table 25.5.

The effect of increasing the value of Δ_1 is to depress the differences in metric relations, that is, to tend toward equally spaced stimuli. This may be seen in Figure 25.2; setting a value for Δ_1 defines a vertical line with Δ_1 as its intercept on the abscissa. Its intersections with the fan of straight lines determine the respective scale values of the stimuli. The two vertical lines for $\Delta_1 = 2$ and $\Delta_1 = 10$ are shown in the figure. We note that *when Δ_1 is given its minimal integral value ($\Delta_1 = 2$) the scale obtained is the same as that obtained with the equal Δ assumption in equation [1].*

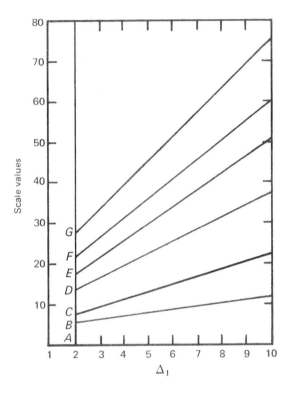

Figure 25.2. Stimulus scale values as a function of Δ_1.

Frank Goode did this for a given ordered metric scale, with the first and last stimuli fixed at the ends of the scale, and allowing the various Δ's to take on all admissible integral values. Fixed boundaries were obtained for each stimulus, with unimodal distributions of admissible scale values for each stimulus. He observed that if the Δ's are all set equal, the scale values of the several stimuli are all close to the modes of their respective distributions.

It is not yet known under what conditions it would be true in general that the equal Δ assumption would yield such a "highly representative" scale. It is correct in this instance, and it is certainly a simple and practical method. It also is a solution in which the metric relations are highly dramatized, as is evidenced by Figure 25.2.

Abelson and Tukey (1959) have developed another method for converting nonmetric information into metric information. Abelson says of their method, in a personal communication, that it is rather too complex to apply often by hand to the ordered metric situation with more than five or six stimuli. On the other hand, he actually did apply it to the ordered metric scale of seven stimuli in this example. Their method seeks that set of scale values for which the minimum correlation (squared) with whatever the true values might be is a maximum. This method insures that the best will be done in the worst possible circumstances.

The set of scale values Abelson obtained, rounded off to a simple integer scale, is as follows, for stimuli A to G respectively: 0, 5, 10, 29, 34, 39, 58. The squared maximum correlation is .959, which means that the "true" scale, no matter how different it is so long as it satisfies the same ordered metric relations, must correlate at least .98 with this one.

He computed the minimum squared correlation for the $\Delta_1 = 2$ solution with constant Δ, as obtained by this simple version of Frank Goode's method, and the value obtained is .944. Hence this solution must correlate at least .97 with the true one.

There is little difference, of course, between .98 and .97, but, on the other hand, there is no assurance that Goode's method with a constant Δ assumption would always do almost as well as the Abelson-Tukey method. The great advantage of the constant Δ method is its simplicity.

Shepard (1962) also has a method, but it is impossible to perform without a computer. There are also uniqueness problems with small n, and partial orders give trouble because they must be converted into simple orders. His method, however, is much more than a method for assigning numbers to stimuli on an ordered metric scale; rather, it is a method for multidimensional scaling of QIVa data, but the method is at least potentially applicable to this simpler case.

The fact that seems to be emerging from these developments is that ordered metric information places rather strong constraints on the admissible scale representations and that there are reasonable criteria for choosing among them. I expect a convention of constructing a numerical representation of ordered metric data to develop very rapidly.

References

ABELSON, R. P., and J. W. TUKEY, 1959, Efficient conversion of non-metric information into metric information, paper presented at the American Statistical Association, Social Sciences Section, December 1959.

COOMBS, C. H., and D. PRUITT, 1960, Components of risk in decision making: probability and variance preferences, *J. exp. Psychol.*, 60, 265–77.

COOMBS, C. H., and D. PRUITT, 1961, Some characteristics of choice behavior in risky situations, *Ann. New York Acad. Sci.*, 89, 784–94.

SHEPARD, R. N., 1960, "Similarity of stimuli and metric properties of behavioral data," Chapter 4 in *Psychological scaling: theory and applications*, edited by H. Gulliksen and S. Messick, John Wiley and Sons, New York.

Chapter 26

SOME APPLICATIONS OF UNFOLDING THEORY TO FERTILITY ANALYSIS

DAVID GOLDBERG AND C. H. COOMBS
University of Michigan

One of the distinguishing features of fertility research during the past decade has been the elaboration of types of data used as the dependent variable. Dimensions of fertility other than behavior have received increasing attention, although the relationship between these dimensions and performance is not completely understood. The extensive use of data such as family size ideals, desires, expectations or values has developed from several factors:

1. Since most studies are limited to one point in time, the dependence on *ex post facto* analysis of completed fertility has often been misleading with respect to some of the more important hypotheses that have been advanced.[1] Because of the problems associated with this type of analysis, it has been argued that we may eventually learn more about behavior by studying expectations or desires.

Reprinted with permission of the author and the publisher from *Emerging Techniques in Population Research*, Proceedings of the 1962 Annual Conference of the Milbank Memorial Fund.

1. Several problems of interpretation resulting from the *ex post facto* analysis of fertility in the Indianapolis Study are described *in* Kiser, Clyde V. and Whelpton, P. K.: "Resume of the Indianapolis Study of Social and Psychological Factors Affecting Fertility," *Population Studies*, 1959, 7: 95–110.

2. For societies characterized by an increasing use of contraception, the non-behavioral dimensions of fertility should become more relevant in guiding performance.[2]

3. Our failure to explain much variance in completed fertility is partially responsible for the treatment of fertility behavior as a complex variable consisting of desired and undesired or error components, as well as components that change with the experience of having different numbers of children.

In combination these factors have resulted in an analysis pattern that treats fertility as a multidimensional phenomenon with the expectation that differences in the relationships between social or psychological factors and the several dimensions of fertility will ultimately be combined in a manner that will lead to a fuller understanding of completed fertility.

Our paper is an attempt to describe a dimension of fertility that is frequently mentioned but, as yet, has not been investigated—the subjective distances between various numbers of children in a family. Up to now, whether investigators have been concerned with fertility expectations, desires, or behavior, distances between the second and third child are taken to be the same as the distance between the fourth and fifth child, for example. If there are large differences in the perceived distances between children, we might expect that variables used to predict additional fertility could vary from one parity to another. This type of problem was explicitly recognized in the design of the Princeton Study in which the sample was restricted to women who had recently given birth to a second child.[3]

There is considerable discussion in the literature to the effect that inter-child differences may be unequal.[4] Similarly, arguments for the existence of a set of cultural norms in each society specifying a preferred range of fertility behavior imply unequal distances between various numbers of children.[5] In the United States, a large body of evidence has been compiled indicating a strong consensus on a particular range of completed family size. Data from the 1955 and 1960 Growth of American

2. A comprehensive discussion of this issue appears *in* Freedman, Ronald: *The Sociology of Human Fertility*. UNESCO. (In press.)

The accuracy of fertility desires and expectations in relation to subsequent performance has been investigated in several studies, including Westoff, Charles, Mishler, Elliot and Kelley, E. Lowell: Preferences in Size of Family and Eventual Fertility Twenty Years After, *American Journal of Sociology*, 1957, 62: 491–497. Goldberg, David, Sharp, Harry, and Freedman, Ronald: The Stability and Reliability of Expected Family Size Data, *Milbank Memorial Fund Quarterly*, 1959, 37: 369–385.

3. Westoff, Charles F., Potter, Robert G., Jr., Sagi, Philip C., and Mishler Elliot G.: *Family Growth in Metropolitan America*. Princeton: Princeton University Press, 1961.

4. *Ibid.,* p. 11.

5. Freedman, *op. cit.*

Families Study show that the overwhelming majority of women prefer, consider ideal and expect to have two, three or four children.[6] It has been suggested that in contrast to previous periods in our history, the present fertility norms result in a complete rejection of childless or one-child families and a strong distaste for families of five or more children.[7] A well-defined range of two to four children indicates that the subjective distances between children in the range two to four are somehow smaller than the other distances.

In this study, the measurement of perceived inter-child distances is developed from two procedures based on unfolding theory.[8] One procedure utilizes the individual ordered preferences for various numbers of children to produce an aggregate scale of distances underlying women's preferences. The second procedure yields a direct measure, for each woman, of the perceived distances or similarities between numbers of children with respect to the problems and pleasures associated with children.

Our data were taken from a study of fertility conducted through the facilities of the Detroit Area Study and the University of Michigan Population Studies Center.[9] The sample consisted of 1,215 women who had recently been married, or had given birth to their first, second or fourth child.[10] Most of the analysis is confined to the 1,148 women who were classified as "probably fecund."

Unfolding theory and two types of scales, the S-scale and the J-scale, are discussed in the next section. The results of the analysis of the data for each of these two types of scales are presented in the two succeeding sections with a final brief section covering conclusions.

6. Freedman, Ronald, "Social Values about Family Size in the United States," in *International Population Conference*. International Union for the Scientific Study of Population, Vienna, 1959, pp. 173–183.

Some preliminary results from the 1960 study are reported by Campbell, Arthur, Whelpton, P. K. and Thomasson, Richard, "The reliability of Birth Expectations of U. S. Wives," paper #70, International Population Union Conference. New York, 1961.

7. Freedman, *op. cit.*

8. Coombs, Clyde H., *A Theory of Psychological Scaling*. Ann Arbor, University of Michigan Press, 1952.

Coombs, Clyde H., *A Theory of Data*. New York, Wiley & Sons., 1964.

9. Ronald Freedman and David Goldberg are the principal investigators of the study.

10. The sample was selected from marriage and birth records to represent on a probability basis white couples in the Detroit metropolitan area who had:
   (a) been married for the first time in July, 1961— (102 couples)
   (b) a first birth in July, 1961— (372 couples)
   (c) a second birth in July, 1961— (372 couples)
   (d) a fourth birth in July, 1961— (369 couples)
Interviews were made during the months of January and February, 1962. More than 92 percent of the selected women were interviewed.

Unfolding Theory

One dichotomy of types of scales is the dichotomy into *J*-scales and *S*-scales. A *J*-scale is a one-dimensional continuum on which *both* individuals and stimuli are scaled, and hence is called a joint scale or *J*-scale. An *S*-scale is a one dimensional continuum on which *only* the stimuli are scaled, and hence is called a stimulus scale or *S*-scale.

A *J*-scale may be obtained from an analysis of the preference orderings of individuals over a set of stimuli, in this case the number of children desired. Each individual may be conceived as corresponding to a point on the continuum which represents his ideal-point or point of maximum preference and each stimulus also corresponds to a point on this continuum. An individual's preferential choice between any two stimuli is assumed to reflect which of the two corresponding stimulus-points is nearer in absolute distance to his ideal-point.

With these assumptions, an individual's preference ordering is the rank order of the stimuli from nearest to farthest. It is as if the *J*-scale were folded at the ideal-point and the preference ordering is the rank order of the stimuli on this folded *J*-scale. This folded *J*-scale or preference ordering is called an *I-scale*. Given a set of *I*-scales representing a variety of preference orderings, the problem is to unfold these and recover the *J*-scale.

It is intuitively evident that individuals whose ideal-points are close together may have the same *rank order* of preferences. Indeed the *J*-scale may be divided into intervals in such a way that all the individuals whose ideal-points are in the same interval have the same rank order *I*-scale. These intervals are generated by the midpoints between stimuli. For example, in Figure 26.1, we illustrate a *J*-scale of 4 stimuli, *A, B, C, D*, divided into seven intervals ($C^m{}_2 + 1$) by their midpoints.

To each of these seven intervals corresponds a unique *I*-scale. For example, any individual whose ideal-point is in I_4 will have the preference ordering *CBAD*. Note that the distance between stimuli *A* and *B*, $\overline{AB}$, is less than that between *C* and *D*, $\overline{CD}$. If $\overline{CD}$ were less than $\overline{AB}$, the midpoints *AD* and *BC* would be reversed in order and the *I*-scale

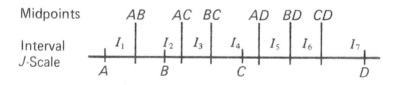

Figure 26.1. A J-Scale of four stimuli.

corresponding to I_4 would then be *BCDA*. In this manner the set of preference orderings may be analyzed to obtain order relations on distances between stimuli, yielding what is called an ordered metric scale.

Note that no other *I*-scales would be changed by this change in the order of those two midpoints, so all the individuals whose *I*-scales place them in one of the other intervals are equally well accommodated by the two differently ordered metric *J*-scales. In any set of real data we may well find that both the *I*-scales *CBAD* and *BCDA* occur. In the case of more than four stimulus-points, there is more metric information, and more alternatively ordered metric *J*-scales may be constructed. If the one which accommodates a maximum number of individuals also accommodates a majority, we speak of it as being the dominant *J*-scale.

While the dominant *J*-scale is of interest in providing a possible and plausible overall structure of the metric relations among the stimuli underlying the preferences of some dominant subset of the respondents, it may also be of interest to know something of the metric relations among the stimuli for *each* individual. This brings us to the *S*-scale, which may be obtained from each individual separately.

To construct an *S*-scale the judgments to be obtained from the respondent are not preferences but judgments about the *differences* between the stimuli. For example, the respondent may be presented with a pair of distances, like $\overline{AB}$ and $\overline{BC}$, and asked which are more alike in some sense relevant to the context of the study.

Such data obtained from each individual may be utilized in a direct manner to construct an ordered metric scale of the stimuli which may well be a different structure for different individuals. This kind of judgments we call "similarity judgments" as distinct from "preferential choices," and the resulting *S*-scale is a scale of the subjective similarity of the stimuli.

In the case of both the *J*-scale or the *S*-scale, this unfolding analysis yields an ordered metric scale. Because of the difficulties of summarizing a series of ordered metric scales and because it is difficult to visualize such a structure and relate it to other variables, a method which seeks a numerical representation of ordered metric scales is employed in the paper.

Similarities Data—The S-Scale

Our data on the perceived distances between children were derived from the question reproduced in Figure 26.2. Respondents were initially asked to distinguish the relative similarity of having no children, or two children, to having one child. Such a judgment is a comparison between the subjective difference between 0 and 1 child $(\overline{01})$ and between 1 and

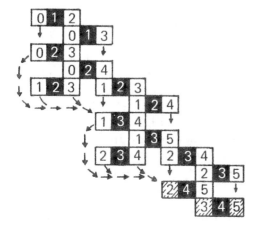

INTERVIEWER: ASK R THE FIRST QUESTION, TAKING THE TRIPLET 0, 1, 2, IN THE PHRASING INDICATED. R WILL ANSWER EITHER 0 OR 2. CROSS OUT (X) THAT NUMBER. ASK THE NEXT TRIPLET IN THE SAME WAY, USING THE TRIPLET INDICATED BY CONTACT WITH THE BOX CROSSED OUT, OR BY ARROW FROM THAT BOX. THE REVERSE NUMBER IS ALWAYS THE NUMBER BEING COMPARED. CONTINUE UNTIL YOU REACH ONE OF THE NUMBERS IN THE SHADED BOXES. THE PATTERN YOU FOLLOW ELIMATES DIFFERENT TRIPLETS IN EACH CASE. YOU NEVER ASK ALL THE TRIPLETS.

Figure 26.2. The "S" Scale: [*In talking to women in Michigan, we find that they have many ideas about how much difference one child more or one child less makes in a mother's satisfactions and problems. (We want to ask you some questions in which you'll have to make some decisions that may seem a little strange. However, it will help the study very much if you just try to give the answers that seem most natural to you.) For example, thinking about the satisfactions and problems you would have in your own life, which is more like having one child, no children or two children?*

For Each Additional Triplet: Which is more like having—child-(ren), — or — ?]

2 children ($\overline{12}$). Such a pair of distances is called a conjoint pair because the two distances have a common terminus, in this case the point corresponding to "1 child." Pairs of distances that do not have a common terminus are called disjoint pairs, as for example, the pair of distances $\overline{01}$ and $\overline{23}$.

The choice made by the subject on each question determined what the next question would be, indicated in Figure 26.2 by an arrow or by the juxtaposition of the next question. For example, if the subject in response to $\boxed{0 \mid 1 \mid 2}$, said "no children" was more like having "one" than "two children" are, she "came out" of that question at the "0" end and, following the arrow, the next question she would be asked would be $\boxed{0 \mid 2 \mid 3}$, i.e., "which is more like having two children, none or three?" If the subject, in response to the initial question, had answered "2", then she would have "come out" of the question at the "2" end and the next question would have been $\boxed{0 \mid 1 \mid 3}$.

A particular sequence of questions is referred to as a "path." In Figure 26.2, depending on the particular path taken by the respondent, a minimum of four and a maximum of eleven questions would be necessary—which then permit determining some of the metric relations throughout the range "no children" to "five children."

Distances between children can be measured in terms of a continuum specified by a given question, such as economic or social considerations. Or, by leaving the continuum unspecified, they can be measured on whatever subjective continuum the individual considers relevant for herself. In this case, the respondents were asked to make their judgments with respect to "a mother's satisfactions and problems," a rather broad but uniformly defined continuum that seemed appropriate in the absence of any previous studies of this kind. Although metric relations can vary from one continuum to another, considerations of the kind of continuum presented to respondents are irrelevant to the formal character of the data.

Since the data consist of judgments as to whether a pair of stimuli are in some sense more alike than another pair, there is a wide variety of ways in which such data may be obtained. We were dealing with five prime intervals, $\overline{01}, \ldots \overline{45}$, and four compound intervals, $\overline{02}, \ldots \overline{35}$, the objective being to obtain a simple order of the nine distances.[11] Compound intervals are included to obtain greater metric information about the prime intervals.

Complete pair comparisons of the distances could have obtained more information but certainly much of it would have been redundant and some of it would have been inconsistent. Also, complete pair comparisons would have taken more time and perhaps reduced rapport with the subject. The data collection procedure used was a compromise device which attempts to minimize redundancy and inconsistency. Interview time for a subject averaged about three minutes. To each path through the figure there corresponds a set of order relations on pairs of conjoint distances. Such a set may be intransitive, implying no consistent metric relations, or the set may be transitive, yielding a simple or a partial order on the distances.

The procedure used here with the 1,215 women yielded 45 percent simple orders, 45 percent partial orders, 7 percent contradictions, and 3 percent not ascertained. The small proportion of contradictions and cases

11. Considering each pair of distances as an individual stimulus, and given nine stimuli, a simple order exists when an order relationship is established among each of the 36 pairs of stimuli. A partial order fails to establish the order relationship for some of the 36 pairs of stimuli.

in which the respondent could not answer the question because it was not comprehended is satisfactorily low in view of the complex series of judgments presented to the women in the sample.

With nine distances involved and with the particular data collection procedure used, there were 284 possible paths that could have been taken by the respondents. One of these paths ($\#180$) is shown on the right (largest interval at the top). The example shown illustrates one of decreasing prime distances in which $\overline{01} > \overline{12} > \overline{23} > \overline{34} > \overline{45}$. The use of compound distances to provide additional information on the relative magnitude of prime intervals is also illustrated. In this case, not only $\overline{01} > \overline{12}$, but $\overline{01} > (\overline{12} + \overline{23})$.

Of the 284 potential paths, 161 different paths were taken by the 1,215 respondents. However, one path dominated the response patterns, nearly 30 percent of the respondents choosing the particular decreasing path illustrated here. The twenty most frequent paths account for 70 percent of the subjects.

Tables 26.1 and 26.2 show, in the qualitative terms, the relative sizes of the prime intervals. It is evident that the prime intervals continually decrease from the largest distance $\overline{01}$ to the smallest $\overline{45}$.

The widths of the prime intervals in the S-scale, which probably represent perceived distances in life style ("satisfactions and problems") associated with the addition of each child, are likely to be important for predictions of future fertility behavior based on statements of fertility values. For example, if the perceived width of the $\overline{45}$ interval is very small and there is a shift in preference from four to five we might expect the changing preference to be followed by a change in fertility performance. If the $\overline{45}$ interval is large, an increase in the preference level may

$\overline{02}$
$|$
$\overline{01}$
$|$
$\overline{13}$
$|$
$\overline{12}$
$|$
$\overline{24}$
$|$
$\overline{23}$
$|$
$\overline{35}$
$|$
$\overline{34}$
$|$
$\overline{45}$

Table 26.1. Proportion of Times Prime Intervals Are Chosen as Largest or Smallest Distance.

| | All Parties | |
Prime Interval	Chosen as Largest Distance	Chosen as Smallest Distance
$\overline{01}$	.814	.021
$\overline{12}$	.095	.073
$\overline{23}$	.016	.164
$\overline{34}$	.009	.169
$\overline{45}$	.066	.573

not be accompanied by higher fertility because the additional child comes at perceived high costs in life style change.

It would, therefore, seem desirable to convert our ordinal data on distances to a set of interval data. A procedure for the conversion of an ordered metric scale to an interval scale has been developed by Frank Goode.[12] A simplified version of his method may be applied to these data as follows.

Table 26.2. Proportion of Times Row Prime Interval Is Greater than Column Prime Interval.

| | All Parities | | | |
| | Prime Intervals | | | |
	$\overline{01}$	$\overline{12}$	$\overline{23}$	$\overline{34}$	$\overline{45}$
$\overline{01}$	—	.878	.910	.946	.901
$\overline{12}$		—	.862	.856	.839
$\overline{23}$			—	.731	.696
$\overline{34}$				—	.627
$\overline{45}$					—

Suppose that, in the example given above, we substitute some positive quantity Δ_1, for the smallest distance $\overline{45}$, given at the bottom. Since $\overline{34}$ is greater than $\overline{45}$, $\overline{34}$ must equal Δ_1 plus some positive quantity, Δ_2. The next interval $\overline{35}$ is determined by adding its components $\overline{34}$ and $\overline{45}$, which results in $\overline{35}$ being equal to $2\Delta_1 + \Delta_2$. Continuing with the same basic principles of: (1) converting inequalities to equalities by adding positive quantities and (2) combining prime interval values to form the values for compound intervals, we arrive at the general form for the entire path:

$$\overline{02} = 8\Delta_1 + 5\Delta_2 + 3\Delta_3 + 2\Delta_4 + \Delta_5$$

$$\overline{01} = 5\Delta_1 + 3\Delta_2 + 2\Delta_3 + \Delta_4 + \Delta_5$$

$$\overline{13} = 5\Delta_1 + 3\Delta_2 + 2\Delta_3 + \Delta_4$$

$$\overline{12} = 3\Delta_1 + 2\Delta_2 + \Delta_3 + \Delta_4$$

$$\overline{24} = 3\Delta_1 + 2\Delta_2 + \Delta_3$$

12. Goode, Frank, "Interval Scale Representation of Ordered Metric Scales," [Mimeograph.] 1962.

$$\frac{|}{23} = 2\Delta_1 + \Delta_2 + \Delta_3$$

$$\frac{|}{35} = 2\Delta_1 + \Delta_2$$

$$\frac{|}{34} = \Delta_1 + \Delta_2$$

$$\frac{|}{45} = \Delta_1$$

The assignment of any positive value to Δ_1 will satisfy the order relations specified by the respondent. This can result in an infinity of solutions. Obviously, an interval scale representation of the ordered metric data becomes meaningful only if the relationships in the data greatly restrict the possible range of solutions. Without going to the limits of the feasible solutions, let us consider two solutions:

1. minimize the smallest interval and maximize each increment thereafter (minimize Δ_1, maximize Δ_2, . . . Δ_5)
2. maximize the smallest interval and minimize each increment thereafter (maximize Δ_1 minimize Δ_2, . . . Δ_5)

Although these solutions seem poles apart and would never be chosen by a researcher using these data, the two solutions are surprisingly similar. If we let the distance $\overline{05}$ equal 1.000 and apply the two conditions to the general solution, we obtain the following results for the prime intervals:

	Solution (1)	Solution (2)
$\overline{01}$	.500	.417
$\overline{12}$	.286	.250
$\overline{23}$	.143	.167
$\overline{34}$	.071	.083
$\overline{45}$	.000	.083

With only ordinal judgments from respondents, Goode argues that there are no grounds for considering parts of the data more valid than others and setting the deltas correspondingly high or low. An "equal delta solution" is recommended. Given an equal delta solution, the general solution for the prime intervals may be written as a row vector times a constant column vector, as follows:

$$\overline{01} = (53211) \cdot \Delta$$
$$\overline{12} = (32110) \cdot \Delta$$
$$\overline{23} = (21100) \cdot \Delta$$
$$\overline{34} = (11000) \cdot \Delta$$
$$\overline{45} = (10000) \cdot \Delta$$

By setting $\Delta = 1$, we arrive at the minimum integer equal delta solution satisfying the order statements of the respondent. The prime distances become: $\overline{01} = 12$, $\overline{12} = 7$, $\overline{23} = 4$, $\overline{34} = 2$, $\overline{45} = 1$ and standardized to a total length of 1.000 are: $\overline{01} = .462$, $\overline{12} = .269$, $\overline{23} = .154$, $\overline{34} = .077$, $\overline{45} = .038$. The scale is illustrated below:

Examples of other types of paths, together with their minimum integer equal delta solutions are given in Figure 26.3. For 78 percent of the responses there is a unique minimum integer equal delta solution. In some of the partial orders, however, amounting to 13 percent of all the responses, there is more than one equal delta solution because the data collection procedure failed to pick up enough of the order relations among the 36 pairs of stimuli. In these cases the "middle" solution was chosen wherever possible (see path 172 in Figure 26.3).

Acceptance of an equal delta solution as a procedure for arriving at perceived distances between children is dependent on its centrality to the distribution of all possible solutions and its effect on analysis. A mathematical description of all possible scale values satisfying the order relations for a selected path has not been developed. But we can explore the relationship between the equal delta solution and a sample from the infinite set of solutions.

In path 180, the total length $\overline{05}$ is: $12\Delta_1 + 7\Delta_2 + 4\Delta_3 + 2\Delta_4 + \Delta_5$. By setting $\overline{05}$ equal to some fixed length, and finding all possible integer solutions for the unknown deltas, the infinite number of solutions ranging over all positive numbers can be approximated. With a total length set at 50, there are 109 integer solutions satisfying path 180, a simple order. Comparable distributions of integer solutions were worked out for representative paths of the two classes of partial orders. The distribution of integer solutions for paths 180, 142, and 172 are shown in Figure 26.4. Each prime interval in the three paths has a unimodal distribution, with the equal delta solution lying fairly close to the mode.

In Table 26.3, the relationship between the equal delta solution and the distribution of feasible solutions is described in several ways. Differences between the average size of prime intervals in the distribution of solutions and the equal delta solution are summarized by the coefficient of dissimiliarity.[13] For the three paths examined, the coefficient

13. The coefficient of dissimilarity is the sum of the positive or negative differences between two proportionalized or percentagized distributions having the same category system. The limits of the coefficient are zero and one. A coefficient of .07, for example, means that seven percent of the cases in one distribution would have to be redistributed to match the second distribution.

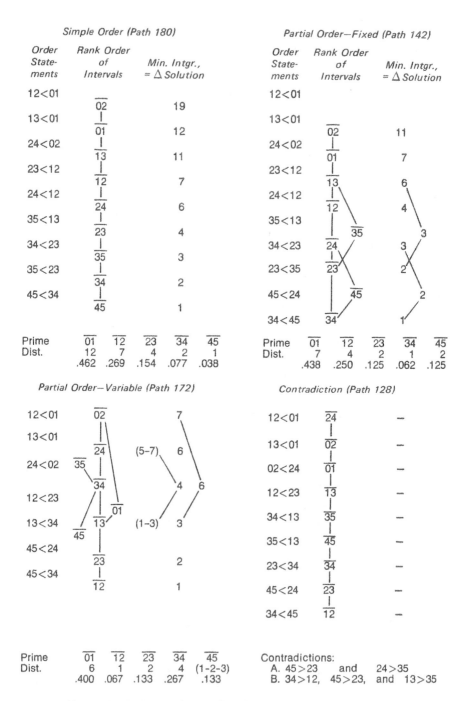

Figure 26.3. Examples of simple orders, partial orders and contradictions from the stimulus scale.

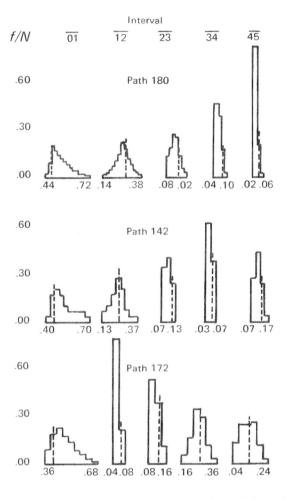

Figure 26.4. Distribution of integer solutions and location of equal delta solution.

is seven or eight percent. The correlations between the scale obtained by the equal delta restriction with each scale in a random sample of 12 scales from the distribution of admissible scales averaged no less than .97 for each of the three paths. This is a rather high correlation, particularly in view of the fact that some of the admissible scales in a distribution are quite extreme. For example, one of the solutions for path 180 weights Δ_5 by the factor of 25, while $\Delta_1 \ldots \Delta_4$ are each given a weight of one. To pick out one of several ordinal judgments that cannot be distinguished in magnitude and argue that is should be given a weight of 25 is an

Table 26.3. Selected Characteristics of the Equal Delta Solution, Distribution of Integer Solutions, and Their Relationship.

Interval	Path 180 Avg. Dist.	$=\Delta$ Dist.	Theor. Lower Bound	Theor. Upper Bound	All solutions (length = 50, solutions = 109) Est. Avg. Corr. = .985 Coef. Dissim. = .073
$\overline{01}$	.535	.462	.417	1.000	Weight 1 to 3 solutions (length = 50,
$\overline{12}$	.261	.269	.000	.500	solutions = 8)
$\overline{23}$	.124	.154	.000	.250	Avg. Corr. = .998
$\overline{34}$	.055	.077	.000	.142	Coef. Dissim. = .004
$\overline{45}$	.024	.038	.000	.083	Theor. Min. Corr. = .856

Interval	Path 142 Avg. Dist.	$=\Delta$ Dist.	Theor. Lower Bound	Theor. Upper Bound	All solutions (length = 30, solutions = 29) Est. Avg. Corr. = .980 Coef. Dissim. = .083
$\overline{01}$	.521	.438	.375	1.000	Weight 1 to 3 solutions (length = 30,
$\overline{12}$	.236	.250	.000	.500	solutions = 8)
$\overline{23}$	.096	.125	.000	.125	Avg. Corr. = .994
$\overline{34}$	.046	.062	.000	.125	Coef. Dissim. = .004
$\overline{45}$	.101	.125	.000	.125	Theor. Min. Corr. = .890

Interval	Path 172 Avg. Dist.	$=\Delta$ Dist.	Theor. Lower Bound	Theor. Upper Bound	All solutions (length = 25, solutions = 100) Est. Avg. Corr. = .970 Coef. Dissim. = .071
$\overline{01}$	.471	.400	.286	1.000	Weight 1 to 3 solutions (length = 25,
$\overline{12}$	.048	.067	.000	.167	solutions = 23)
$\overline{23}$	.102	.133	.000	.250	Avg. Corr. = .976
$\overline{34}$	.254	.267	.000	.500	Coef. Dissim. = .018
$\overline{45}$	.124	.133	.000	.333	Theor. Min. Corr. = .839

unusual demand to make upon the data. If the range of weights given the deltas is restricted to a maximum value of, say, 3 to 1 in contrast to the 25 to 1 example, debate concerning the delta weights is a waste of time because all solutions turn out to be virtually identical. For perhaps 80 percent of our respondents the equal delta solution will be within one percent of any other solution involving delta weights of one to three. We feel, therefore, that the similarities data describe perceived distances

Table 26.4 Perceived Distances Between Numbers of Children by Parity.

	Parity							
	0		1		2		4	
Distance	Mean	Std. Dev.	Mean	Std. Dev.	Mean	Std. Dev.	Mean	Std. Dev.
$\overline{01}$	.381	.099	.383	.089	.381	.099	.366	.111
$\overline{12}$	.247	.072	.231	.069	.246	.075	.253	.079
$\overline{23}$	.136	.046	.150	.050	.150	.055	.163	.062
$\overline{34}$	.114	.050	.114	.053	.114	.052	.110	.056
$\overline{45}$	.122	.098	.122	.094	.109	.089	.108	.085
N	89		304		315		321	

between children at the interval level of measurement with a very modest degree of measurement error.

Means and standard deviations of inter-child distances are given in Table 26.4. Interestingly, the average prime interval distances do not vary from one parity to another. Recently married women expect to find a set of "satisfaction and problem" distances associated with having children that are almost identical to the experience of women who have already had four children. A limited number of studies have shown that aggregate predictions of future fertility are surprisingly accurate for both short and long periods of time. These data suggest that the accuracy of prediction may derive from the fact that at the aggregate level, women are also capable of predicting the general consequences of having children. The experience of having children apparently does not result in some overall shock effect requiring revision of subjective "satisfaction and problem" distances.

Anticipated and experienced distances related to family growth follow a pattern of decreasing increments from $\overline{01}$ to $\overline{34}$, with $\overline{45}$ being roughly comparable in size to $\overline{34}$. The $\overline{24}$ interval is much smaller than the perceived changes in life style associated with having a first child and is approximately equal to $\overline{12}$. These data supplement previously collected materials dealing with the normative or preferred range of fertility. With the first two children taking up more than 60 percent of the $\overline{05}$ "satisfactions and problems" space, it would appear that families of less than two children are perceived as being distinctly different from larger families. Within the $\overline{25}$ range, changes in life style associated with family growth are small and the slope of decreasing size for average prime intervals becomes flat at $\overline{34}$, $\overline{45}$. Together with the relatively high variance observed in $\overline{45}$, this suggests a lack of consensus with respect to the upper limits of the normative range. In data not presented here, it was found that 55 percent of the women had S-scales with consistently decreasing prime intervals from $\overline{01}$ to $\overline{45}$, whereas only one percent viewed the child rearing experience as being represented by an increasing set of prime intervals. The high relative variance in the $\overline{45}$ interval exists because about one-fourth of the respondents have S-scales in which the interval $\overline{45}$ is large relative to the prime intervals in the intermediate $\overline{24}$ range.

The relationship between the S-scale values and other independent variables has not been carried through as yet. Some preliminary runs indicate that the decreasing path is more characteristic of Catholics than non-Catholics and is found more frequently among college educated women than among those with less education. There seems to be no relationship between S-scale values and age or duration of marriage.

The marked relationship between expected fertility and perceived distances is shown in Table 26.5. Women who perceive the distance between their present parity and five children as being represented by a small space are more likely to continue having children. For example, one parity women who have a relatively small subjective distance between one and five children expect to complete their

Table 26.5. Relative Distance to Fifth Child by Parity by Minimum and Maximum Expected Number of Children.

Parity	Distance		Minimum Expectations (Percent)					Maximum Expectations (Percent)					N
		1–2	3	4	5+	Total	1–2	3	4	5+	Total	
0	($\overline{35}$) Small[a]	31	13	41	15	100	6	13	43	38	100	48
0	($\overline{35}$) Large	52	17	29	2	100	22	27	36	15	100	41
1	($\overline{15}$) Small[b]	24	30	33	13	100	8	28	39	25	100	162
1	($\overline{15}$) Large	46	31	15	8	100	25	37	28	10	100	141
2	($\overline{25}$) Small[c]	34	25	29	12	100	9	32	36	23	100	146
2	($\overline{25}$) Large	49	22	24	5	100	22	32	32	14	100	169
4	($\overline{45}$) Small[d]	—	—	75	25	100	—	—	36	64	100	197
4	($\overline{45}$) Large	—	—	84	16	100	—	—	48	52	100	124

[a] Distance between 3 and 5 less than 20% of the distance from 0 to 5 in S-scale.
[b] Distance between 1 and 5 less than 60% of the distance from 0 to 5 in S-scale.
[c] Distance between 2 and 5 less than 30% of the distance from 0 to 5 in S-scale.
[d] Distance between 4 and 5 less than 10% of the distance from 0 to 5 in S-scale.

families with more children than one parity women whose subjective space, $\overline{15}$, is large. Among zero parity women, the $\overline{35}$ distance was used, since $\overline{05}$ is unity. The strength of the relationship is most pronounced in the lower parities because our scale of distance was limited to five children, resulting in some of the critical distances being unexplored for the higher parity women and because relevant distances are considerably smaller at the higher parities, making prediction a more difficult task. The fact that the relationship is in the observed direction indicates that anticipated "problems" rather than "satisfactions" dominate future growth patterns if we can assume that the fertility model is rational. This speaks well for conceptual schemes emphasizing the conflict between child rearing and other activities.

The Joint Scale

In the preceding section we were concerned with scaling the "satisfaction and problem" space associated with the growth of families. We believe the similarities question tapped a space roughly equivalent to the perceived effect of children on the mother's life style. This space is not necessarily equivalent to her *preference* space which may include several unknown factors which determine desires for specific numbers of children.

The analysis of preferences, contained here, focuses on implied distances between various numbers of children desired. We are not attempting to describe the typical number of children desired, but rather the subjective continuum or latent attribute mediating individuals' choices among a set of alternatives including 0, 1, 2, 3, 4, 5, or 6 children.

Our source of data for the *J*-scale analysis was the following question:

> The number of children people expect and want aren't always the same. If you could choose and have just the number of children you want by the time your family is completed, how many children would that be?

This question has been asked in several previous fertility studies, including the 1955 and 1960 Growth of American Families Study. The change introduced in this study was to follow the first question with:

> Suppose you couldn't have that number, but had to choose between———
> and—————. Which would you choose?
> and
> If you couldn't have that, would you choose————— or —————?

The numbers put in the blanks depended on the previous response and were chosen so as to yield the individual's preference ordering on number of children. For example, if a respondent initially chose four children as the number most preferred, the next question asked her to indicate her preference between three and five. If she chose three, then she was asked to choose between two and five, etc. The interviewer had a set of "trees," one for each ideal number, as a guide to the proper succession of questions and to simplify recording. The "tree" when the ideal number was four children, was as shown below:

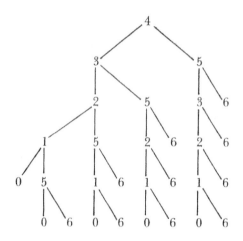

A respondent's rank order of preferences, say, 2 3 4 5 6 1 0, is her I-scale. With seven stimuli, there are $64 = (2^{n-1})$ continuous orders ending with the preferences zero or six. If a respondent preferred more than six children, it was assumed that her preference ordering was 6 5 4 3 2 1 0.

For any given J-scale with n stimuli there are $C^n_2 + 1$ class intervals formed between the midpoints of the stimuli. Each interval has a unique I-scale. A particular J-scale with the seven stimuli 0, 1 . . . 6, contains 22 intervals corresponding to a set of 22 I-scales. With 64 I-scales available and only 22 being combined to form a J-scale, there are many incompatible I-scales. Consider, for example, two individuals, A and B, having the preference ordering 2 3 4 1 0 5 6 and 3 2 1 0 4 5 6, respectively, A's I-scale reveals that she is to the right of the 14 midpoint but to the left of the 23 midpoint, hence 14 comes before 23 on the J-scale which implies that the distance $\overline{12}$ is greater than the distance $\overline{34}$. For individual B, the reverse is true. For her the 23 midpoint precedes 14 on the J-scale, so $\overline{34} > \overline{12}$.

The 1,215 respondents in the study were distributed across virtually all of the 64 I-scales. Therefore, the objective of the unfolding analysis is to recover the particular J-scale which is most representative of the underlying attribute. The J-scale chosen to represent the underlying attribute is the one from which a set of 22 I-scales may be derived and which contains the largest number of respondents. A J-scale cannot be

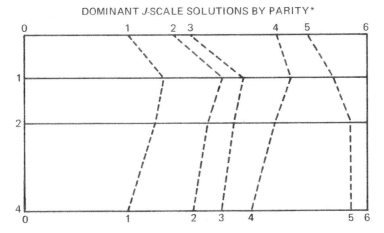

DOMINANT J-SCALE SOLUTIONS BY PARITY*

* The percentage of respondents whose preference orderings may be derived from the *J*-Scale is 71%, 61%, 62%, 74% for zero, one, two and four parity.

Figure 26.5. Dominant J-scale solutions by parity.

forced upon all the data. In fact, the proportion of respondents included in the dominant *J*-scale is an indication of the degree of consensus in the population being studied.

Selection of the 22 compatible *I*-scales with the maximum number of respondents is a fairly complex process described elsewhere.[14] Once the *I*-scales have been selected, their metric implications are combined and a partial order of distances is made. An equal delta solution can then be applied to the partial order to obtain an interval scale representation of the set of subjective distances underlying the preference orderings.

The dominant *J*-scales are shown in Figure 26.5 for the four parities in juxtaposition to facilitate comparisons. The proportion of respondents whose *I*-scales are consistent with the dominant *J*-scale solution varies from 61 percent among the first parity women to 74 percent among the fourth parity women. Proportions of this magnitude may be interpreted to mean that there is a moderately strong level of consensus. If all 64 preference orderings were equally likely, the "dominant" solution would contain about 34 percent of the cases.

Certain features of the *J* space are very similar to the average values of the *S* space. Both share the characteristic $\overline{01} > \overline{12} > \overline{23}$. The $\overline{01}$ distance in both cases is very large, comprising between 30 and 40 percent of the $\overline{06}$ distance on the *J*-scale. Other features clearly distinguish the *S*-

14. Coombs, *op. cit.*

and J-scales. Whereas the S-scales were stable from parity to parity, the J-scales vary considerably in the critical range, three children to five children. It is critical in the sense that the $\overline{34}$ and $\overline{45}$ distances underlying the most common preference orders must be viewed as at least partial determinants of the upper range of fertility performance.

The $\overline{34}$ distance is relatively large for recently married couples, in fact so large that it is impossible to argue that the two-child to four-child norm has emerged. It appears to be more like a two- to three-child norm. This pattern at the early stages of marriage, prior to the birth of children, may have no significant effect on performance if the process of family growth itself has an impact on preferences for size of family, as it apparently does.

Although we do not have cohort data, a cohort interpretation of the data may be legitimate because of the small differences in age among the parities. Mean ages of the four parities are 21, 23, 25 and 29 years. The $\overline{34}$ distance becomes relatively smaller and $\overline{45}$ larger with the introduction of each additional child. In contrast, subjective distances measured by the S-scale were, most often, continuously decreasing. Relative reduction of the $\overline{34}$ J-space and increase of $\overline{45}$ may reflect revision of preferences based on the experience of caring for children.

In the J-scale, $\overline{24}$ is relatively small, amounting to less than 20 percent of the $\overline{06}$ distance for one, two, and four parity women. Coupled with the large distance $\overline{45}$ and the large distance $\overline{02}$, the dominant J-scale lends additional weight to the existence of a two- to four-child normative range.

Conclusions

Our exploration of the application of unfolding theory to the analysis of fertility has been limited to a rather brief examination of the so-called normative range. The J-scale yields more direct information on what most demographers would consider the norm, whereas the S-scale, based on the experience women have with children, may be more appropriate for understanding fertility performance, given a particular norm. Very clearly the majority expression of the subjective distance between various numbers of children, as measured by the S-scale, is a decreasing function. There are several popular expressions in the language which imply the existence of such a function. Typically, the normative range is thought to be different by demographers—large distances between $\overline{01}$, $\overline{12}$, and $\overline{45}$. This kind of distance pattern is basically what was found in the J-scale analysis although the space seems to be subject to revision by the family building process. Since the S-space and J-space need not be compatible, both the demographers and the people may be correct in their evaluations of fertility behavior.

Although the S- and J-scales are different, their interaction may give us useful information for understanding performance. In the S-scale we are told that the $\overline{02}$ space has great consequences for the life style of the mothers. The dominant J-scales also indicates that the $\overline{02}$ space is large, which in conjunction with the preference orders means that no children or one child is to be avoided. The consequences of spanning the $\overline{02}$ space are desirable. Having reached two children, both the S- and J-scales suggest that the $\overline{24}$ distance is relatively small, whether we consider experienced changes or preferences. Performance clusters within this range and our ability to distinguish differential behavior within the range may be difficult. A fifth child would not present many problems according to the S-scale, but for some unidentified reasons it is not desired according to the J-scale. The interaction of S and J in the $\overline{45}$ space is likely to produce a substantial minority of children beyond the two to four range.

Some of the potential applications of the S- and J-scales to fertility analysis can be briefly enumerated:

1. The S-scale seems particularly useful as a device for measuring the utility of various numbers of children. It is often hypothesized that differences in fertility between societies, within societies, and in societies over time, can be attributed to differences in the utility of children.

2. The shape of the S-scale can be used as a control variable for relating other variables to fertility. If a very large proportion of the $\overline{05}$ space is consumed by the first two or three children ("satisfactions and problems" nearly exhausted), the fourth or fifth child is a luxury item that should occur or not, depending on economic circumstances. If the space between certain numbers of children is large, we should be concerned with variables that measure components of life style which may be in conflict with the child rearing process. For example, it has been shown that economic variables are directly related to fertility for males and inversely related to fertility for females. The S-space is probably perceived differently by men and women, with additional children having a greater consequence for women than for men. For men the only distance that may have real significance is $\overline{01}$. Similar differences in urban-rural and Catholic-non-Catholic populations may be a function of differences in the subjective space underlying different numbers of children.

3. The apparent inconsistency between family size ideals and performance in several underdeveloped countries may be comprehended, given certain S- and J-scales. Relatively low ideals are perfectly compatible with S- and J-scales that are characterized by rapidly decreasing functions. Under these conditions there is nothing to prevent high fertility even with the low ideals.

4. One of our immediate uses for the S-scale will be in the follow-up

to the present study.[15] We will intend to examine its predictive utility together with expectation data as we periodically gather information about the fertility performance of our respondents.

15. The first follow-up study was in the field in October–November, 1962, about eight or nine months after the original interview. Respondents are being questioned about their fertility history during the intervening period and about their future fertility expectations.

PART VI

Related Materials

The final section of this volume contains a set of articles that is clearly related to scaling and to the materials included in the earlier sections. However the selections are sufficiently unique, or in some cases, tangential to the methods discussed earlier; thus, it seems better to locate them in this separate section.

Chapter 27, by Edwards and Kilpatrick, presents a "scale discrimination technique," a procedure one can employ to create attitude scales. This is a technique that uses Thurstone scaling procedures, Likert methods of evaluating the ability of an item to discriminate, and Guttman's coefficient of reproducibility. This method is a synthesis of these techniques of item evaluation. The approach involves obtaining equal-appearing scale scores of a large pool of items. By using this set of scores and measuring the spread of judgments, it is possible to screen the items. The selected items are then prepared in Likert form, administered and item analyzed to aid in the final selection of items. Guttman scaling procedures are then employed. The total method provides an objective approach to item selection.

The second article in this section, Chapter 28, introduces a process called H-technique. This method involves the determination of cutting points in Guttman scales by using several items instead of only one. This procedure creates a "contrived item" which uses a greater amount of the information available in the basic data. Because of this maximization of information, scale error is reduced and confidence in the scalability of the content area is increased.

The paper by Leik and Matthews (Chapter 29) describes developmental processing scaling. This type of scale is shown to be a general case of which the Guttman scale is a special case. A Guttman scale satisfies developmental scale criteria, but the reverse is not necessarily true. The distinction between the two types of scales involves the fact that a developmental scale allows for the acquisition of traits in a cumulative manner, as does a Guttman scale, and it also allows for the dropping of traits that a Guttman scale does not. Therefore, presence of a trait or the endorsement of an item must continue once it has occurred in Guttman scaling, but this is not so in the developmental process scaling. The process recognizes that, for example, in a scale of card games, the game of Rummy may not be enjoyed or even played by someone who has progressed developmentally to Bridge. Leik and Matthews provide a discussion of the process of developmental process scaling, a description of the method, some illustrative examples, an index of scalability, and an appropriate significance test.

Chapter 30, by Sanford Labovitz, is a consideration and examination of the effect of treating ordinal data as interval data. The author takes up the position presented by Burke in the first section of this book and provides a demonstration that under certain conditions ordinal variables can be treated as if they were interval variables. He demonstrates this by generating and assigning random and nonrandom numbers to rank orders and then examining the effects of this transformation. In this way, Labovitz suggests, we obtain advantages that include the retention of knowledge, the use of more powerful statistics, and an increase in versatility in statistical manipulation.

The position presented in the preceding article and endorsed in the subsequent one is not a resolution of the debate and issues that exist between the measurement-oriented and the measurement-independent position. These articles serve rather to point up the nature of the two sides of the question. There are serious criticisms which are continually leveled at the measurement-independent position. Other authors, for example, Wilson[1] indicate persuasively that ordinal variables do not allow any but very weak inferences to be made in regard to the fit between data and theoretical models which are formulated in terms of interval variables and that less than interval variables cannot figure directly in the formultion of substantive theoretical models.

The 31st and final paper is a short parable written by Frederic Lord. This charming story takes a practical position and points out the potential dangers of taking the whole business too seriously. An informed aware researcher can, after all, employ a great variety of scaling methods and still not be driven in every instance, panic stricken, into a fear of parametric statistics and the advantages they allow.

1. Thomas P. Wilson (1971) "Critique of Ordinal Values," *Social Forces*, Vol. 49, No. 3, pp. 432–444.

Lord, Labovitz, and Burke are clearly less concerned with scales of measurement than are many authors whose works are included in this book. This work is included since it can serve as an obvious complement to the other articles, which are inclined generally to a measurement-oriented position. Burke, Labovitz, and Lord would encourage researchers to use all scales as if they had generated interval data, taking what was described early in this volume as a measurement-independent position.

The goal of this volume is not to rule in this debate but simply to make students aware of the development and the procedures of scaling and also to make them aware of the existence and nature of a debate regarding the importance of the levels of measurement.

Chapter 27

A TECHNIQUE FOR THE CONSTRUCTION OF ATTITUDE SCALES

ALLEN L. EDWARDS
University of Washington

AND FRANKLIN P. KILPATRICK
University of Delaware

Earlier articles (3, 6) have reviewed the various methods which have been used in the construction of attitude scales: the method of equal appearing intervals developed by Thurstone (16), the method of summated ratings developed by Likert (14), and the method of scale analysis developed by Guttman (9). The method of equal appearing intervals and the method of summated ratings are similar in that both provide techniques for selecting from an initial large number of items, a set of items which constitutes the measuring instrument. Scale analysis differs from these two methods in that it is concerned with the evaluation of a set of items, *after* the items have been selected in some fashion or another.

In the method of equal appearing intervals, items of opinion are sorted by a judging group into 9 or 11 categories constituting a continuum ranging from unfavorable to favorable. The scale value of each item is found by locating the point on the continuum above which and below which 50 percent of the judges place the item. The spread of the judges' rating is measured by Q, the interquartile range. A high Q value

Reprinted with permission of the author and the publisher from the *Journal of Applied Psychology*, Vol. 32, No. 4, 1948. Copyright 1948 by the American Psychological Association.

for an item indicates that the judges are in disagreement as to the location of the item on the continuum and this, in turn, is taken to mean that the item is ambiguous. Both Q and scale values are used in selecting items for the attitude test. Approximately 20 items with scale values equally spaced along the continuum and with low Q values are selected for the test. Scores on the test are determined by finding the median of the scale values of the items with which a subject agrees.

In the method of summated ratings, items are selected by a criterion of internal consistency. Subjects check whether they strongly agree, agree, are undecided, disagree, or strongly disagree with each item. Numerical weights are assigned to these categories of response using the successive integers from 0 to 4, the highest weight being consistently assigned to the category which would indicate the most favorable attitude. A high and low group are selected in terms of total scores based upon the sum of the item weights. The responses of these two groups are then compared on the individual items and the 20 or so most discriminating items are selected for the attitude test. A subject's score on this test is determined by summing the weights assigned to his responses to the 20 items.

In scale analysis, a complete set of items is tested to determine whether they, as a group, constitute a scale in the sense that from the rank order score it is possible to reproduce a subject's response to the individual items. The degree to which this is possible is expressed by a coefficient of reproducibility.[1] Although ordinarily Guttman uses 10 to 12 items, to give a simple explanation of this coefficient let us suppose that we have but 3 items, each with but 2 categories of response, agree and disagree. We shall assume that the agree response, in each instance, represents a favorable attitude and the disagree response an unfavorable attitude. A weight of 0 is assigned to the disagree response and a weight of 1 is assigned to the agree response. Let us also suppose that for the first item we have in our sample 10 subjects with weights of 1 and 90 with weights of 0; for the second item we have 20 subjects with weights of 1 and 80 with weights of 0; and for the third item we have 40 with weights of 1 and 60 with weights of 0.

In the case of perfect reproducibility, the 10 subjects with weights of 1 on the first item will be the 10 subjects with the highest rank order scores. These 10 subjects will also be included in the 20 who have weights of 1 on the second item and these 20, in turn, will be included in the 40 who have weights of 1 on the third item. It would also be true that only 4 patterns of item response would occur, if the set of items were perfectly reproducible. For the sample at hand, these patterns and the scores associated with them would be: AAA-3; DAA-2; DDA-1; DDD-0. Since all responses could be perfectly predicted from the scores, the coefficient of reproducibility, in this instance, would be 100 percent. Perfect reproducibility is seldom found, however, and in practice a coefficient of 85 percent or higher is believed satisfactory for judging a set of items to be a scale.[2] Various techniques for computing the coefficient of reproducibility have been developed and are described in the articles by Festinger (7), Clark and Kreidt (2), and Guttman (11, 12).

Scale analysis, in the sense mentioned above, thus becomes a technique *secondary* to the problem of item selection.[3] The important problem is to obtain a set of items which the investigator may have some assurance will scale when a particular technique of testing for scalability is applied. Up to the present time, the problem of item selection in scale analysis seems to have been left largely to the intuition and experience of the investigator. The only practical rules suggested are that one should simply rephrase the same question in slightly different ways (7, p. 159) or that one should look for items with as homogeneous content as possible (12, p. 461). This latter suggestion indicates that if we are interested in the problem of attitude toward the Negro, we should break this universe of content down into sub-universes constituting perhaps such areas as attitude toward the Negro in public eating places, attitude toward the Negro as a resident in the community, attitude toward the Negro as a voter, attitude toward the Negro as an employer, attitude toward the Negro in public conveyances, and so on. But even here, we find that attitude toward the Negro, let us say, in public conveyances can be broken down into areas of content even more homogeneous by enumerating the specific conveyances: street cars, busses, trains, planes, and so on. Each of these areas of content might possibly be broken down into still more homogeneous areas. Eventually, we may end up, as Festinger suggests, with multiple rephrasings of the same question, and our two rules are thus but one (7, p. 159).

Obviously, any technique which enables us to select a set of items from the large number of possible items, with some assurance that the set of items selected will, in turn, meet the requirements of scale analysis

2. There are other criteria to be applied in determining whether a set of items constitutes a scale in addition to the coefficient of reproducibility (10, 12). Little has been published, however, in which these criteria have been applied empirically to a concrete set of data. This may be remedied in the forthcoming volumes by Guttman and his associates which are to be published by the Social Science Research Council. So far, however, the application of these criteria has simply been mentioned along with the theoretical and practical implications. The coefficient of reproducibility has been stressed in all of Guttman's publications, perhaps for the reason that it is considered a primary and necessary condition but an insufficient condition for a scale.

3. This is not to deny the importance of the theory underlying scale analysis.

would be of great value. In this paper, a technique which has proved successful in doing this is described. For reasons which will become clear as we proceed, we have called this technique the scale-discrimination method of attitude scale construction (5).

The Scale-Discrimination Technique

The scale-discrimination method is based upon preliminary investigations which showed that the cutting point[4] of an item is related to the Thurstone scale value of the item and that the reproducibility[5] of an item is related to the discriminatory power of the item (6). The discriminatory power of an item, it has also been shown, is not, as might seem at first glance, merely a function of the item's scale value. It can easily be demonstrated that items with comparable Thurstone scale and Q values may differ tremendously in their power to differentiate between those with favorable and those with unfavorable attitudes.[6]

Statements of opinion concerning science were collected from a variety of sources. Books and essays were consulted. Individuals were asked to express their opinions in brief written statements. We eventually collected 266 statements of opinion about science. In editing these items, particular attention was paid to eliminating those items which: (1) were liable to be endorsed by individuals with opposed attitudes; (2) were factual or could be interpreted as such; (3) were obviously irrelevant to the issue under consideration; (4) appeared likely to be endorsed by everyone or by no one; (5) seemed to be subject to varying interpretations for any reason; (6) contained a word or words not common to the vocabularies of college students. Also, due to emphasis upon the matter during both the collecting and editing of the statements, most of the 155 statements finally selected expressed a clear-cut favorable or unfavorable opinion about science.

Thirteen other items, which might be called "control" items, were added to the original 155. These 13 items were added to determine how they would fare at various stages of the scale-discrimination method. Of the 13 items, we judged that 7 were "neutral" items in the Thurstone

4. The cutting point of an item marks the place in the rank order scores of the subjects where the most common response shifts from one category (agree) to the next (disagree). Between cutting points, in a *perfect* scale, all responses would fall in the same category.

5. The reproducibility of an item is measured by degree to which responses to the item can be reproduced from the rank order scores of the subjects.

6. For example, the extreme item: "All Republicans should be executed" would undoubtedly show a scale value at one extreme of the continuum and a definitely low Q value. But this item will not differentiate between those with favorable and unfavorable attitudes toward Republicans for the obvious reason that both groups would probably react in the same fashion to the item.

sense; 2 were items which could possibly be interpreted as factual; 1 was believed to be too extreme for many endorsements; 1 was judged ambiguous because the words "scientific holiday" could be interpreted as meaning a moratorium or as meaning a celebration; 1 was judged ambiguous because more than one dimension was involved; and 1 was judged irrelevant. Thus there were 168 items in all which were used in testing the scale-discrimination method of scale construction.[7]

Determining Scale and Q Values of the Items

Envelopes numbered 1 through 110 were prepared. In each envelope we placed a set of 3 $\times$ 5 cards lettered A, B, C, D, E, F, G, H, I, and a pack of slips of paper approximately 2 $\times$ 4 inches in size. On each slip of paper one of the 168 items was printed along with the number of the item. In each case the pack of slips was shuffled so that the items would be arranged in no set order. The envelopes were given to an elementary psychology class along with a set of instructions describing the Thurstone sorting procedure and the members of the class were asked to sort the items in accordance with the instructions.

The item sortings of each subject were examined and we discarded those subjects whose sortings showed obvious reversals of the continuum or failure to carry out instructions. On this basis we were left with 82 completed sets of judgments.

Frequencies of judgments in each of the 9 categories for each item were tabulated, translated into cumulative frequencies, and then into cumulative proportions.[8] An ogive was plotted for each item with cumulative proportions on ths axis of ordinates and scale values on the axis of abscissas. Scale values were read to two decimal places (the second decimal place being merely an approximation) by dropping a perpendicular to the baseline of scale values at the point where the cumulative proportion curve crossed the 50 percent mark. In a similar fashion Q values were determined by dropping perpendiculars at the 25th and 75th percent levels, Q being the scale distance between these two points or the interquartile range.[9]

7. It should be emphasized that the inclusion of the "control" items mentioned is not to be considered part of the scale-discrimination procedure.

8. This task was most laborious. Almost 14,000 slips of paper had to be sorted and then tabulated. Some judging technique similar to that used by Ballin and Farnsworth (1) or Seashore and Hevner (15) would reduce much of this labor, but even here the task is not simple. Various methods which simplify the judging process are now being tried and will be reported upon in another paper.

9. This operation was simplified by setting up a master chart with the cumulative proportions on the Y axis and the scale values on the X axis. This chart was then taped to a ground-glass plate which fitted over an enclosed wooden box containing a 100 watt bulb. Tracing paper could then be placed over this chart and the ogives for the individual items quickly drawn.

The 168 items were then plotted in a bivariate distribution according to scale and Q values, the scale values being plotted on the baseline. The distribution of scale values was bimodal in shape. There were very few items in the "neutral" section (none at all in between 5.0 and 5.9), the modal categories being 1.0 to 1.9 and 7.0 to 7.9. The Q values of the 7 items which did fall in the "neutral" scale interval (4.0 to 4.9) were quite low, 6 of the 7 falling well below the median Q value for all 168 items. All 7 of these items were "control" items, described previously.

A line was drawn through the distribution at approximately the median Q value of all the items, 1.29. All items with Q values above this point were rejected. We worked from here on with the remaining 83 items or with approximately the 50 percent of the initial set of items with the least degree of ambiguity as measured by Q. One of the "neutral" control items was eliminated by this standard and 6 were acceptable. These 6 items all had scale values between 4.0 and 4.9. No items at all had been found in the scale interval 5.0 to 5.9 and the Q criterion eliminated all items in the interval 3.0 to 3.9. One of the 2 factual items was rejected by the Q criterion and the ambiguous item with the words "scientific holiday" was also eliminated. The remaining 10 "control" items would have to be judged acceptable by the Q criterion.

Item Analysis

The 83 items were prepared in a form suitable for Likert type reactions. Each item was followed by a 6 point forcing scale (strongly agree, agree, mildly agree, mildly disagree, disagree, strongly disagree). Subjects were instructed to check for each item the one expression which most nearly described their own attitude with respect to the item. In all, 355 subjects filled out the questionnaire: 245 from sociology, psychology, and speech classes at the University of Washington; 60 from a local junior college; and 50 from a police school. Of these 355 papers, 346 were usable, 9 of them being incomplete or having more than one answer for a single item.

Scoring was done in the usual Likert fashion, weights of 0 through 5 being assigned to the 6 response categories, the weight of 5 being given to the strongly agree response in the case of items expressing a favorable opinion about science, and to the strongly disagree response in the case of items expressing an unfavorable opinion about science. For the 6 items in the scale interval, 4.0 to 4.9, the direction of the weights was assigned on the basis of whether the scale value of the item was larger or smaller than 4.5. Response weights on the 83 individual items were summed for each subject and a frequency distribution plotted for the resulting scores. The obtained range of scores was only 64 percent of the possible range (140–405 obtained, 0–415 possible) with considerable bunching at the upper (favorable) end of the distribution.

Two criterion groups were chosen, approximately the upper and lower 27 percent, in terms of total scores. The range of scores for the lower 94 papers was from 140 to 300 and the upper 94 papers had scores ranging from 343 to 405. The 83 items were then subjected to item analysis. For each item, frequencies in each of the response categories for the high and for the low group were tabulated. The 6 categories were then reduced to 2 by combining categories 0, 1, 2, 3, and 4.[10] From the resulting 2 × 2 tables, phi coefficients were calculated.[11] The phi coefficients ranged in size from .16 to .78.

Next the 83 items were plotted in a bivariate distribution with phi values on the Y axis and scale values on the X axis.[12] The 4 items with the highest phi coefficients were selected from each half-scale interval; due to the previously mentioned gaps in the scale continuum, this involved only the intervals from .5 to 2.5 and from 6.5 to 8.0. No items were selected from the "neutral" control items in the scale interval 4.0 to 4.9. The 28 items thus selected were assigned to Forms A and B of the questionnaires by alternating scale values between the two forms.

The final scales then consisted of 14 items each, with the items very closely equated as to Thurstone scale values, Q values, and phi values. For Forms A and B, respectively, the mean scale values of the 14 items were 3.85 and 3.91; the mean Q values were .90 and .92. Phi coefficients of the items in Form A ranged from .58 to .78 with a median value of .65; for Form B they ranged from .58 to .76 with a median value of .66. Only 1 of the remaining 10 "control" items had a phi value above .58. This was one of the 6 "neutral" items and it had a phi value of .61. The other "control" items would be rejected by the phi criterion.

Reliability and Reproducibility of the Scale

The reliability coefficient of the two forms of the scale, 14 items versus 14 items, based upon the responses of 248 new subjects was .81, uncor-

10. This grouping was necessary because our subjects gave predominantly favorable responses to the items. If our universe of content had been attitude toward labor unions, we would expect a more symmetrical distribution of responses and consequently a different grouping of categories.

11. The nomographs of Guilford (8) or the tables prepared by Jurgensen (13) make these calculations quite simple.

12. A plot of phi values against Q values indicated no discernible relationship, the variability within columns being approximately the same as the total variability. This would indicate that in the procedure followed here, the scale-discrimination procedure, the phi analysis adds to the process of item selection when items with comparable Q values are used. We have, it may be recalled, already eliminated the 50 percent of the items with the highest Q values. The relationship between the discriminatory power of an item and Q value when this is not the case is described in another paper (4).

rected. For both forms of the test the range of scores was quite restricted, 30 to 70 in each case with possible ranges from 0 to 70. Within this restricted range, bunching at the upper, or favorable, end was present. The mean score for Form A was 58.22 and the standard deviation was 7.33. For Form B the mean was 57.20 and the standard deviation was 7.79.

Scale analysis based upon the performance of a sample of 87 subjects drawn from the larger group of 248 subjects was carried out with both forms of the test by the Cornell technique (11). A coefficient of reproducibility of 87.5 percent was obtained for Form A and a coefficient of reproducibility of 87.2 percent was obtained for Form B. Response categories in each instance were dichotomized. Cutting points were established and we observed Guttman's rule that "no category should have more error than non-error" (11, p. 17). The range of modal response categories was from .51 to .82 for Form A. The mean value of the modal categories, .57, which is the minimum value [13] of the coefficient of reproducibility for this set of items with the sample at hand, may be compared with the observed coefficient of reproducibility of 87.5 percent. For Form B the range of the modal categories was from .52 to .67. The mean value, which again is the lower limit of the coefficient of reproducibility, was .57, whereas the observed value of the coefficient of reproducibility was 87.2 percent.

The two observed values of the coefficient of reproducibility are sufficiently high to constitute evidence that but a single dominant variable is involved in the sets of items or that, in other words, uni-dimensionality is present. Such sets of items are said to be scalable or to constitute a scale. The coefficients of reproducibility also mean that it is possible to reproduce item responses from rank order scores with the accuracy indicated by the value of the coefficients.

The error of reproducibility which is present is simply 1.00 minus the observed coefficient of reproducibility. If the error of reproducibility can be assumed to be random, then these sets of items possess an important property: the simple correlation between rank order scores and an external criterion will be equal to the multiple correlation between the items and the external criterion (10). This, in turn, means that efficiency of prediction is maximized by the simple correlation.

It would also be true in the case of sets of items which meet the criteria demanded of scales[14] that the interpretation of the rank order scores is unambiguous and that it is possible to make meaningful statements about one subject being higher (more favorable) than another on the variable

---

13. This is the lower limit because the reproducibility of any single item cannot be less than the frequency in the modal category. The method of computing the minimum value of the coefficient assumes independence of the items. See Guttman (12) .

14. See footnote 2.

in question.[15] This would not be true of a test involving more than one variable. Suppose, for example, a test involves two variables. Then a subject might obtain a given score by being high on one variable and low on the other. Another subject might obtain the same score by being high on the second variable and low on the first. From the rank order scores alone it would be impossible to tell the relative positions of the subjects on the two variables, and the interpretation of the composite score is ambiguous. Statements of "higher and lower than" might be made, but we would not know what the "higher and lower than" referred to, for by increasing or decreasing the number of items related to either variable, the rank order scores of the subjects could be altered.[16] This would not be true of a test in which the items all belong on a single continuum, that is, a test which is uni-dimensional. In such a test, increasing the number of items would not shift the rank order scores of the subjects.

Summary

The method of scale construction described in this paper has been called the scale-discrimination method because it makes use of Thurstone's scaling procedure and retains Likert's procedure for evaluating the discriminatory power of the individual items. Furthermore, the items selected by the scale-discrimination method have been shown, in the case described, to yield satisfactory coefficients of reproducibility and to meet the requirements of Guttman's scale analysis. The scale-discrimination method is essentially a synthesis of the methods of item evaluation of Thurstone, Likert, and Guttman. It also possesses certain advantages which are not present in any of these methods considered separately.

The scale-discrimination method, for example, eliminates the least discriminating items in a large sample, which Thurstone's method alone fails to do. The unsolved problem in the Thurstone procedure is to select from within each scale interval the most discriminating items. Items within any one scale interval may show a high degree of variability with respect to a measure of discrimination. For example, we found within a single interval items with phi values ranging from .24 to .78. That Thurstone's criterion of Q does not aid materially in the matter of selecting discriminating items is indicated by the plot of phi values against Q values, *after* the 50 percent of the items with the highest Q values had

15. In the case of perfect scales, where the coefficient of reproducibility is unity, it also follows that an individual with a low rank order score will not have given a more favorable response to any item than any person with a higher rank order score.

16. We do not mean to imply by this discussion that multi-dimensional scales are without value.

already been rejected. Under this condition, items with Q values from 1.00 to 1.09 had phi coefficients ranging from .32 to .76. Thurstone's method also, by the inclusion of "neutral" items, tends to lower reliability and to decrease reproducibility of the set of items finally selected (6).

Thus when selecting items by Thurstone's technique alone, we have no basis for making a choice between items with comparable scale and Q values, and yet these items are not equally valuable in the measurement of attitude. By having available some measure of the discriminatory power of the items, the choice becomes objective as well as advantageous as far as the scale itself is concerned.[17]

The advantage of the scale-discrimination method over the Guttman procedure lies essentially in the fact that we have provided an objective basis for the selection of a set of items which are then tested for scalability. It may happen that not always will the scale-discrimination method yield a set of items with a satisfactory coefficient of reproducibility. But this is not an objection to the technique any more than the fact that not always will a set of intuitively selected items scale. Rather, it seems that the scale-discrimination method offers greater assurance of scalability than any intuitive technique such as applied by Guttman. Furthermore, the set of items selected by the scale-discrimination technique provides a wider range of content than do the intuitive Guttman items. In the scale-discrimination method, we obtain items which are not essentially multiple phrasings of the same question as is often true when the selection of a set of items to be tested for scalability is left to the experience of the investigator (7, p. 159).

Several different areas of content are now being studied by variations of the scale-discrimination method and the results of these researches should provide additional evidence concerning the relationship between the scale-discrimination method and scale analysis.

References

1. BALLIN, M., and FARNSWORTH, P. R. A graphic rating method for determining the scale values of statements in measuring social attitudes. *J. soc. Psychol.*, 1941, 13, 323–327.
2. CLARK, K. E., and KREIDT, P. H. An application of Guttman's new scaling techniques to an attitude questionnaire. Unpublished paper, 1947. *Educ. psychol. Measmt.*, 1948, 8, Summer, No. 2.
3. EDWARDS, A. L., and KENNEY, K. C. A comparison of Thurstone and Likert techniques of attitude scale construction. *J. appl. Psychol.*, 1946, 30, 72–83.

17. Additional research may indicate that the Thurstone scaling procedure is not necessary. See, however, the articles by Edwards and Kilpatrick (6) and Clark and Kreidt (2).

4. EDWARDS, A. L. A critique of "neutral" items in attitude scales constructed by the method of equal appearing intervals. *Psychol. Rev.*, 1946, 53, 159–169.

5. EDWARDS, A. L., and KILPATRICK, F. P. The scale-discrimination method for measuring social attitudes. *Amer. Psychol.*, 1947, 2, 332.

6. EDWARDS, A. L., and KILPATRICK, F. P. Scale analysis and the measurement of social attitudes. *Psychometrika*, 1948, 13, June.

7. FESTINGER, L. The treatment of qualitative data by "scale analysis." *Psychol. Bull.*, 1947, 44, 149–161.

8. GUILFORD, J. P. The phi coefficient and chi square as indices of item validity. *Psychometrika*, 1941, 6, 11–19.

9. GUTTMAN, L. A basis for scaling qualitative data. *Amer. sociol. Rev.*, 1944, 9, 139–150.

10. GUTTMAN, L. *Questions and answers about scale analysis.* Research Branch, Information and Education Division, Army Service Forces, Report D-2, 1945.

11. GUTTMAN, L. The Cornell technique for scale and intensity analysis. Mimeographed, 1946.

12. GUTTMAN, L. On Festinger's evaluation of scale analysis. *Psychol. Bull.*, 1947, 44, 451–465.

13. JURGENSEN, C. E. Table for determining phi coefficients. *Psychometrika*, 1947, 12, 17–29.

14. LIKERT, R. A technique for the measurement of attitudes. *Arch. Psychol., N. Y.*, 1932, No. 140.

15. SEASHORE, R. H., and HEVNER, K. A time-saving device for the construction of attitude scales. *J. soc. Psychol.*, 1933, 4, 366–372.

16. THURSTONE, L. L., and CHAVE, E. J. *The measurement of attitude.* Chicago: Univ. Chicago Press, 1929.

Chapter 28

A TECHNIQUE FOR IMPROVING CUMULATIVE SCALES

Samuel A. Stouffer, Edgar F. Borgatta
Queens College, City University of New York
David G. Hays and Andrew F. Henry (deceased)
Suny, Buffalo

This paper introduces a simple new procedure for obtaining a cumulative-type scale which should have properties of high reproducibility, high test-retest reliability, and high stability from sample to sample in rank order of cutting points. Moreover, none of these properties need be obtained at the cost of restricting the scale to content of too narrowly limited specificity or to questions with too uniform a format. We shall call the new procedure the H-technique.

Actually, the H-technique produces a Guttman scale or a Lazarsfeld latent distance scale with one important modification. Instead of using only one item to determine a given cutting point on the scale, as in the conventional procedure, the H scale uses two or more items.

Reprinted from the *Public Opinion Quarterly*, Summer, 1952, pp. 273–91, with permission of Columbia University Press and the senior serving author.

This research, carried out at the Harvard Laboratory of Social Relations, was supported in part by the United States Air Force under Contract AF33 (038) –12782 monitored by the Human Resources Research Institute. Permission is granted for reproduction, translation, publication and disposal in whole and in part by or for the United States Government.

The basic idea of the cumulative scale is that all items have a structure such that a person who answers "Yes" to any item to which p proportion of respondents answers "Yes" will tend to answer "Yes" to all other items to which larger than p proportion of the respondents answers "Yes."[1] In the case of a perfect Guttman m-item scale, respondents can be ordered without error into m + 1 classes. Each item serves to define the limits of one class or rank group.

In practice, however, perfect Guttman scales are not likely to occur. There will be errors in response to any particular item. Hence some of the respondents will fall into non-scale types—that is, they will say "No" to one or more items which are more frequently approved than an item to which they say "Yes." How many errors are permissible before the hypothesis of scalability of a set of items is rejected is an arbitrary matter. Guttman sets a minimum standard of 90 percent "reproducibility"—that is, at least nine times out of ten, on the average, if we know a respondent's rank we should be correct in specifying his response to any particular item. There are further requirements: (a) that the errors thus made with respect to any single item must not exceed the number of correct calls of either positive or negative response, whichever are fewer, and (b) that the errors be at random, such that not too many individual respondents have identical scale patterns.

Hundreds of scales have been constructed which approximate these minimum standards. But sometimes these standards have been attained under less than happy circumstances.

Most commonly, perhaps, the errors have been held down by keeping the number of items few—say four or five—and the fraction used in the scale of all the information initially available may be very small. A questionnaire may contain a dozen items, each allowing as many as four or five possible categories of response. Item analysis may show that most or all of these items are correlated with the total score, yet it may be possible to use only a few of these items in a single scale. Furthermore, scales ordinarily use only dichotomous responses; trichotomies have been used, but the necessary conditions are rarely met. Hence information is lost in two ways—through elimination of items and consolidation of responses.

If out of a dozen or so initial questions only four or five form a

1. For general orientation in cumulative scales see Samuel A. Stouffer, et al., *Measurement and Prediction*, Princeton University Press, 1950. Chapter 1 by Stouffer gives an overview of the problems; Chapters 2 to 9 by Louis Guttman and Edward A. Suchman deal with Guttman scale theory; Chapters 10 and 11 by Paul F. Lazarsfeld deal with the theory of latent structures.

scale, there is, as Guttman and others have warned, great risk of over-capitalization on chance unless the scale is thoroughly replicated. Even more serious, perhaps, is the possibility that the four or five items hold together merely because they have something highly specific in common—either in phrasing of content or in format—and lack the generality of meaning which the author has been seeking. This is one of the more serious charges directed at conventional Guttman scaling.

These considerations have led to an insistence on the desirability of requiring at least ten or a dozen items to hang together in an initial scale—even if, once scalability is established, a smaller number of items, perhaps only four or five, may be selected for practical eventual use in applying the scale.

But a ten or twelve item cumulative-type scale is easier to talk about than to accomplish. As we increase the number of items, and, as almost always happens, are confronted with pairs or triplets of items with about the same frequency of favorable response, we almost invariably increase the number of non-scale types. With a four-item scale which has a reproducibility of .90, as many as 40 percent of the respondents may be non-scale types. With a 10-item scale which has a reproducibility of .90, it is possible for nearly 90 percent of the respondents to be non-scale types—that is, to have an error on at least one item. Thus we have the somewhat paradoxical situation that increasing the number of items may strengthen our confidence in the scalability of an area under consideration and in the generality of the dimension which the scale is defining, at the same time that it creates more non-scale types and thus introduces more ambiguity in ordering respondents. And the closer some of the original items are to each other in frequency the less likely that the rank order of items will remain invariant from sample to sample.

Consider two items with 50 percent and 55 percent favorable, respectively. Suppose the fourhold table from the items looks as follows:

ITEM 2

	Unfavorable	Favorable	
Favorable	5	45	50
Unfavorable	40	10	50
	45	55	100

ITEM 1

There are ony five apparent errors (in the upper left-hand corner). But, unless we have a very large sample we can by no means be confident that in replications Item 2 will continue to have a larger favorable frequency than will Item 1.[2] Consequently, scale patterns classified as correct in one sample would be classified as incorrect in another.

The obvious remedy, namely, to use a very large initial sample in determining the scale types, is not always feasible and is no panacea. It is quite possible that a particular item will have a special significance for some segment of the population not adequately covered in the initial sample and hence will have a larger or smaller frequency through such a non-chance factor. Hence, reversals are not at all unlikely, even if the items are spaced relatively far apart.

Only by spacing the cutting points of our scale quite widely can we guard against reversals. This limits us to a four or five item scale. It may indeed be that a small number of rank groups is sufficient for the practical task of this particular scale in ordering individuals. If so, is there some way by which we can utilize more of the available information and can build greater precision into the cutting points than is usually feasible if each cutting point depends on the responses to one item alone?

This is the task of the H-technique. Instead of using one item to determine a given cutting point we use two, three, or even more. In effect, what we are doing is to convert the responses to two or more observed items into a response to a "new" item, which we call a *contrived* item.

Suppose that the following three observed items have approximately the same frequencies of "Yes often" responses:

1. Do you have sick headaches?   Yes often; Yes sometimes; No.
2. Do you have trouble sleeping?   Yes often; Yes sometimes; No.
3. Do you have backaches?   Yes often; Yes sometimes; No.

Now, provided further, that each of these items at the cutting point used satisfies the condition of correlating with a provisional scale based on all observed items, we combine the responses to the three into a new contrived item.

There are at least two ways of scoring this new contrived item. If an odd number of items is used, such as three as in the present example, an individual can be arbitrarily scored as positive if he answers "Yes often" to the majority of questions. Or, the number of simultaneous "Yes often" responses to the three items can be recorded and the "best" cutting point of the contrived item be determined, in conjunction with other contrived items, by the conventional Guttman procedures.

Here we have treated "Yes often" as a positive response and others

2 This is easily seen, for example, by testing a null hypothesis as to the differences between the frequencies favorable in the two items. We have $x^2 = (10 - 5 - 1)^2/15 = 1.1$, with 1 degree of freedom.

as negative. We might, however, have treated "Yes sometimes" as a positive response also and could proceed exactly as above, if the three items had about the same positive frequency by this definition and if at this cutting point there was a satisfactory correlation with a provisional scale.

Actually, it might be that the "Yes often" response to one item had about the same frequency as the "Yes often" plus the "Yes sometimes" responses to the other items. Then these three items, in spite of different cutting points, could be converted into a contrived item.

Furthermore, Item 1, cut at "Very often," might be used with Items 2 and 3, and at the same time Item 1, cut at "Very often" plus "Sometimes" might be used over again with two other items. Thus, different response categories of the same item may contribute to more than one contrived item. This tends to maximize the information obtainable from a single set of observed items each of which has multiple response categories.

A Theoretical Illustration of the Advantages of the H-Technique

The perfect Guttman scale is a limiting case of the Lazarsfeld latent distance model which postulates that respondents are ordered into latent classes. The Lazarsfeld model makes it possible to compute the probability that a particular response pattern to individual questions will be given by a member of a given latent class. (In the perfect Guttman case these probabilities are either unity or zero.)

Let us now examine a special case of five latent classes equally populated. Let us assume that we have four observed items, and that a favorable response to a given item has a probability of occurrence from members of a given latent class according to the schedule which follows:

	ITEM			
Latent Class	1	2	3	4
I	.9	.9	.9	.9
II	.1	.9	.9	.9
III	.1	.1	.9	.9
IV	.1	.1	.1	.9
V	.1	.1	.1	.1

For example, the probability that Item 3 will be endorsed by a member of Latent Class III is .9; the probability that it will be endorsed by a member of Latent Class IV is .1.

Consider now a particular response pattern to the four items simultaneously:

Item 1+, Item 2—, Item 3+, and Item 4+

The probability that this response pattern $+ - + +$ would be produced by a member of Latent Class I is $.9 \times .1 \times .9 \times .9 = .0729$. The probability that it would be produced by a member of Latent Class II is $.1 \times .1 \times .9 \times .9 = .0081$.

Or take response pattern $- - + +$. The probability that it would be produced by a member of Latent Class III is $.9 \times .9 \times .9 \times .9 = .6561$, while the probability that it would be produced by a member of Latent Class I is $.1 \times .1 \times .9 \times .9 = .0081$.

There are sixteen response patterns in all and the probability that each will be given by a member of a given latent class is shown in Table 28.1. The following observations may be made from Table 28.1:

1. The theoretical probability that members of each latent class will fall into a unique perfect scale type ($+ + + +$, $- + + +$, $- - + +$, $- - - +$, or $- - - -$) is .6561.

2. The proportion of all cases falling into perfect scale types, irrespective of latent class from which recruited is found, from the right-hand column of Table 28.1, to be $.14762 + .16218 + .16362 + .16218 + .14762 = .78322$. Hence, the proportion of non-scale types is $1 - .78322 = .21678$.

3. In computing Guttman reproducibility two non-scale types ($+ + - -$ and $+ - + -$) must be counted twice. The proportion of cases in these types is $.00362 + .00522 = .00884$. The sum $.21678 + .00884 = .22562$. The Guttman reproducibility coefficient is $1 - .22562/4 = .94360$.

4. For ranking respondents by assigning non-scale types to the nearest perfect scale type, Table 28.1 is helpful. It can be used in a number of ways. A simple way is to allocate the non-scale type to the perfect scale type associated with the latent class which has the greatest probability of producing this non-scale type. For example, in Table 28.1, the type $+ + + -$ would be assigned to Latent Class I. There are five non-scale types whose assignment would be ambiguous by this method ($+ - + +$, $- + - +$, $- - + -$, $+ + - -$, and $+ - + -$), constituting $.03258 + .03402 + .03258 + .00362 + .00522 = .10802$ of all the cases.

Now, let us examine the gains achieved when the H-technique is used. Instead of using one item to determine a particular cutting point, we shall use three items, each with the same theoretical proportion "positive" and each with the same probability of "error" as the single item used initially, namely ten percent. We shall assume the errors uncorrelated. Hence, the new contrived item will have four categories, with probabilities as follows:

3 responses without error .... $.9 \times .9 \times .9 = .729$
2 responses without error, 1 with error ..... $3 \times .9 \times .9 \times .1 = .243$
1 response without error, 2 with error ...... $3 \times .9 \times .1 \times .1 = .027$
0 response without error .... $.1 \times .1 \times .1 = .001$

$$1.000$$

If we decide to call persons who make either two or three positive responses as positive on our contrived item, the latter will have an "error" term of only .028, a considerable improvement over our initial error term of .1. Here lies the fundamental basis of the theoretical advantage of the new contrived item.

How the advantage actually works out is shown by Table 28.2, which is based on four contrived items (each in turn derived from triplets) with an error of .028. The table was computed precisely in the same manner as Table 28.1 except that the error term is .028 throughout instead of .1.

Table 28.1. Probability That Response Patterns Will Be Given By Members of the Classes, and Recruitment

Response Patterns	Latent Class					Sum	Proportion of all cases within a specified response pattern (Sum ÷ 5)
	I	*II*	*III*	*IV*	*V*		
+ + + +	.6561	.0729	.0081	.0009	.0001	.7381	.14762
+ + + −	.0729	.0081	.0009	.0001	.0009	.0829	.01658
+ + − +	.0729	.0081	.0009	.0081	.0009	.0909	.01818
+ − + +	.0729	.0081	.0729	.0081	.0009	.1629	.03258
− + + +	.0729	.6561	.0729	.0081	.0009	.8109	.16218
− + + −	.0081	.0729	.0081	.0009	.0081	.0981	.01962
− + − +	.0081	.0729	.0081	.0729	.0081	.1701	.03402
− − + +	.0081	.0729	.6561	.0729	.0081	.8181	.16362
− − − +	.0009	.0081	.0729	.6561	.0729	.8109	.16218
− − + −	.0009	.0081	.0729	.0081	.0729	.1629	.03258
+ − − +	.0081	.0009	.0081	.0729	.0081	.0981	.01962
− − − −	.0001	.0009	.0081	.0729	.6561	.7381	.14762
− + − −	.0009	.0081	.0009	.0081	.0729	.0909	.01818
+ − − −	.0009	.0001	.0009	.0081	.0729	.0829	.01658
+ + − −	.0081	.0009	.0001	.0009	.0081	.0181	.00362
+ − + −	.0081	.0009	.0081	.0009	.0081	.0261	.00522
Σ	1.0000	1.0000	1.0000	1.0000	1.0000	5.0000	1.00000

Table 28.2. Probability That Response Patterns Will Be Given By Members of the Classes, and Recruitment (Triplets)

Response Patterns	Latent Class					Sum	Proportion of all respondents with specified response patterns (Sum ÷ 5)
	I	II	III	IV	V		
+ + + +	.892617	.025713	.000741	.000021	.000001	.919093	.183819
+ + + −	.025713	.000741	.000021	.000001	.000021	.026497	.005299
+ + − +	.025713	.000741	.000021	.000741	.000021	.027237	.005447
+ − + +	.025713	.000741	.025713	.000741	.000021	.052929	.010586
− + + +	.025713	.892617	.025713	.000741	.000021	.944805	.188961
− + + −	.000741	.025713	.000741	.000021	.000741	.027957	.005591
− + − +	.000741	.025713	.000741	.025713	.000741	.053649	.010730
− − + +	.000741	.025713	.892617	.025713	.000741	.945525	.189105
− − − +	.000021	.000741	.025713	.892617	.025713	.944805	.188961
− − + −	.000021	.000741	.025713	.000741	.025713	.052929	.010586
+ − − +	.000741	.000021	.000741	.025713	.000741	.027957	.005591
− − − −	.000001	.000021	.000741	.025713	.892617	.919093	.183819
− + − −	.000021	.000741	.000021	.000741	.025713	.027237	.005447
+ − − −	.000021	.000001	.000021	.000741	.025713	.026497	.005299
+ + − −	.000741	.000021	.000001	.000021	.000741	.001525	.000305
+ − + −	.000741	.000021	.000741	.000021	.000741	.002265	.000453
Σ	1.000000	1.000000	1.000000	1.000000	1.000000	5.000000	1.000000

We can place side by side the salient comparisons from the two tables:

	Conventional technique using four individual items	H-technique: using four contrived items, each based on triplets of individual items
1. Probability that members of each latent class will fall into a unique perfect scale type	.6561	.8926
2. Proportion of all cases falling into perfect scale types, irrespective of latent class from which recruited	.7832	.9346
3. Guttman coefficient of reproducibility	.9436	.9835
4. Proportion of cases which cannot be ranked unambiguously	.1080	.0327

A further important comparison is possible by looking at test-retest reliabilities. These can be computed, in each case, by assuming that a respondent does not change in the latent class to which he belongs, but that his scale score on the "first" test does not alter the possibilities of being assigned to various scale scores on the "second" test.[3] The theoretical product-moment correlations between test and retest are as follows:

For scale based on four observed items .... r = .8085
For scale using H-technique based on four triplets r = .9565

The theoretical example discussed has been for a special case where each original item is subject to a ten percent error. The situation is generalized somewhat further in Figure 28.1. Assuming, as before, equal errors for all four equally spaced items, we can observe in Figure 28.1 the gains expected from the H-technique over the entire possible range of error. This figure deserves careful examination, since it demonstrates, on the one hand, that the expected gains are very substantial, and, on the other hand, that the H-technique *cannot be expected to salvage a situation where the original item error is much larger than about 20 percent*. In other words, the H-technique is definitely not a procedure for lifting one's self with one's own bootstraps.

Earlier in this exposition it was suggested that an original item can be used more than once. One cutting point can be used to contribute toward a given contrived item; another cutting point of the same original item can be used to contribute toward a second contrived item. The theoretical effect on reproducibility, test-retest reliability etc. appears

---

3. To compute test-retest reliability, response patterns first are grouped and weighted as follows:

				x: weight
+ + + +,	+ + + −,	+ + − +		+ 2
− + + +,	+ − + +,	− + + −		+ 1
− − + +,	− + − +,	+ + − −,	+ − + −	0
− − − +,	− − + −,	+ − − +		− 1
− − − −,	− + − −,	+ − − −		− 2

Within each latent class, the sums of squares and sums of cross products are computed. For example, on Latent Class I, the frequency for X = 2 as seen from Table 28.1 is .8019. The contribution toward ΣX^2 is $2^2 \times .8019$. Since *within* a latent class, the probabilities of a given response on test X_1 and retest X_2 are independent, the contribution toward $\Sigma X_1 X_2$ made when X_1 and X_2 in Latent Class I both equal 2, is $(2 \times .8019)^2$. For $X_1 = 1$ and $X_2 = 2$ we have $(1 \times .1539)$ $(2 \times .8019)$, etc. The process is repeated for each of the five latent classes and the totals for all classes summed for utilization in the conventional formula for Pearson product-moment correlation. (An alternative and more laborious procedure for computing theoretical test-retest reliability, which gives response patterns a separate value based on assigning a separate weight to the frequency within each latent class, yields only slightly different results from the method used above.)

to be almost the same as when no original item is used more than once. See Table 28.3 for the special case of equally spaced items with equal errors. If two original items are used with different cutting points in the same two contrived items, the reproducibility is substantially unaffected, but the test-retest reliability drops somewhat, depending on which contrived items are involved.

If the original number of items is few, one way of cutting corners might be to use only two original items to form the extreme contrived

Table 28.3. Effect on Reproducibility and Test-Retest Reliability of Using the Same Original Item in Two Different Contrived Items

	Coefficient of Reproducibility		Test-Retest Reliability	
	Average item error .1	Average item error .2	Average item error .1	Average item error .2
Four original items only	.9436	.8949	.8085	.5443
Four contrived items, each based on three unduplicated original items	.9835	.9415	.9565	.7990
Four contrived items, each based on three original items, but with an original item used twice in the pair of contrived items specified:				
Items I and II	.9830	.9409	.9511	.7777
I and III	.9836	.9424	.9470	.7845
I and IV	.9845	.9456	.9550	.7954
II and III	.9846	.9457	.9526	.7859
II and IV	.9852	.9474	.9547	.7926
III and IV	.9861	.9501	.9566	.7982
Four contrived items, each based on three original items but with two original items used twice in the pair of contrived items specified:				
Items I and II	.9822	.9397	.9353	.7365
I and III	.9831	.9420	.9422	.7589
I and IV	.9852	.9491	.9529	.7880
II and III	.9854	.9492	.9405	.7565
II and IV	.9864	.9515	.9502	.7845
III and IV	.9880	.9572	.9545	.7925

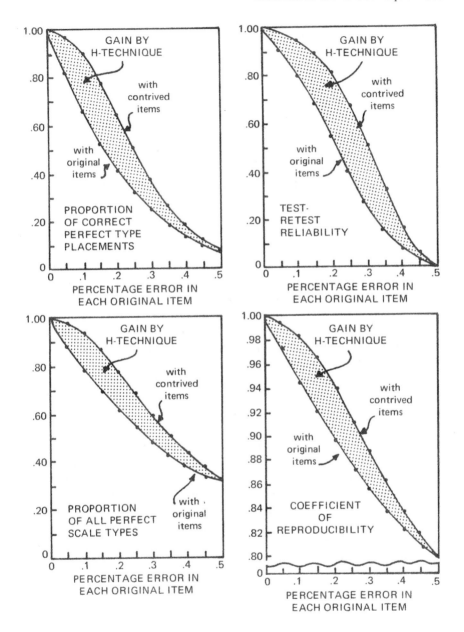

Figure 28.1. Gains by H-Technique, for Varying Amounts of Error in the Original Items

items I and IV, while using three original items to form the middle contrived items II and III. Contrived Item I would be scored as positive if the respondent is positive on *both* of the original items involved. Contrived Item II would be scored as positive if two out of three of the original items involved were positive; the same with contrived Item III. Contrived Item IV would be scored as positive if at least *one* of the two items involved were positive. The effect of such an operation is to produce a slightly higher reproducibility than would be obtained if three items were used in each case, but at a small and noticeable cost in lowered test-retest reliability. In the example given above, when the average error in original items is ten percent, the use of doublets instead of triplets at the two extremes changes reproducibility from .9835 to .9888 but cuts reliability from .9565 to .9298. When the average error is 20 percent, the use of doublets changes reproducibility from .9415 to .9597 but cuts reliability from .7990 to .7630. This kind of result ought to be noted carefully by students of scaling theory, because, incidentally, it illustrates one of the dangers in relying on reproducibility alone as a measure of success.

In the examples given, several restrictions were introduced to facilitate computation and exposition. No general mathematical treatment is at present available. Further preliminary theoretical explorations have shown that the assumption of equal error of all items in a triplet is not necessary to obtain a substantial reduction in over-all scale error. Moreover, there seems to be no special virtue in the use of three responses to form the new contrived item as in the example cited; more responses should do even better.

One suggested line of inquiry should be explored in greater detail than has been possible as yet. This is to treat a set of individual items making up a given contrived item as contributing to a Lazarsfeld latent dichotomy. (At least four original items probably would be needed to form such a contrived item.) The final cumulative scale would then consist of a series of ordered cutting points, each of which has first been established by latent dichotomy analysis. Computation might be too heavy to justify this procedure in practice, but its theoretical implications are of much interest.

An Empirical Example, with Computing Procedures

In connection with the Harvard-Air Force research project, 633 officers at six air bases filled out an eleven-item questionnaire designed to order respondents along the single dimension "sensitivity to sanctions."

There are four response categories for each question, which are arbitrarily scored 1, 2, 3, 4, respectively, a score of 1 going to the response intended to indicate the most independence from pressure. As in a conventional item analysis procedure or in the Guttman Cornell technique, a provisional total score was computed for each person based on his answers to all eleven questions. This total score, together with the individual item responses, was punched on IBM cards.

The remaining machine job, done on an IBM Electronic Statistical Machine, Type 101, required less than four hours, including time spent in wiring a special board. The job also could have been done easily on the more common type of counter sorter, but would have taken more time.

The steps were as follows:

1. Tabulating the correlation tables of each of the eleven items with the provisional total score. These are shown in Table 28.4.

2. Selecting the cutting points for each item which correlated with the total score high enough to form a four-fold table in which neither "error" cell had a frequency higher than the smaller of the two frequencies on the principal diagonal. For example, if Item 1 is dichotomized with 3 and 4 combined as positive we can form from the data in Table 28.4 the following acceptable four-fold table:

Table 28.4. Responses to Individual Items as Related to Provisional Total (Air Force Data)

Provisional Total Score*	Item 1				Item 2				Item 3			
	1	2	3	4	1	2	3	4	1	2	3	4
43–44	—	—	—	46	—	—	3	43	—	—	2	44
39–42	—	—	18	23	—	—	9	32	—	3	13	25
35–38	1	4	38	29	2	3	16	51	1	4	36	31
33–34	—	3	59	7	1	2	36	30	—	5	58	6
30–32	—	25	42	5	2	6	37	27	—	19	47	6
28–29	—	43	25	1	1	5	53	10	1	29	33	6
26–27	4	40	21	2	—	11	42	14	5	37	19	6
24–25	3	48	6	1	2	14	37	5	2	38	17	1
21–23	11	59	4	—	—	22	49	3	3	62	8	1
11–20	33	30	2	—	7	28	28	2	31	30	4	—
	52	252	215	114	15	91	310	217	43	227	237	126

Table 28.4.—(Continued)

Provisional Total Score*	Item 4				Item 5				Item 6			
	1	2	3	4	1	2	3	4	1	2	3	4
43–44	—	—	2	44	—	—	1	45	—	—	2	44
39–42	1	1	18	21	—	—	7	34	—	2	21	18
35–38	—	9	46	17	—	—	33	39	—	22	38	12
33–34	—	16	49	4	—	3	58	8	1	16	49	3
30–32	2	50	20	—	—	4	62	6	2	49	21	—
28–29	5	61	3	—	1	14	54	—	4	60	5	—
26–27	13	51	3	—	2	26	38	1	13	51	2	1
24–25	23	32	3	—	—	38	20	—	19	35	4	—
21–23	42	32	—	—	1	65	8	—	29	45	—	—
11–20	58	7	—	—	21	40	3	1	58	7	—	—
	144	259	144	86	25	190	284	134	126	287	142	78

Provisional Total Score*	Item 7				Item 8				Item 9			
	1	2	3	4	1	2	3	4	1	2	3	4
43–44	—	—	—	46	—	—	—	46	—	—	—	46
39–42	—	—	13	28	—	2	10	29	—	—	9	32
35–38	—	11	31	30	2	12	29	29	1	1	40	30
33–34	1	5	54	9	6	6	49	8	1	4	55	9
30–32	—	16	52	4	1	21	45	5	2	10	54	6
28–29	—	22	47	—	4	22	39	4	—	29	37	3
26–27	2	28	35	2	9	31	24	3	2	40	22	3
24–25	1	45	12	—	6	35	17	—	1	40	16	1
21–23	3	66	5	—	7	58	9	—	8	64	2	—
11–20	25	38	2	—	30	32	2	1	38	21	5	1
	32	231	251	119	65	219	224	125	53	209	240	131

Provisional Total Score*	Item 10				Item 11			
	1	2	3	4	1	2	3	4
43–44	—	—	—	46	—	—	1	45
39–42	—	—	2	39	—	—	7	34
35–38	—	1	22	49	1	1	39	31
33–34	1	—	51	17	1	2	55	11
30–32	—	2	48	22	1	12	52	7
28–29	—	5	54	10	1	31	34	3
26–27	1	10	44	12	5	41	19	2
24–25	—	19	38	1	2	45	11	—
21–23	1	46	24	3	12	53	8	1
11–20	10	41	13	1	33	32	—	—
	13	124	296	200	56	217	226	134

* Grouped to form class intervals approximating deciles. N = 633.

	1, 2	3, 4
	+ 33	267
Provisional Score		
	− 271	62

From the same item, by dichotomizing with 4 as a positive answer, we can also form an acceptable table:

	1, 2, 3	4
	+ 18	69
Provisional Score		
	− 501	45

Also, from the same item, by dichotomizing with 2, 3, 4 as positive answers another acceptable table is obtained (although it just barely meets the minimum requirements):

	1	2, 3, 4
	+ 19	549
Provisional Score		
	− 33	32

In each of the examples, it will be noted, the sum of the errors is well below 20 percent of the total frequency. If the errors go much above 20 percent the item ordinarily should be discarded, although an occasional exception may be tolerated.

3. Ordering all acceptable cutting points from the largest positive to the smallest, as shown in Table 28.5.

4. Selecting sets of triplets to constitute four new contrived items. The original items actually used are indicated in Table 28.5 by Roman numerals which designate the contrived item they served to form. Thus, Contrived Item I is formed from Original Items 4 and 6 (with categories 2, 3, 4 treated as positive), and Original Item 10 (with categories 3, 4 treated as positive). It will be noted in Table 28.5 that several cuts which met the test of acceptable correlation were not used (designated by **). For example, Items 1, 9, and 11 (with categories 2, 3, and 4 treated as positive) were not used, for two reasons: (1) Cutting points so close to the end of the scale were not desired; and (2) it was preferred to use two of the same items (with different cuts) in forming Contrived Item II. The main objective is to select acceptable items with approximately the same frequency for a given triplet and to space these sets of triplets as evenly as possible over the range.

5. Scoring each individual on each contrived item by calling him positive on a given item if he were positive in either two or three of the component individual responses.

Table 28.5. Items and Cutting Points Used in Construction of Contrived Items (Air Force Data)

Item	Positive Response Categories	Frequency Positive	Contrived Item in which Original Item Is Used
10	2, 3, 4	620	—*
2	2, 3, 4	618	—*
5	2, 3, 4	608	—*
7	2, 3, 4	601	—*
3	2, 3, 4	590	—*
1	2, 3, 4	581	—**
9	2, 3, 4	580	—**
11	2, 3, 4	577	—**
8	2, 3, 4	568	—*
2	3, 4	527	—*
6	2, 3, 4	507	I
10	3, 4	496	I
4	2, 3, 4	489	I
5	3, 4	418	—**
9	3, 4	371	II
7	3, 4	370	—**
3	3, 4	363	II
11	3, 4	360	II
8	3, 4	349	—**
1	3, 4	329	—**
4	3, 4	230	III
6	3, 4	220	III
2	4	217	III
10	4	200	—**
5	4	134	IV
11	4	134	—**
9	4	131	—**
3	4	126	—**
8	4	125	IV
7	4	119	—**
1	4	114	IV
4	4	86	—**
6	4	78	—**

*Item at this cutting point does not correlate satisfactorily with provisional total score.
**Item at this cutting point satisfies the criterion of adequate correlation with the provisional total, but was not used in the scale.

6. Tabulating the 16 response patterns. Actually operations 5 and 6 were performed together in five minutes by a single pass of the cards through the 101 machine, after a special board had been wired. This would require considerably more time on the older machine, but actually not much more time than is required for a conventional Guttman scale analysis. The 16 response types are exhibited in Table 28.6.

From Table 28.6 we see that only 45 of the 633 respondents fall into non-scale types. This is an almost unbelievably satisfactory result, as compared with what experience with conventional scales has led us to expect. Because, to some extent, artificial restrictions on error are introduced by using more than one cutting point for some of the same items, there is need for caution in reporting a coefficient of reproducibility (just as in the case when trichotomies are used in a conventional Guttman analysis). Even if the computed reproducibility coefficient of .982 is a little inflated, a conservative estimate would place it well above .95, since there are only seven percent of non-scale types altogether.

Because of the high precision with which each cutting point on the cumulative scale is marked off by the H-technique, the rank order of the contrived items, if they are relatively widely spaced, should tend to remain invariant from sample to sample.

A good test of this is provided by sample data made available by the American Telephone and Telegraph Company which has asked a set of morale questions to respondents in 25 different operating units, varying in size from 116 to 963 persons, with an average of 388. A six-step scale based on five contrived items held up in all 25 units with only four instances of reversals of rank order of adjacent contrived items out of 100 possibilities, in spite of the fact that we found it impossible to make up conventional scales based on single items without frequent reversals. The average proportion of non-scale types was only ten percent (highest among the 25 operating units was fifteen, lowest was eight) and the average coefficient of reproducibility, perhaps slightly overestimated, was .98 with no individual coefficient below .97. A five-step four-item scale also was constructed by the H-technique, from the same collection of original questions, with even better results. The average proportion of non-scale types was five percent (highest among the twenty-five operating units was seven, lowest was three). The average reproducibility was above .98. The rank order of the contrived items was invariant in all twenty-five units.

The H-technique has been used on other sets of data with equal success, where the original item error, as indicated in the correlation tables with the provisional total score, does not exceed about 20 percent. Further experience with it may reveal shortcomings not now seen or may suggest better procedures for handling particular steps.

Table 28.6. Frequency Assigned to Each Response Pattern by the H-Technique (Air Force Data)

Response Patterns	Frequency
++++	92
+++−*	1
++−+*	1
+−++*	22
−+++	103
−++−*	—
−+−+*	8
−−++	157
−−−+	106
−−+−*	12
+−−+*	—
−−−−	130
−+−−*	—
+−−−*	—
++−−*	—
+−+−*	1
	633

* Non-scale types.

Chapter 29

A SCALE FOR DEVELOPMENTAL PROCESSES

ROBERT K. LEIK AND MERLYN MATTHEWS
University of Washington

One of the major orientations to the study of social phenomena is "developmental," which is to say that the focus is on a sequence of stages through which an individual, an interpersonal relationship, a group, an organization, or perhaps a social institution passes.[1] The concept of a developmental sequence has usually been an intuitive one, however, with little explicit concern for the form that data must take if they are to support a developmental model. The purposes of this paper will be to examine briefly the implications of a developmental process, to present a

Reprinted by permission from the American Sociological Association and the author from the *American Sociological Review*, Vol. 33, No. 1, 1968, pp. 62–75.

Revised version of a paper read at the Pacific Sociological Association meetings, Long Beach, California, March, 1967.

1. Developmental conceptualizations are common to such diverse areas as *personality theories*, see Calvin S. Hall and Gardner Lindzey, *Theories of Personality*, New York: John Wiley and Sons, Inc., 1957, particularly regarding the question of early developmental experiences, p. 23; *small group problem-solving*, see Robert F. Bales. "The Equilibrium Problem in Small Groups," in A. Paul Hare, *et al.* (eds.), *Small Groups*, New York: Alfred A. Knopf, 1965, pp. 467–470; *the family*, see Reuben Hill and Roy H. Rodgers. "The Developmental Approach," chap. 5 in Harold T. Christensen (ed.), *Handbook of Marriage and the Family*, Chicago: Rand McNally, 1964; and *general social change*, many developmental versions being apparent in Charles P. Loomis and Zona K. Loomis, *Modern Social Theories*, Princeton: Van Nostrand, 1965. It is not possible to document all developmental orientations due to the pervasiveness of this type of conceptualization in social science.

scaling procedure appropriate for assessing certain aspects of developmental processes, and finally, to illustrate the procedure with data pertaining to the family.

A Developmental Process

Inherent in some uses of a developmental conceptualization is an assumption that a particular sequence of stages is a functional necessity.[2] In its most stringent interpretation, this usage would imply that stage k cannot occur unless stage j has occurred, assuming that j precedes k in the sequence. A less stringent functional interpretation is that stage j *should* precede stage k if certain favorable outcomes are to be realized, although it is possible but not very probable for k to occur without j having been present.[3] Regardless of the question of a functional ordering, it may be that a sequence is hypothesized as the norm, implying that it is modal in the statistical sense.[4] As with the less stringent functionalist form, this

2. In the terminology of causal relations, this position is equivalent to the assertion that each stage is a *necessary* condition of the next stage. Usually other conditions are part of the sufficiency requirement for such a sequence. Thus orthodox Freudian theory would assert that each stage of development is a necessary step toward a complete adult personality, but that simply passing through a given stage does not guarantee attainment of the next. (Hall and Lindzey, *op. cit.*, chap. 2.) Much of the developmental approach to the family is of this character, stating that certain "developmental tasks" are prerequisites of development. It is frequently not clear, however, whether the intention is that these developmental tasks are functional necessities or serve simply to make more likely a desirable family pattern. Cf. Hill and Rodgers, *op. cit.*

3. A clear example of this type of conceptualization appears in Sjoberg's discussion of industrialization of preindustrial societies. "Apart from any personal political preference, it appears necessary to recognize that if a preindustrial civilized society is to industrialize *rapidly* it can do so only in the context of a one-party state exercising a high degree of totalitarian or authoritarian rule. (Sjoberg, 1963). Only charismatic totalitarian leaders can give direction to the uneducated mass populace, and their ideological fervor is apparently required to break the 'cake of custom' that survives from the preindustrial past." Gideon Sjoberg, "Rural-Urban Dimension in Preindustrial, Transitional, Industrial Societies," chap. 4 in Robert E. L. Faris (ed.), *Handbook of Modern Sociology*, Chicago: Rand McNally, 1964, p. 149. Udy's scale of administrative rationality in nonindustrial organizations leads him to suggest that "if specialization should be the first rational characteristic to develop in an organization [not the first item of the scale], the scale implies that such an organization if it is to be stable must immediately develop a centralized management and compensatory rewards." S. H. Udy, "Administrative Rationality, Social Setting and Organizational Development," *American Sociological Review*, 68 (1962), pp. 299–308.

4. This type of formulation is evident in Zetterberg's presentation of axiomatic theory. Hans L. Zetterberg, *On Theory and Verification in Sociology*, Stockholm: Almqvist and Wiksell, 1954, chap. 2. Similarly, Loomis and Loomis highlight such a conceptualization by quoting portions of Robin Williams' presidential address to the American Sociological Association as follows: "Let us begin by noting an empirical tendency for larger size of . . . [formal organizations] to lead to greater specialization of function . . . increased differentiation of interests, of status-ranking, of rewards, and of control. . . . The high degree of interdependence . . . tends in turn to lead

usage would imply that stage k would more frequently follow stage j than the reverse ordering.

Common to the above conceptualizations is the conclusion that the order in which a particular pair of stages occurs is a consequence of the developmental process, and that, therefore, it will be unlikely that those stages will appear in reverse order. In its pure form, this conclusion is similar to one aspect of a perfect Guttman scale. If the items in a Guttman scale form a perfect order, then no case will occur in which item k is endorsed unless item j has been endorsed, assuming that j precedes k in the scale. This similarity between a Guttman scale and the concept of a developmental sequence is not sufficient to warrant the use of Guttman procedures for assessing the presence of a developmental process, but it will provide a point of departure for establishing appropriate procedures. Two questions need to be considered: (1) Is the notion of a stage in a developmental process the same as a single item in a Guttman scale? (2) Does an advanced stage in a developmental sequence contain all the characteristics of earlier stages, only adding new characteristics to the pattern? Let us turn briefly to an example from the area of the family as a guide to answering these questions.

Recently the authors attempted to apply Guttman scale criteria to data on the independence of wives from their husbands. Conceiving of independence as being acquired in developmental stages, and following the usual scalogram approach, we assumed that as a wife becomes independent in new spheres, such as the economic, she retains whatever earlier independent behavior she had exhibited, such as independence in family decision-making. The data, unfortunately, did not share our enthusiasm for the hypothesized pattern.

Careful examination of individual response patterns suggested that one aspect of the Guttman approach was inappropriate: many of the women evidently ceased exhibiting earlier types of independence once they had attained more "advanced" independence. Reflection on a variety of similar developmental processes led us to the conclusion that a new scaling procedure was needed—a procedure which would allow individuals to drop traits on one side of the scale pattern as they acquired traits on the other side.

Note that a second aspect of a developmental sequence has been indicated, namely that, as development proceeds, earlier traits are dropped and new traits are acquired. Developmental stages may thus be

---

to a recognition [for need] of preserving the existing order of relationships, in whole or in part [in which case] differentiation will lead to increased *formality* in communication . . . [as a] means of controlling tension. . . ." Loomis and Loomis, *op. cit.,* p. 587.

thought of as points in time during which certain sets of traits are present. Furthermore, the acquisition of traits *and the dropping of traits* occurs in the same ordered sequence. There is no implication that acquiring and dropping traits occurs at the same rate. Consequently, some stages may be characterized by possession of only one, or a very few traits. Other stages may contain many traits. It is likely, for example, that husband-wife interaction will show rapid acquisition of traits in early stages of marriage, with a much more gradual dropping of "young-married" traits as the marriage matures. Similarly, some individuals may quickly drop immediately preceding traits on acquisition of new ones, whereas others may retain a wide set of traits as they progress over the scale.

A graphic comparison of the model of a developmental process being described here with a Guttman scale should prove useful. To use more common scaling terminology, let us refer to endorsement of items rather than possession of traits.

Table 29.1a shows a perfect Guttman pattern if only the endorsement of each item is listed.[5] All items must be endorsed from the left-most item of the scale through the right-most item which is endorsed in any given set of responses, or scale type. Consequently, a triangular area of endorsements occurs, with a complementary triangular area of items not endorsed.

Table 29.1b shows all possible perfect response types in the type of scale being developed here. Note that the only criterion which the graphic pattern need satisfy is that all endorsed items in any response type are in adjacent columns. Conversely, any nonendorsed item should not have endorsed items both preceding and following it. Endorsements need not extend back to the left-hand item of the scale. At first glance such a simple requirement appears too easily satisfied by chance. For three items, for example, seven of the eight possible response types are acceptable. For four items, 11 out of 16 are acceptable. As the number of items increases, however, the number of acceptable response types becomes a much smaller proportion of the total. For ten items, only 56 out of a possible 1024 types will fit the scale requirements.

As will be shown, appropriate criteria can be developed to assess the extent to which data are (1) not distributed by chance, and (2) accurately predicted by the perfect scale pattern. It should be noted that, contrary to the Guttman type of scale, the *number* of items endorsed, i.e., the num-

---

5. The so-called parallelogram pattern in a Guttman scale is produced by listing each answer category, and hence each question, twice for dichotomous questions. Although the present scaling procedure could be developed for polytomies, this presentation will be concerned only with dichotomies.

Table 29.1. Two Types of Perfect Scales
for Five Items

Type	a. Guttman Scale Endorsements				
I					
II	X				
III	X	X			
IV	X	X	X		
V	X	X	X	X	
VI	X	X	X	X	X

Type	b. Developmental Scale Endorsements				
I					
II	X				
III	X	X			
IV	X	X	X		
V	X	X	X	X	
VI	X	X	X	X	X
VII		X			
VIII		X	X		
IX		X	X	X	
X		X	X	X	X
XI			X		
XII			X	X	
XIII			X	X	X
XIV				X	
XV				X	X
XVI					X

NOTE: The numerals assigned to scale types in the developmental scale do not necessarily imply order. The ordering of types must depend upon whether initial point, terminal point or center of the endorsement pattern is salient to the purposes of the analysis.

ber of traits in a given stage, will not indicate *which* items have been endorsed, even in a perfect scale.

Ordering of Scale Items

It is possible to place items in an appropriate order by use of the frequency of joint endorsement of each pair of items. Assume that no items (traits) are skipped in the developmental process and consider only perfect scale response types. The joint frequency of endorsement of items j and j', denoted $f_{jj'}$, will depend upon the frequency of each response type and whether that type includes both j and j'. Each response type may be specified by a double subscript indicating the first item endorsed and the number of items endorsed.[6] Designate the first endorsed item in the scale type according to its placement in relation to the item j, the first in the pair for which a joint frequency is to be defined. Thus $k=j$ would mean that the first item endorsed is the jth item, $k=j-1$ would mean that the first item endorsed immediately precedes the jth item, etc. Let i denote the number of items endorsed in a given pure scale response type.

Any unique response type, i(k), occurs with frequency $f_{i(k)}$. Then, for m items, if $j'=j+1$, i.e., j and j' are adjacent items,

$$f_{j,j+1}= \quad f_{2(j)}+f_{3(j)}+\cdots+f_{m-j+1(j)}+f_{3(j-1)}+f_{4(j-1)}+$$
$$\cdots+f_{m-j+2(j-1)}+\cdots+f_{j+1(1)}+f_{j+2(1)}+\cdots+f_{m(1)},$$

$$= \sum_{k=1}^{j} \sum_{i=j-k+2}^{m-k+1} f_{j(k)}. \qquad [1]$$

Similarly, for two items separated by a third, i.e., for $j'=j+2$,

$$f_{j,j+2}= \quad f_{3(j)}+\cdots+f_{m-j+1(j)}+f_{4(j-1)}+\cdots$$
$$+f_{m-j+2(j-1)}+\cdots+f_{j+2(1)}+\cdots+f_{m(1)},$$

$$= \sum_{k=1}^{j} \sum_{i=j-k+3}^{m-k+1} f_{i(k)}. \qquad [2]$$

Note that $f_{j, j+1}$ includes all entries in $f_{j, j+2}$ plus the first entry in each row of equation [1]. Since, in general, these frequencies are nonzero, we may employ similar procedures to assert that $f_{jj'}$ is maximal for adjacent items and monotonic decreasing as items j and j' are farther apart in the scale.[7] Furthermore, the marginal frequencies of endorsement, f_j, must be greater than or equal to any joint frequency involving that item. As a consequence, a matrix showing joint frequencies in the off-diagonal cells, and marginal frequencies in the main diagonal, will contain entries which decrease monotonically along rows and columns as one departs from the main diagonal toward the borders of the matrix, if rows and columns are correctly ordered.

6. As noted earlier, the number of endorsements does not determine *which* endorsements, in a perfect developmental scale, even though it will for a perfect Guttman scale. Two subscripts are therefore needed to indicate origin and width of the band of endorsed items. This fact poses questions about respondent ordering which will be discussed later.

7. By contrast, in a perfect Guttman scale, the probability that item j' is endorsed, given that item j is endorsed, equals unity if item j is more stringent than item j'; hence joint frequencies equal marginal frequencies of the more stringent item.

Because the joint frequency matrix is related to the intercorrelation matrix, which is common to most scaling and factor analytic procedures, a few words about the intercorrelation matrix are in order. Following the previous approach, it may be demonstrated that, if all response types are perfect scale types, the matrix of point correlations has unit entries in the major diagonal and is monotonic decreasing away from that diagonal in both rows and columns. This property obtains for a Guttman scale as well. In what way are the matrices different? The developmental matrix differs from the Guttman model in that it includes negative correlations as well as positive ones, whereas the latter is restricted to positive correlations. It would appear, then, that this is a more general model of which the Guttman procedure is a special case. That is evident, also, in the fact that any Guttman scale completely satisfies the developmental scale criteria, but the reverse is not true.

Returning to the problem of item ordering, an initial trial ordering may be obtained from either the joint frequencies or the correlations by placing the smallest frequency, or lowest (largest negative) correlation, in the upper right cell of an item-by-item matrix. This determines the first and last items of the scale. Table 29.2 shows hypothetical responses to five items, and indicates this first step in ordering items. By arranging the first row (hence the first column) so that it is monotonically nondecreasing[8] as cells depart from the upper right corner, the entire trial ordering of items is obtained, again as illustrated in Table 29.2. If the first row does not determine the ordering of a specific pair of items, such as d and e in the illustration, then their joint frequencies with remaining items can be used to determine their initial placement.

Subsequent potentially better orderings may be obtained by summing over rows, for each pair of adjacent columns, the difference between entries in a given row *if* those entries are inverted with respect to the hypothetical monotonic form of the matrix. Such a sum, if large, implies that the positions of those items should be reversed in the scale. Reversal of a pair of items necessarily implies that both rows and columns representing those items must be interchanged. Few reversals should be needed if the items scale well, since the initial trial should be close to the correct order. In any event, final ordering is a question answerable only in terms of scale error, which does not bear a one-to-one relationship to the joint frequency matrix or the intercorrelation matrix.[9]

8. Note that tied joint frequencies are acceptable. To be technically correct, the joint frequency criterion should be stated as *nondecreasing*, rather than *increasing*.

9. Without a computerized routine, establishing order among a large number of items may be extremely tedious, especially if considerable scale error exists even under optimal ordering. A program for ordering of items is being developed but is not yet operational.

Table 29.2. Illustrative Ordering of Items: Hypothetical Data

Assume the following response patterns:

Item a	b	c	d	e	Frequency
X		X	X	X	12
X		X			14
X			X		8
X	X		X		5
	X			X	11

Original Joint Frequency Matrix

	a	b	c	d	e	
a	39	5	26	25	12	
b	5	16	0	5	11	Smallest entry to
c	26	0	26	12	12	be placed in upper
d	25	5	12	25	12	right-hand cell, as
e	12	11	12	12	23	below.

Initial Ordering of Joint Frequency Matrix

	c	a	d	e	b
c	26	26	12	12	0
a	26	39	25	12	5
d	12	25	25	12	5
e	12	12	12	23	11
b	0	5	5	11	16

NOTE: No pairs of frequencies are contrary to monotonic requirements; hence, trial ordering is determined to be c, a, d, e, b. This procedure does not determine which end of the scale represents the earliest developmental stage.

Computing Errors

With standard Guttman scales, it is not clear whether an endorsement or an omission should be considered an error under certain circumstances, making the placement of an error pattern dependent upon convention. For the scale presented here, the respondent ideally endorses all items (or possesses all traits) between the two most extreme items endorsed. Endorsement is therefore not considered erroneous; only lack of endorsement of an item within the set bounded by extreme endorsements can be considered in error. To compute the number of errors, therefore, requires only counting the number of items, within the range of endorsement, which have not been endorsed. The sum of such omissions across all subjects is the number of errors in the total scale.

Table 29.3. Illustrative Scale: Hypothetical Data

c	a	Item d	e	b	f	Errors per Pattern	Total Errors
X	X				14	0	0
	X	X			8	0	0
X	X	X	X		12	0	0
	X	X		X	5	1	5
			X	X	11	0	0

Mean errors per respondent $= \dfrac{5}{50} = .100$

Again referring to the hypothetical data of Table 29.2, re-record the answer patterns according to the established ordering; *c, a, d, e, b*. This produces the scale shown in Table 29.3. Only one answer pattern contains blank entries between the first and last X's for that pattern. Total scale error is easily computed to be five errors in this example. Note that the fact that the third pattern in Table 29.3 extends farther left than does the second pattern does not result in attributing pattern error. All that need be said about such a fact is that the second pattern represents a narrower range of endorsement, which does not extend as far in either direction as the third pattern. Both are entirely consistent with the scale criteria.

Ordering of Response Types

As may have been evident from the preceding remark and the double subscripting used in discussing joint frequencies, it is not possible to order pure response types in the unambiguous manner provided by scalogram analysis. This fact is a consequence of variation of both width and origin of the band of endorsed items for any given type. Consequently, although a single ordering of items is possible and can be used to check hypotheses about developmental sequences, a single ordering of response types will require a decision that some aspect of the scale is more important for the purposes at hand than are other aspects. Thus one might order by first item endorsed, i.e., the most regressive item, then by increasing number of endorsements within each set of types beginning with the same item, and thus by increasing extent of development beyond the most regressive trait. This procedure produces the ordering shown in Table 29.1, given five items with the left-hand item as first in the developmental sequence.

An alternative would be to order initially by last (most progressive) item endorsed, then by increasing number of endorsements within each set of types ending with the same item, and thus by increasing retention of regressive traits. This procedure also produces the ordering shown in Table 29.1 if the left-hand item is the most progressive item. These two procedures differ in their emphasis on "not yet rid of trait j" versus "already progressed to trait j." Clearly these emphases are differentially useful for varying analytical purposes.

Yet another procedure would be to order by median trait endorsed, with ordering of types showing the same median to be determined by, e.g., most regressive trait or most progressive trait endorsed. No doubt other bases of ordering response types could be suggested. The point, however, is that analytic intent will determine ordering criteria for response types. Item order will not be subject to the same variation by purpose of analysis. Perhaps this suggests that the scaling procedure will find greater utility in examining processes than in determining specific ordering of persons within such processes.

Relationship to Other Scaling Procedures

Two sources of scaling methods appear particularly pertinent for assessing the newness of a proposed procedure. These are the works of Coombs[10] and Torgerson.[11] The latter author mentions the possibility of a scale of this type, but dismisses it with the statement that "in what might be called the Guttman spirit, this model would not seem very worthwhile."[12] Hopefully, this paper will demonstrate that, to the contrary, such a scale would indeed be worthwhile.

To use Coombs' terminology, the type of scale proposed here is a combination of his categories QIIa and QIIb. "In these data, then, the relation may be either an order or a proximity relation on a pair of points from distinct sets. Such data could be obtained if an individual in responding to a statement of opinion were permitted to say (1) he endorsed it, (2) he did not endorse it, it is too radical, (3) he did not endorse it, it is too conservative. The first response is an instance of a proximity relation, the latter two are instances of order relations."[13] Having indicated such a possibility, however, Coombs makes no further use of this combination of relations. Consequently he recognizes but does not treat the problem we pose.

10. Clyde H. Coombs, *A Theory of Data*, New York: John Wiley, 1964. This work incorporates extensive earlier work by Coombs on matters of measurement and scaling.

11. Warren S. Torgerson, *Theory and Methods of Scaling*, New York: John Wiley, 1958.

12. *Ibid.*, p. 315.

13. Coombs, *op. cit.*, p. 24.

The developmental procedure appears similar to Coombs' parallelo-gram analysis, but it differs in that his procedure requires what he terms "pick k/n" data. (These are preferential, or QI data in his taxonomy.) Our procedure, on the other hand, might be labeled "pick any num-ber/n." Returning to Coombs' QII data, we find that the works of Gutt-man, Lazarsfeld, Torgerson and Thurstone are examples. These works are variations on a theme which gives fixed properties to the scale points and treats individually variability of response to an item in terms of a (normal) curve of response probability, given the distance between the respondent's actual scale location and the scale point in question. All respondents at a given location on the scale are identical within response error. Our scale treats respondents as variable on scale location and response width (number of endorsements); hence the probability of response to a given item is not a single function for all respondents. A consequence of this fact is that precise trace lines for items are not specifiable in general form, but only in terms of particular response types. Therefore equations for the underlying scale model cannot be stated without specific consideration of the probability, $p_{i(k)}$, of each response type. It would seem that any assumptions about this probability distribu-tion would be population-specific, making general equations unobtain-able.

Need the Scale be "Developmental"?

The above references to other scaling literature should indicate that a developmental (or time-based) conceptualization is not essential to the formal properties of the scale being presented. It would appear, in fact, that this procedure is an extension of the general scaling literature with-out reference to a time variable. We will not pursue nondevelopmental applications, however, for two reasons. First, such applications constitute a whole separate topic of exploration. Second, we feel that the intuitively most compelling use of such a scale is with time-based, and thus develop-mental, data. Perhaps subsequent usage will extend the sphere of appli-cation into more traditional scaling realms.

Significance and Utility

Simply counting the number of errors is inadequate for assessing the utility of the scale. Two questions should be raised: Could the observed pattern reasonably have occurred by chance? Is the observed scale error small enough to permit fairly accurate measurement by the scale? As is well known from Guttman scaling procedures, setting observed errors in ratio to the number of responses made, and subtracting from unity, will produce an index which can be spuriously high due to marginal response probabilities. Departure from minimal marginal reproducibility has been used to assess the extent of patterning beyond marginally induced reproducibility. A similar procedure will be employed for the present type of scale, except that random or chance error will be computed according to observed distributions, and then actual error compared with random error.

Two aspects of observed data will be treated as fixed marginal conditions. First, the proportion of all respondents who endorse any given item will be important. Let this proportion be designated $p_j = f_j/n$. Second, consider the number of respondents who have endorsed a total of i items out of the set of m items, regardless of which i items. This number will be designated n_i Treating the distributions of p_j and n_i as given, what is the expected number of errors across all respondents if response to each item is independent of responses to other items? This question is the same as asking what expected error would be if the items did not form a scale.

Although the correct answer to the question posed will appear somewhat complicated initially, it can be greatly simplified. Each set of n_i respondents will be treated separately. Letting Y_i indicate the number of errors contained in a response pattern which has i endorsements, we will first compute the expected value of Y_i, designated $E(Y_i)$. Then these expected values need only be weighted by n_i, added, and the sum divided by n, to produce the expected error over all values of i. Formally:

$$E(Y) = \frac{1}{n} \sum_i n_i E(Y_i). \qquad [3]$$

For any arbitrary ordering of items, it is possible to specify all arrangements of i X's and $m-i$ blanks. Let each of these arrangements be called a pattern. For any i, there will be $m!/i!(m-i)!$ patterns. The probability of any given pattern being observed is obtainable from the product of the p_j and $q_{j'}$, if item endorsements are independent. Also, the number of errors for any given pattern can be determined by inspection. It can be demonstrated (proofs available on request) that systematically examining all orderings, and assuming that they are equally probable, leaves $E(Y_i)$ independent of the marginal endorsement probabilities p_j and $q_{j'}$. The general equation for $E(Y)$ is

$$E(Y) = \frac{1}{n} \sum_i \frac{n_i}{\binom{m}{i}} \sum_{Y_i}$$

$$Y_i (m-i-Y_1+1) \binom{i+Y_i-2}{Y_i}. \qquad [4]$$

Equation 4 provides a general solution to the expected number of errors, given the marginal distributions of endorsements and of number of items endorsed. It must be remembered, of course, that $Y_i = 0$ for $i = 0$, 1 and m, taking integer values between 1 and $m - i$ for $2 \leqslant i \leqslant m$. Although the equation looks rather forbidding, it appears to be equal to a shorter formula:[14]

$$E(Y_i) = \frac{(i-1)\ (m-i)}{i+1} \qquad [5]$$

To make computation quick and simple, all values of $E(Y_i)$ for m from 2 through 10 are provided in Table 29.5. These values are the upper entry in each cell representing a given m and i. To use Table 29.5 for computing $E(Y)$, when m is not greater than 10, just multiply the appropriate $E(Y_i)$ values by n_i, then sum and divided by n, as indicated in Equation 1. By this procedure, $E(Y)$ for the data in Table 29.2 would be computed as shown in Table 29.4.

Table 29.4. Computation of E(Y) for Data from Table 29.2

i	n_i	$E(Y_i)$ from Table 29.5	$n_i E(Y_i)$
0	0	0	0
1	0	0	0
2	33	1.000	33.000
3	5	1.000	5.000
4	12	.600	7.200
5	0	0	0
	50		45,200

$$E(Y) = \frac{45.200}{50} = .905$$

NOTE: Values of $E(Y_i)$ are found in the row headed m = 5, since Figure 29.2 contains 5 items.

Of what use is $E(Y)$? Following Goodman,[15] our procedure, like other scaling procedures, systematically assigns a number to each specific pattern of endorsements and nonendorsements which indicates how different that pattern is from an ideal one. He designates the tth pattern number

14. Although the equivalence of Equations 4 and 5 can be demonstrated for $i = 2$ and 3, a general proof has not yet been developed for all m. Equation 5 correctly produces all entries in Table 29.5 for $E(Y_i)$.

15. Leo A. Goodman, "A Coefficient of Reproducibility," *Psychometrika*, 24 (March, 1959).

(or error score) as X_t, noting that all perfect patterns will be assigned a score of zero. Dividing X_t by the number of scale items provides a variable, Y_t, representing that pattern's relative error. A coefficient of reproducibility is then definable as

$$C = 1 - \sum_t P_t Y_t, \qquad [6]$$

where P_t is the proportion of the population in category t. Thus, treating Y_t as a proportional error score per pattern, C is the mean proportion of responses which are "correct" in the population. In our notation, where m indicates number of items and Y denotes number of errors,

$$C = 1 - E\left(\frac{Y}{m}\right) = 1 - \frac{E(Y)}{m} \qquad [7]$$

under an hypothesis of independence.

Goodman goes on to note the population reproducibility would be estimated from a sample of n respondents by an analogous formula equivalent to one minus the mean observed error over number of items. Letting sample reproducibility be designated c, then the expected value of c is C and the variance of c, translated from his notation to ours, is

$$\sigma^2_c = \frac{1}{n}\left\{ E\left(\frac{Y}{m}\right)^2 - \left[E\left(\frac{Y}{m}\right)\right]^2 \right\} = \frac{E(Y^2) - [E(Y)]^2}{nm^2} \qquad [8]$$

"When $\sigma^2_c > 0$, using the usual central limit theorem, the distribution of $z = (c - C)/\sigma_c$ will be approximately normal with zero mean and unit variance, when $N \to \infty$."[16] Therefore a normal-based test of the scalability of items in a developmental sense may be defined if $E(Y^2)$ can be computed. All that is required to obtain $E(Y^2)$ is to substitute Y^2 into Equation 4 for Y. Again, the computation can be tedious, but Table 29.5 contains values $E(Y^2_i)$ for each set of values of m and i. These values are the lower entries of each cell. Parallel to computations for $E(Y)$, the value of $E(Y^2)$ for data in Table 29.2 may be computed as $1/50\ [33(2.0) + 5(1.6) + 12(.6)] = 1.624$.

Note that the constant m will appear in both numerator and denominator for z, and thus may be eliminated. Letting $\bar{Y}$ stand for the sample mean error per pattern, the normal deviate may be computed by

$$z = \frac{E(Y) - \bar{Y}}{\sqrt{\dfrac{1}{n}\left\{ E(Y^2) - [E(Y)]^2 \right\}}} \qquad [9]$$

Using Equation 9 we can determine that, for the data in Table 29.2,

16. *Ibid.,* p. 37.

Table 29.5. $E(Y_i)$ and $E(Y^2_i)$ for i Endorsements among m Items

Number of Items ($m =$)	Number of Endorsements ($i =$)										
	0	1	2	3	4	5	6	7	8	9	10
2	0 / 0	0 / 0	0 / 0	...	...	...	...	...	...	...	...
3	0 / 0	0 / 0	.333 / .333	0 / 0	...	...	...	...	...	...	...
4	0 / 0	0 / 0	.667 / 1.000	.500 / .500	0 / 0	...	...	...	...	...	...
5	0 / 0	0 / 0	1.000 / 2.000	1.000 / 1.600	.600 / .600	0 / 0	...	...	...	...	...
6	0 / 0	0 / 0	1.333 / 3.333	1.500 / 3.300	1.200 / 2.000	.667 / .667	0 / 0	...	...	...	...
7	0 / 0	0 / 0	1.667 / 5.000	2.000 / 5.600	1.800 / 4.200	1.333 / 2.286	.714 / .714	0 / 0	...	...	...
8	0 / 0	0 / 0	2.000 / 7.000	2.500 / 8.500	2.400 / 7.200	2.000 / 4.857	1.428 / 2.500	.750 / .750	0 / 0	...	...
9	0 / 0	0 / 0	2.333 / 9.333	3.000 / 12.000	3.000 / 11.000	2.667 / 8.381	2.143 / 5.357	1.500 / 2.667	.778 / .778	0 / 0	...
10	0 / 0	0 / 0	2.667 / 12.000	3.500 / 16.100	3.600 / 15.600	3.333 / 12.857	2.857 / 9.286	2.250 / 5.750	1.556 / 2.800	.800 / .800	0 / 0

NOTE: The upper entry in each cell is $E(Y_i)$, the lower entry is $E(Y^2_i)$. The value of $E(Y_i)$ may be calculated by the short formula: $E(Y_i) = \dfrac{(i-1)\ (m-i)}{i+1}$. No comparable short formula has been found for $E(Y^2_i)$. [See editor's note below]

[Editor's Note: Robert K. Leik recently informed me that Neil W. Henry has provided a general proof of the short equation for $E(Y_i)$ above. More importantly, he has worked out a similar general form for $E(Y^2_i)$, which is:

$$E(Y_i^2) = \frac{(i-1)\ (m-i)\ (mi - i^2 + 2)}{(i+1)\ (i+2)}$$

These make the use of the scale much simpler for larger numbers of items. Also, Thomas Steinburn of the University of Washington has written a computer program for the scale.]

$z=6.337$, which is less probable than one in a million under the hypothesis of random ordering.

The question of measurement utility is not answered by statistical significance. Although a pattern may contain significantly less error than would be expected if no order among items existed, it may still contain too much error for adequate measurement. Because the coefficient c is subject to minimum marginal restrictions, similar to Guttman coefficient restrictions,[17] let us define what Costner has termed a P-R-E measure.[18]

$$\text{Scalability} = \frac{\text{Error based on chance} - \text{Scale error}}{\text{Error based on chance}} \qquad [10]$$

For the illustration used thus far,

$$\text{Scalability} = \frac{E(Y) - \overline{Y}}{E(Y)} = \frac{.905 - .100}{.905} = .890 \qquad [11]$$

This is analogous to the computation of scalability for Guttman scales, and is interpretable in the same way.[19] In this instance there is high scalability as well as statistical significance. In general, both properties should obtain for the scale to be useful.

Two Illustrations

Now that the presentation of the scaling procedure is completed, there remains the question of possible uses to which such a scale might be put. Let us return to consideration of wife independence as a developmental process. The data for this illustration are from a nonrandom sample of 48 midwest and west coast women, predominantly middle-aged and middle class, and about equally divided between urban and rural residence.[20] Questionnaires included sets of questions on each of five areas of wife

17. The restrictions on c for a developmental scale pertain to the frequency with which i items have been endorsed, whereas the restrictions on c for a Guttman scale pertain to the frequency with which each item has been endorsed.

18. Herbert L. Costner, "Criteria for Measures of Association," *American Sociological Review*, 30 (June, 1965), pp. 341–353.

19. Scalability for Guttman scales represents proportional improvement over minimal marginal reproducibility. See H. Menzel, "A New Coefficient for Scalogram Analysis," *Public Opinion Quarterly*, 17 (1953–4), pp. 268–280. Proportional improvement over random error is therefore analogous but more stringent.

20. Cases reported here do not represent the total original sample. Due to various nonresponses, nearly half of the original questionnaires were not usable for this analysis. Since our intent is illustrative rather than to test hypotheses, nonresponse bias is not important. For the original study, see Merlyn Matthews, "Autonomy and the Marital Roles of Women," unpublished M.A. thesis, University of Washington, 1967.

independence. These were, briefly: (1) attitude toward wives working under varying conditions of financial need and with children of varying ages; (2) attitude toward public and private disagreement with husband on matters of opinion; (3) attitude toward wives taking part in a social or working group, mixed sex or all male, with or without spouse; (4) actual disagreement and attempt to argue with husband regarding expectations for wife's behavior; and (5) actual behavior in disregard of husband's contradictory preferences.[21]

Treating the five areas of independence as dichotomous, we can describe each case as displaying either high or low independence in each area, where low implies dependence. Using the Goodenough criteria,[22] the best Guttman scale of these five dichotomies had a coefficient of reproducibility of only 0.82 and an accompanying coefficient of scalability of 0.51, implying that the pattern was by no means undimensional and additive. The scale accounts for about half of the responses which could have possibly been errors *under the worst conceivable patterning of responses.*

Application of the developmental scale criteria produced much more satisfying results. Given the marginal conditions of the observed responses, a random ordering of items would produce mean error of 0.77 responses per respondent. Observed mean error under the best ordering was only 0.34, providing scalability of 0.56. This figure indicates that the developmental scale correctly accounts for over half of the responses which would appear as errors given a *random* ordering of items. The value of z for the difference between the expected error under randomness (0.77) and the observed mean error (0.34) is 4.3. Statistically there is a highly significant departure from randomness. Table 29.6 shows the observed response patterns, their frequency of occurrence, and the accompanying errors.

The developmental scale procedure placed the items in the following order: (1) attitude toward opinion difference; (2) actual arguing with husband; (3) attitude toward social independence; (4) attitude toward working; and (5) actual behavior in opposition to husband's preference.[23]

---

21. Attitudinal dimensions were assessed by scaling responses to standard kinds of attitude questions. Actual behavior dimensions involved asking first for the respondent to list three instances in which her husband had expected her to behave in a way contrary to her own expectations. Then she was asked whether she argued with and whether she disregarded the wishes of her husband. For specific questions see Matthews, *op. cit.*

22. W. H. Goodenough, "A Technique of Scale Analysis," *Educational and Psychological Measurement,* 4 (1944), pp. 179–190.

23. This was somewhat different from optimal ordering for the Guttman scale, which was: attitude toward working, attitude toward social independence, actual arguing with husband, attitude toward opinion difference, actual behavior in opposition to husband's preferences.

Table 29.6. Developmental Scale of Wives' Independence

Attitude re Opinions	Actual Arguing	Attitude re Social Independ.	Attitude re Working	Actual Disregarding	f	Errors
..	..	..	..	..	3	0
X	..	..	..	..	1	0
X	X	..	..	..	2	0
X	X	X	..	..	2	0
X	..	X	..	..	1	1
..	X	X	..	..	4	0
X	X	X	X	..	8	0
X	X	..	..	..	2	2
X	..	..	X	..	1	2
..	X	X	X	..	4	0
..	..	X	X	..	6	0
..	..	..	X	..	1	0
X	..	X	X	X	2	2
X	..	..	X	X	2	4
X	..	X	..	X	1	2
X	X	..	X	X	1	1
..	X	X	X	X	1	0
..	X	X	..	X	1	1
..	..	X	..	X	1	1
..	..	..	X	X	2	0
..	..	..	..	X	1	0
					47	16

CS = .56, z = 4.3, Pr. <.001.

Such an ordering makes sense in terms of increasing independence, but does it make sense that, as occurred, and as the developmental model implies, women at the more independent end of the scale no longer endorsed items associated with early independence?

Two indications from the family literature tend to support such a notion. Hoffman has found that working mothers showed less participation in household tasks and made fewer decisions about routine matters than did nonworking mothers.[24] Similarly, Blood's review of the literature led to the conclusion that compared to nonworking wives, ". . . the working wife's (a) task participation decreases, and (b) activity control decreases."[25] Although these references are to working vs. nonworking wives,

24. Lois W. Hoffman, "Parental Power Relations and the Division of Household Tasks," *Marriage and Family Living*, 22 (February 1960), pp. 27–35.
25. Robert O. Blood, "The Husband-Wife Relationship," chap. XX in F. Ivan Nye and Lois W. Hoffman (eds.), *The Employed Mother in America*, Chicago: Rand-McNally, 1963.

which represents only one type of indepedence, they do indicate the plausibility of a decrease in one area of independence or autonomy as another area is activated.

The second illustration concerns patterns of family recreation. Specifically, a sample of adults from 600 households in the Puget Sound area was asked an extensive set of questions about the participation of household members in a variety of outdoor activities.[26] The mean annual participation in each of 18 activities was computed for all residents of all households.

Our interest for the purpose of this paper lies in whether the family recreation patterns are in any way related to family life cycle, a developmental concept. From the total sample, those households were selected which represented intact nuclear families (two parents plus children, with no other adults living in the house). There were 307 such families. For each parent separately, and for the family children on the average, it was determined whether participation in a given activity was above or below mean annual participation for the total sample. There are three dichotomous variables represented here, whether (1) father, (2) mother and (3) children of a given family are above or below the mean.

Two variables which are related to family life cycle were added to the above three. They were whether the oldest child was over twelve years of age, and whether the youngest child was over twelve years of age. Thus there were families with young children only, with both young and teen-aged children, and with teen-aged children only, representing successive stages from young to middle-aged families.

Application of the logic of either a Guttman scale or a developmental scale to these data would imply that there should be a specific chronological pattern of becoming active. For example, perhaps one parent is active, induces the other to be, and, as soon as the children are old enough, they too are taught to enjoy the activity. Following Guttman procedures, it would be necessary that the parents remained active, however, if they once were. The developmental scale, on the other hand, makes possible a subsequent reduction in parental activity, e.g., with increasing age, such that only the children might remain active.

Each of seventeen activities was examined for both a Guttman and a developmental scale pattern.[27] To avoid a problem of spurious ordering, which will be discussed later, only those families in which someone was above mean participation for a given activity were included in the scale

26. These data are from L. K. Northwood, Robert K. Leik and Robert Reed, *Outdoor Recreation in the Puget Sound Region—1963*, research report to the Puget Sound Governmental Conference, August, 1964.

27. One of the original 18 activities, sailing, had so low a mean rate of participation that the dichotomy between participation above or below the mean became meaningless.

for that activity. Activities scaled were (1) swimming or going to the beach, (2) water skiing, (3) power boating, (4) other boating, (5) hunting, (6) fishing, (7) camping, (8) hiking, (9) bicycling, (10) golfing, (11) snow skiing or other winter sports, (12) horseback riding, (13) playing outdoor games, (14) picnicking, (15) walking for pleasure, (16) driving for pleasure, and (17) attending outdoor sports.

Guttman scales proved in general to be inadequate. The mean coefficient of reproducibility was 0.81, ranging from 0.68 to 0.95, and the mean coefficient of scalability was only 0.41, ranging from 0.08 to 0.81. Only two scales had adequate reproducibility and scalability. It is interesting that they were both passive activities, for which age would not be a particular deterrent: picnicking (CR=0.93, CS=0.72) and driving for pleasure (CR=0.95, CS=0.81).

In contrast, all activities provided good developmental scales. Coefficients of scalability averaged 0.90, ranging from 0.80 to 0.99. It should be emphasized that scalability in this usage means proportional improvement over a *random* pattern. Scalability as applied to a Guttman pattern means proportional improvement over the lowest coefficient of reproducibility which is mathematically possible. Since a random pattern would show reproducibility well above the minimum possible value, the coefficient of scalability for a Guttman scale must be considerably higher than that for a developmental scale in order to represent the same proportional improvement over randomness.

The values of the statistic z which are associated with the coefficients of developmental scalability are all highly significant. They average 8.1, ranging from 4.8 to 10.9, with degrees of freedom ranging from 74 to 168. Clearly there is a nonrandom ordering of the three activity variables and the two life-cycle stage variables. That ordering, the same for all activities, is mother's activity, father's activity, oldest child over twelve, children's activity, youngest child over twelve. Table 29.7 shows the Guttman scale and developmental scale data for the 17 activities.

The fact that mother's activity precedes father's activity need not be indicative of the wife inducing the husband into outdoor recreation. In fact, it appears from the data as if both spouses begin marriage with high activity rates, or both with low activity rates. The ordering is apparently a consequence of the father's *maintaining* activity longer than his wife. Since order of dropping is as important to the developmental sequence as is order of acquisition, when one of these aspects does not particularly determine item ordering, the other may, as in this instance.

It is not clear from the data and the nature of the developmental scale whether those families in which parents are inactive but children are active represent cases of parental "retirement" from earlier activity which

*Table 29.7. Guttman and Developmental Scale Values for
Seventeen Activities*

Activity	Guttman scales		Developmental scales	
	C.R.	C.S.	C.S.	t
Swimming or Beach	.77	.32	.95	10.9
Water Skiing	.72	.25	.99	8.1
Power Boating	.84	.52	.84	7.0
Other Boating	.79	.40	.89	7.6
Hunting	.88	.47	.82	4.8
Fishing	.85	.47	.90	8.0
Camping	.89	.61	.91	8.2
Hiking	.72	.21	.88	8.1
Bicycling	.76	.21	.83	5.4
Golfing	.84	.44	.88	9.2
Skiing, Winter sports	.79	.38	.95	7.8
Horseback Riding	.75	.33	.94	6.7
Playing outdoor games	.68	.08	.95	10.7
Picknicking	.93	.72	.96	10.1
Walking for Pleasure	.86	.53	.80	6.5
Driving for Pleasure	.95	.81	.95	8.8
Attending Outdoor Sports	.75	.26	.91	9.8
Means	.81	.41	.90	8.1

C.R. = Coefficient of Reproducibility
C.S. = Coefficient of Scalability
t = Student's *t*

induced the children to be active. Equally compatible with the scale would be cases in which the children became active but the parents never were active and were not instrumental in the children's induction to outdoor recreation. The extent to which either pattern obtains is a historical question which the scaling procedure cannot answer.

The scale does indicate, however, that young children do not in general become active (above average) unless they do so with older siblings. No case was observed of children being above average activity when the oldest was twelve or younger, but frequently both parents were active even though the children were too young to be similarly active. Thus the plausible pattern emerges, as suggested by the scale, of early parental activity, increasing child activity as the children approach teen-age, and decreasing parental activity as the younger children become teen-agers or older. Whether the fact that these are outdoor recreations alters the pattern compared to, e.g., attending movies, playing indoor sports and games, or engaging in hobbies, is not evident. It is apparent, however, that outdoor activities, at least in the Puget Sound area, show a distinct family life-cycle pattern.

Naturally-Ordered Items

Two of the items in the recreation activity scales are what might be termed "naturally ordered." It is necessarily the case that the oldest child will reach twelve years of age before the youngest child does. When such natural ordering exists for a pair (or more) of scale items, certain considerations are in order. First, the time-ordering of a developmental process is obviously determined by the placement of naturally-ordered pairs of items. Thus there is no question which stage of the family life cycle is earlier or later in the scales presented. Secondly, the use of Table 29.5 to compute $E(Y)$ and $E(Y^2)$ may be invalid when some items are naturally ordered.

To make intuitively clear the problem of natural ordering as it affects computation of means and variances of error, suppose that all m items represent points on a single naturally-ordered continuum. Assuming accurate data, it would be impossible for any gaps between responses to appear in response patterns. Because this in no way limits the number of items endorsed, however, it is still possible to have i range from 0 to m. According to Table 29.5, $E(Y_i)$ will be non-zero for $2 \leq i \leq m-1$, but according to the preceding discussion, $E(Y_i)$ will be zero for all i.

It is possible, when confronted with natural ordering of some but not all of the items in the scale, to examine systematically the number of errors for a pattern containing i responses when all acceptable orderings of items are considered and to compute the needed $E(Y_i)$ and $E(Y_i^2)$. This was done for the reaction scale (two naturally-ordered items out of five), and it was found that Table 29.5 values were still valid. As yet there has been no attempt to find a general solution for $E(Y_i)$ and $E(Y_i^2)$ for varying numbers of naturally-ordered items in scales which also contain non-naturally-ordered items.

A final point about natural ordering concerns the decision, mentioned earlier, to scale only families in which someone participated above average. The two "ages of children" items, of course, would always be endorsed for older families. If a majority of families did not participate at all in a given activity, such as horseback riding, then the order would be greatly affected by the large number of cases with only the age items endorsed. These would have to be adjacent in the scale. The consequences would be that the nonparticipant cases would obscure the pattern of development of activity, which was, after all, the focus of the scale construction.

Implied in the preceding comments is the fact that use of naturally ordered and naturally occurring items as part of a larger set of items may arbitrarily produce misleading results if large numbers of cases do not endorse the other items. The natural items will be endorsed regardless of the pattern of the others. They may overshadow that pattern, therefore unless caution is used in determining what cases are appropriate for inclusion in the scale. Obviously this is a question concerning the purpose of the scale, not a way of getting around low coefficients of scalability.

Developmental Scales and Theoretical Advancement

A new technique is by no means automatically desirable. The proof of the pudding is not, in this case, in the eating, but in the digesting—digesting the multitude of minutiae that research accumulates. If the concept of ordered change is theoretically meaningful (and we are convinced that it is), then there must be some way of assessing whether order is evident in the observable cases. Particularly in the area of the family, developmental conceptualizations have been explicitly stated. In other areas similar notions are used, such as personality development, childhood and adult socialization, the development of organizational structure, or the orderly process of ecological change.

To validate these conceptualizations completely requires longitudinal evidence. It is possible, however, to invalidate such a conceptualization by showing that cross-sectional data do not fit any of the time-slice patterns that a longitudinal model could produce. This is the essence of the developmental scale we have proposed. If cross-sectional data scale well, then the plausibility of a developmental process is established. If the scale is poor, a developmental model is challenged. These are the best assessments available short of extensive longitudinal analysis.

Chapter 30

THE ASSIGNMENT OF NUMBERS TO RANK ORDER CATEGORIES

SANFORD LABOVITZ

University of Southern California

Empirical evidence supports the treatment of ordinal variables *as if* they conform to interval scales (Labovitz, 1967).[1] Although some small error may accompany the treatment of ordinal variables as interval,[2] this is offset by the use of more powerful, more sensitive, better developed, and more clearly interpretable statistics with known sampling error. For example, well-defined measures of dispersion (variance) require interval

Reprinted by permission from the American Sociological Association and the author from the *American Sociological Review*, Vol. 35, No. 3, 1970, pp. 515–24.

I am grateful to Robert Hagedorn, Harvey Marshall, Ross Purdy, and the referees of *ASR* for their helpful comments and critical reading of an earlier draft.

1. Labovitz demonstrates the utility of treating ordinal variables as interval for a hypothetical problem relating two types of therapy to four subjective responses: it made me worse $(-)$; it had no effect (0) ; it helped a little $(+)$; and it helped quite a bit $(++)$. The four ordinal responses are assigned scores ranging from highly skewed (e.g., 0, 1, 2, 10) to equidistant systems (e.g., 0, $3\frac{1}{3}$, $6\frac{2}{3}$, 10). The monotonic scoring systems produce largely similar point-biserial coefficients, t-tests, and critical ratios. Furthermore, the divergent scoring systems are highly interrelated. The r's between the two types of therapy and the four subjective responses are somewhat higher (averaging about .20) than the correlation coefficients in this study (averaging about .12).

2. Small error may result because the difference between two adjacent ranks may not be the same as the difference between two other adjacent ranks.

or ratio based measures. Furthermore, many more manipulations (which may be necessary to the problem in question) are possible with interval measurement, e.g., partial correlation, multivariate correlation and regression, analysis of variance and covariance, and most pictorial presentations. The arguments presented below are general enough to apply to any ordinal scale, and perhaps with even greater confidence they apply to variables that fall between ordinal and interval, e.g., I.Q. scores and formal education (Somers, 1962:800).

To determine the degree of error of results when treating ordinal variables as if they are interval, the relation between occupational prestige and suicide rates is analyzed. Prestige rankings obtained by NORC in its 1947 survey are related to suicides by occupation for males in the United States in 1950. The list of occupations, taken from Duncan's comparisons of occupational categories used in the survey, are matched to the detailed occupational classification in the U.S. Census of 1950 (Reiss *et al.*, 1961). Because suicides are not reported for all of these occupations and sometimes the reported suicides are for two or three occupations grouped into one, 36 occupations were selected which contain the necessary data used in this study (see Table 30.1). Measurement of occupational prestige is based solely on the principle of ordinal ranking. In the survey, respondents were given occupations to rank by the method of paired comparisons; consequently, the resulting prestige scores indicate merely the rank of one occupation relative to the others (Reiss *et al.*, 1961:122–123).[3]

The rank correlation (*rho*) between occupational prestige and suicide is .07. The scatter diagram of the NORC prestige ratings and suicide rates suggests that the relation is roughly linear, although the plotted points are widely scattered. The Pearsonian correlation coefficient (*r*) on the same data is slightly larger (.11). The .04 discrepancy between the two measures is due to the magnitude of the differences between adjacent scores which are not considered in *rho*, but do influence the value of *r*.

Assignment of Scoring Systems to Ordinal Categories

Twenty scoring systems are used on NORC's occupational prestige values. One scoring system is the actual prestige ratings resulting from the study (the NORC Prestige Rating Scale in Table 30.1). A second scoring system is the assignment of equidistant numbers (i.e., an equal distance between assigned numbers) to the occupational categories (Table 30.2). The remaining scoring systems in Table 30.2 were generated from

---

3. Duncan's socioeconomic index, based upon the income and educational levels of each occupation, correlates highly with the NORC prestige scale.

Table 30.1. Prestige, Income, Education, and Suicide Rates for 36 Occupations, United States, Males, Circa, 1950

Occupation	NORC Prestige Rating Scale [a]	Male Suicide Rate [b]	Median Income [c]	Median School Yrs. Completed [d]
Accountants and auditors	82	23.8	3,977	14.4
Architects	90	37.5	5,509	16+
Authors, editors and reporters	76	37.0	4,303	15.6
Chemists	90	20.7	4,091	16+
Clergymen	87	10.6	2,410	16+
College presidents, professors and instructors (n.e.c.)	93	14.2	4,366	16+
Dentists	90	45.6	6,448	16+
Engineers, civil	88	31.9	4,590	16+
Lawyers and judges	89	24.3	6,284	16+
Physicians and surgeons	97	31.9	8.302	16+
Social welfare, recreation and group workers	59	16.0	3,176	15.8
Teachers (n.e.c.)	73	16.8	3,465	16+
Managers, officials and proprietors (n.e.c.) —self-employed—manufacturing	81	64.8	4,700	12.2
Managers, officials and proprietors (n.e.c.) —self-employed—wholesale and retail trade	45	47.3	3,806	11.6
Bookkeepers	39	21.9	2,828	12.7
Mail-carriers	34	16.5	3,480	12.2
Insurance agents and brokers	41	32.4	3,771	12.7
Salesmen and sale clerks (n.e.c.), retail trade	16	24.1	2,543	12.1
Carpenters	33	32.7	2,450	8.7
Electricians	53	30.8	3,447	11.1
Locomotive engineers	67	34.2	4,648	8.8
Machinists and job setters, metal	57	34.5	3,303	9.6
Mechanics and repairmen, automobile	26	24.4	2,693	9.4
Plumbers and pipe fitters	29	29.4	3,353	9.3
Attendants, auto service and parking	10	14.4	1,898	10.3
Mine operatives and laborers (n.e.c.)	15	41.7	2,410	8.2
Motormen, street, subway, and elevated railway	19	19.2	3,424	9.2
Taxicab-drivers and chauffeurs	10	24.9	2,213	8.9

[a] Albert J. Reiss, Jr., et al., 1961 : 122–123. The scale is based on a 1947 survey.

[b] Males, aged 20–64. National Office of Vital Statistics, *Vital Statistics—Special Report,* Vol. 53, No. 3 (September, 1963) .

[c] 1949 Median income. *United States Census of Population, 1950. Occupational Characteristics* (Special Report, P-E No. 1B) , Table 19.

[d] 1950 Median school years completed. *Ibid.,* Table 10.

Table 30.1.—(Continued)

Occupation	NORC Prestige Rating Scale [a]	Male Suicide Rate [b]	Median Income [c]	Median School Yrs. Completed [d]
Truck and tractor drivers, deliverymen and routemen	13	17.9	2,590	9.6
Operatives and kindred workers, (n.e.c.), machinery, except electrical	24	15.7	2,915	9.6
Barbers, beauticians and manicurists	20	36.0	2,357	8.8
Waiters, bartenders and counter and fountain workers	7	24.4	1,942	9.8
Cooks, except private household	16	42.2	2,249	8.7
Guards and watchmen	11	38.2	2,551	8.5
Janitors, sextons and porters	8	20.3	1,866	8.2
Policemen, detectives, sheriffs, bailiffs, marshals and constables	41	47.6	2,866	10.6

a computer according to the following conditions: (1) the assigned numbers lie between the range of 1 and 10,000, (2) the assignment of numbers is consistent with the monotonic function of the ordinal rankings, (3) any ties in the ordinal rankings are assigned identical numbers, and (4) the selection of a number is made on the basis of a random generator in the computer program. To be consistent with the monotonic function, any subsequent randomly selected numbers must be higher than previous ones (except for ties). The resulting largely random scoring systems vary among themselves (sometimes to a large extent) on the actual values assigned to each rank, the range of values, and the size of the differences between adjacent values. Although all are necessarily consistent with the monotonicity of the ordinal rankings, they vary widely among themselves. In fact, some of the scoring systems show definite curvilinear patterns—logarithmic, exponential or higher order curves (two or more inflection points).

Because this computer approach to assigning numbers to rank order data partially is based on a random selection of numbers, the generality of the findings is somewhat limited. It is possible that some systematic selection of numbers will not yield such consistent results as those reported herein.

The similarity among the scoring systems can be assessed by their matrix of intercorrelations (Table 30.3). By assuming, in turn, that each scoring system is the "true" one, the intercorrelations (Pearson product-moment coefficients) indicate the extent of "error" of using one of the other 19 scoring systems. For example, if (4) is the "true" system and (7) has been used in its place, then .97 (the correlation between the two

Table 30.2. NORC Prestige Ratings, Linear Scoring, and Five Monotonic Random Generated Scoring Systems[a]

Linear (1)	NORC (2)	Monotonic Random Generated Scoring Systems				
		(3)	(5)	(9)	(13)	(18)
1.0	7	13	79	52	849	418
2.0	8	34	105	109	909	585
3.5	10	99	233	380	923	648
3.5	10	99	233	380	923	648
5.0	11	248	389	518	1152	820
6.0	13	407	580	557	1167	869
7.0	15	727	605	799	2300	1271
8.5	16	1824	771	2167	2343	1478
8.5	16	1824	771	2167	2343	1478
10.0	19	1897	1042	2790	2845	1647
11.0	20	2021	1287	2796	2876	1789
12.0	24	2470	1374	3209	3107	2112
13.0	26	2978	1713	3558	3159	2627
14.0	29	2995	2083	3598	3231	2628
15.0	33	3330	2595	3808	3409	2777
16.0	34	3412	2715	3945	3760	2921
17.0	39	3535	2751	4087	4238	3077
18.5	41	3952	2861	4094	4898	3156
18.5	41	3952	2861	4094	4898	3156
20.0	45	4082	3003	4745	5336	3209
21.0	53	4485	3266	4885	5903	3600
22.0	57	4865	4013	4892	6016	4304
23.0	59	5091	4267	5044	6106	4323
24.0	67	5146	4449	5300	6242	4762
25.0	73	5349	5318	5819	6270	5020
26.0	76	5775	6330	5876	6681	5528
27.0	81	5995	6547	5923	6787	5797
28.0	82	6304	6810	5932	6915	6027
29.0	87	6356	6974	5976	7118	6388
30.0	88	6644	7660	5995	7229	6471
31.0	89	6742	8145	6060	7652	6560
33.0	90	7657	9085	6231	7926	6911
33.0	90	7657	9085	6231	7926	6911
33.0	90	7657	9085	6231	7926	6911
35.0	93	7841	9108	6458	8283	6972
36.0	97	8164	9461	7094	8472	7588

[a]See text for an explanation of the scoring systems. The five random scoring systems are indicative of the eighteen used in the study.

scoring systems) indicates the degree to which the two systems vary together. On the other hand, r^2 (the values below the diagonal in Table 30.3) indicates "error" in terms of the amount of variance in the assigned

Table 30.3. Intercorrelations (r) Among Twenty Scoring Systems[a]

Scoring Systems	(1)	(2)	(3)	(4)	(5)	(6)	(7)	(8)	(9)	(10)	(11)	(12)	(13)	(14)	(15)	(16)	(17)	(18)	(19)	(20)
(1) [b]	..	.98	1.00	.99	.97	.99	.98	.99	.97	.98	.99	.99	.99	.99	1.00	.98	.99	.99	.99	1.00
(2) [c]	.98	..	.97	.96	.97	.98	.95	.97	.96	.94	.97	.97	.96	.98	.93	.97	.94	.99	.98	.96
(3)	1.00	.94	..	.99	.97	.99	.98	.98	.98	.98	.99	.99	.99	.99	.99	.98	.99	.99	.99	.99
(4)	.98	.92	.98	..	.98	.99	.97	.99	.95	.99	.98	.98	.98	.99	.99	.99	.99	1.00	.98	.99
(5)	.94	.94	.94	.96	..	.99	.93	.99	.91	.99	.96	.95	.96	.98	.97	.99	.98	.99	.94	.96
(6)	.98	.96	.98	.98	.98	..	.97	.99	.94	.99	.97	.98	.98	.98	.99	.99	.99	.99	.97	.98
(7)	.96	.90	.96	.94	.86	.94	..	.96	.98	.95	.96	.98	.98	.98	.98	.94	.96	.96	.99	.98
(8)	.98	.94	.98	.98	.98	.98	.92	..	.93	.99	.97	.97	.98	.98	.99	.99	.98	.99	.96	.98
(9)	.94	.92	.96	.90	.83	.88	.96	.86	..	.93	.97	.98	.97	.97	.96	.92	.96	.95	.99	.98
(10)	.96	.88	.96	.98	.98	.98	.90	.98	.86	..	.97	.97	.97	.98	.98	.99	.99	.98	.96	.98
(11)	.98	.94	.98	.96	.92	.94	.92	.94	.94	.94	..	.99	.98	.99	.99	.98	.99	.98	.98	.99
(12)	.98	.94	.98	.96	.90	.96	.96	.94	.96	.94	.98	..	.98	.99	.98	.97	.98	.98	.99	.99
(13)	.98	.92	.98	.96	.92	.96	.96	.96	.94	.94	.96	.98	..	.98	.99	.97	.98	.98	.99	.99
(14)	.98	.96	.98	.98	.92	.96	.96	.96	.94	.96	.96	.98	.98	..	.98	.97	.98	.99	.98	.99
(15)	1.00	.86	.98	.98	.94	.98	.96	.98	.92	.96	.98	.96	.98	.98	..	.98	.99	.99	.98	.99
(16)	.96	.94	.96	.98	.98	.98	.88	.98	.85	.98	.96	.94	.94	.94	.96	..	.99	.99	.96	.97
(17)	.98	.88	.98	.98	.96	.98	.92	.96	.92	.98	.98	.96	.96	.96	.98	.98	..	.99	.98	.99
(18)	.98	.98	.98	1.00	.98	.98	.92	.98	.90	1.00	.96	.96	.96	.98	.98	.98	.98	..	.97	.98
(19)	.98	.96	.98	.96	.88	.94	.98	.92	.98	.92	.96	.98	.98	.96	.96	.92	.96	.94	..	.99
(20)	1.00	.92	.98	.98	.92	.96	.96	.96	.96	.96	.98	.98	.98	.98	.98	.94	.98	.96	.98	..

[a] r above the diagonal; r^2 below.
[b] linear scoring system.
[c] NORC prestige ratings.

scoring system accounted for by the variation in the "true" scoring system (Abelson and Tukey, 1959).[4] In this instance, between scoring systems (4) and (7), 94% of the variance in (7) is accounted for by the variation in (4).

The r and r^2 values in Table 30.3 are consistently and substantially high, indicating a high degree of interchangeability among the 20 scoring systems. Out of 190 correlation coefficients, all are above .90 (a few even reach unity), and 157 are .97 and above. Therefore, even without a rationale concerning the differences between ranks, by using a nearly random method of assigning scoring systems (consistent with the monotonic function), it is possible that under specific conditions the selected scoring system will deviate from the "true" system by a near zero or negligible amount. The r^2 values are slightly lower than the r values, but still exceedingly high. For example, only nine of the 190 are below .90, and none are below .83. (Since r^2 is the square of a decimal fraction, it is necessarily smaller than r.)

Note that if the equidistant (linear) scoring system is always selected (no matter what the "true" scoring system may be), the expected error is smaller than the larger errors cited above. Almost all the r's and r^2's for the linear system (1) are near unity, with the lowest r being .97 and the lowest r^2 being .94. The linear scoring system lies midway between the other scoring systems (in correlational terms), which by definition excludes the most extreme scoring systems in each direction. The correlations between the extremes are lowest, and, therefore, selecting the linear scoring system eliminates the lowest r's and the highest potential "errors" in selecting a scoring system different from the "true" one.

Possessing some knowledge about the amount of differences between ranks can reduce the small error even further, if the linear scoring system

---

4. Abelson and Tukey also use r^2 as the criterion for assessing the adequacy of numerical assignments and, in addition, present a "maximin" r^2 to assess the largest possible error in a scoring system. Briefly, an assigned scoring system X is correlated with a "true" system Y so that the minimum possible r^2 between X and Y achieves its maximum value. Their analysis, instead of leading to an average error rate (in which the "true" r^2 may be equally above or below the rate), results in a conservative lower limit estimate. This lower limit estimate is based on a sequence called "corners," which is consistent with the inequalities (i.e., it follows the monotonic or equality functions) and is based on a set of dichotomized values. For example, given the following relations $Y_1 \leq Y_2 \leq Y_3 \leq Y_4$, a set of corners is $(0, 0, 0, 1)$, $(0, 0, 1, 1)$, and $(0, 1, 1, 1)$. One of these corner sequences yields the maximum r^2. There are three problems with Abelson and Tukey's analysis: (1) an average error rate is more indicative of a representative error (i.e., the most likely error in assigning a scoring system) and, therefore, is more useful to the researcher, (2) the corner sequence is based on dichotomies which is a highly unlikely occurrence and a waste of information, and (3) they analyze only "greater than" and "equal to" models in combination $(Y_3 \geq Y_2 \geq Y_1)$, while most frequent ordinal cases are "greater than" between most ranks $(Y_3 > Y_2 > Y_1)$. The "greater than" model leads into a dichotomous analysis only if there are two ranks (a trivial case).

has been assigned to the ordinal categories. Perhaps, the best strategy, if there is some knowledge of the differences between ranks, is to modify the linear scoring system accordingly. For example, in the relation $X_1 > X_2 > X_3$, X_2 is assumed to be closer to X_3 than to X_1. Consequently, the linear scoring system of 10, 20 and 30 (as values for X_1, X_2 and X_3) can be modified to 10, 25 and 30 to account for this additional knowledge. It should be stressed that without prior knowledge or theory such score assignments are not likely to prove useful for analysis.

Table 30.4 offers further evidence that ordinal data can be treated as if they are interval by assigning scoring systems to the ordered categories. In this instance, the predictive ability of each scoring system is assessed in terms of its relation to suicide rates. As indicated previously, the *rho* value between the NORC prestige scale and 1950 suicide rates for males in 36 occupations is .07; for the same data, *r* is .11. Table 30.4 reports the *r* and r^2 values between the 20 scoring systems and the suicide rate. (The last two columns in Tables 30.4 are *r* values for 20 and 10 occupations respectively and will be discussed later in the paper.) The

Table 30.4. Correlation Coefficients (r) between Suicide Rates and Twenty Scoring Systems of Occupational Prestige[a]

Scoring System		*r* (*N*=*36*)	*r*²	*r* (*N*=*20*)	*r* (*N*=*10*)
(1)	(linear)	.13	.02	.35	.28
(2)	(prestige ratings)	.11	.01	.35	.25
(3)		.13	.02	.31	.24
(4)		.11	.01	.32	.30
(5)		.10	.01	.30	.21
(6)		.11	.01	.35	.18
(7)		.14	.02	.28	.33
(8)		.12	.01	.38	.34
(9)		.14	.02	.26	.15
(10)		.09	.01	.29	.24
(11)		.13	.02	.30	.24
(12)		.11	.01	.28	.22
(13)		.15	.02	.41	.35
(14)		.14	.02	.37	.25
(15)		.13	.02	.35	.32
(16)		.09	.01	.33	.18
(17)		.12	.01	.37	.34
(18)		.11	.01	.30	.33
(19)		.15	.02	.33	.25
(20)		.14	.02	.38	.41

[a] Partially based on the data in Tables 1 and 2. Scoring systems 3–18 are randomly generated.

similarity in predicting an outside variable is extremely high. The r's vary between .09 and .15, and the r^2 values are either .01 or .02.[5] Given some degree of unreliability in occupational prestige and suicide data, and the rather crude measurement procedures, these results substantiate the point that different systems yield interchangeable variables. Each indicates a quite low positive (statistically nonsignificant) relation between occupational prestige and suicide. These results are consistent with a previous study (Labovitz, 1967), which also found the relations to be very similar; however, in the previous study, the relations are somewhat higher and statistically significant.

As indicated by the results in Table 30.4, the greater the number of ranks (N), the greater the confidence in assigning an interval scoring system to ordinal data (Labovitz, 1968a; Morris, 1968). The last two columns (Table 30.4) report the correlations for the first twenty and for the first ten occupational groups (between suicide rates and the 20 scoring systems). That the correlation coefficients based on smaller N's are appreciably higher than for $N = 36$ is not a major concern. A statistical explanation for the higher r's with smaller N's is that in this case and for whatever reason, the error variance diminishes more rapidly than the total variance as N increases. By restricting the range in a systematic manner, i.e., taking the first twenty and first ten, some occupations with rather high suicide rates and low prestige levels are eliminated. The net effect is an increase in the positive correlation. The major concern is not with the magnitudes but with the similiarity of coefficients within each of the three groups ($N = 10$, $N = 20$, and $N = 36$). The standard deviations among the correlation coefficients decrease as N increases: (1) for $N = 10$; $SD = .07$; (2) for $N = 20$; $SD = .04$; and (3) for $N = 36$; $SD = .02$.

Note the similarity between the equidistant correlation coefficient (scoring system 1) and and the mean correlation coefficient for all 20 scoring systems for each of the three groups. For 10 occupations, the mean correlation is .27 and the equidistant correlation is .25; for twenty occupations, the mean is .33 and the equidistant is .35; and finally, for all 36 occupations, the mean is .12 and the equidistant is .11. This lends further support to the suggested strategy of imposing an equidistant scoring system on ordinal categories.

It may be argued that these results do not hold for "extreme" nonlinear monotonic transformations of ordinal measures. Admittedly, there is a point beyond which the transformation will not yield interchangeable measures. This is the case where the assigned scoring system has essentially dichotomized the ranks. For example, if the true or assigned

5. It should be noted that the usual purpose of a transformation in correlation work is to raise the correlation. However, the stability of the correlations in this study is not inconsistent with this general principle.

system is 1, 2, 3, 4, and the assigned or true system is 1, 9,996, 9,998, 10,000, then we are in essence scoring the categories as 1, 10,000, 10,000, 10,000. Under such conditions, it is obvious that treating ordinal categories as if they are interval is not an aid in data analysis, unless the "dichotomy" is recognized. The problem of dichotomizing becomes increasingly serious (in terms of faulty interpretations) as the number of ranks increases. Under the condition of a true equidistant scoring system and exponential power function or logarithmic power function assigned scoring systems, as k increases the true and assigned systems increasingly diverge in values.

To illustrate this last point, consider an equidistant scoring system for X, e.g., 1, 2, 3, 4. Suppose X' (assigned scoring systems) is set equal to X^k and scores are generated. For given values of k the scoring systems are: (1) 1, 2, 3, 4 (k = 1); (2) 1, 4, 9, 16 (k = 2); (3) 1, 8, 27, 64 (k = 3); (4) 1, 16, 81, 256 (k = 4); and (5) 1, 32, 243, 1024 (k = 5). As noted above, as k increases the new scoring systems come increasingly closer to dichotomizing (polarizing) the values. The equidistance between numbers is progressively lost; the fourth number becomes large more rapidly than the first, second, or third; the third number increases more rapidly than the second; and the second more rapidly than the first. Consequently, the most deviant result in this process (i.e., the farthest from the true scoring system) is polarization into an essential dichotomy for very large k's.

Although we generally can partition a variable into more than two intervals, it is useful to consider the correlation between the linear system (1, 2, 3, 4) and the dichotomies (0, 0, 0, 1), (0, 0, 1, 1), (1, 1, 0, 0), and (1, 0, 0, 0). The correlations are respectively .77, .83, −.83, and −.77. The correlation of .77 (or −.77) represents the lower limit (the worst possible situation) for the four number sets compared to the equidistant scoring system (1, 2, 3, 4). With regard to an "outside" variable, "extreme" transformations yielding a dichotomy do make a difference. The equidistant system X (1, 2, 3, 4) and the exponential system X' (1, 8, 27, 64) are related to Y (2, 1, 3, 5) as follows: $r_{xy} = .83$ and $r_{x'y} = .94$. In this case, X and X' are not interchangeable variables.

Perhaps the most important reason for treating an ordinal variable as if it conforms to an interval scale lies in the opportunity it provides for applying well-developed and interpretable multivariate techniques. Although partials can be applied to ordinal measures—e.g., partial *tau*, partial *gamma*, or partial *rho*—these are often difficult to interpret. Further, a multiple relationship measure is not defined for ordinal variables, unless they are assumed to be the counterpart of the correlation coefficient.

To illustrate the utility of using multivariate analysis, education and income are combined with the equidistantly (linearly) scored prestige scale to account for the variance in suicide rates. Although occupational prestige is not highly related to suicide (in a zero-order correlation), when combined with an additive combination of income, the multiple R is .31. The zero-order correlation between income and suicide is .26. An additive combination of occupational prestige, income, and education results in an R of .55. This represents an increase of .05 over the R of .50 between suicide and the independent variables of income and education. Over 30% of the variance in suicide rates is accounted for by an additive combination of the three major independent variables. Treating occupational prestige as ordinal would not have permitted this analysis, although this variable adds the least amount of explained variance to the multiple R.

Partial correlations also result in significant findings. When partialed on income, the relation between occupational prestige and suicide is negative (−.16). Although the relation between the two is still quite low, a reasonable interpretation of the partial correlation is that income is determining the positive association between the two (by its positive effect on both variables). An implication is that conflicting results among several studies between prestige and suicide may be resolved by controlling for variations in income (Powell, 1958; Hirsh, 1959; Dublin, 1963; Breed, 1963, Labovitz, 1968b; Maris, 1967).

Consistent with the above partial of −.16, the relation between income and suicide increases when the effects of occupational prestige are partialed out. It appears that occupational prestige acts as a suppressor variable in relating income to suicide, and, therefore, since prestige and income are differentially related to suicide (one positive and one negative), they are to some extent canceling out each other's effects.

With regard to multivariate analysis, the treatment of occupational prestige as an interval variable has several advantages. First, a rather small N (in this instance the number of occupations is only 36) can be used with most intervally based multivariate techniques. Partialing by modes of elaboration techniques (cross-tabulation) may require an extremely large N. Second, these techniques are well-developed for interval data, but are either not developed or poorly developed for ordinal data. Consequently, the degree of latitude or versatility (Anderson, 1961)[6] in statistical analysis is increased substantially by using interval statistics. Finally, and in summary, the multivariate analysis led to some highly suggestive conclusions that would have been overlooked if prestige was treated as an ordinal variable. For example, (1) income may be determining the positive relation between prestige and suicide, (2)

---

6. This is Anderson's basic reason for selecting parametric over nonparametric statistics.

prestige may be suppressing or depressing the relation between income and suicide, and (3) the predictive model of an additive combination of prestige, income, and education accounts for a moderate part (30%) of the variance in suicide.

The researcher should be warned that the similarity among the scoring systems in terms of intercorrelations and predictive ability should not be uncritically generalized to regression problems. The bivariate regression coefficients (slopes) of the scoring systems on suicide rates for 36 occupations range from .04 to 32.8. This substantial variation is largely due to the linear and NORC scoring systems which used smaller numbers than those generated by the computer. The computer generated systems are closer is slope values, ranging from 18.2 to 32.8. The range of the standardized zero-order (gross) regression coefficients stand in sharp contrast to these results. The bivariate standardized slopes (which are the zero-order correlation coefficients in the standard score regression line) range from the lowest of .09 to the highest of .15.

These results indicate that if scores are assigned to an ordinal system (or if there is uncertainty regarding the magnitude of the differences between adjacent scores), regression coefficients should be standardized because of their greater stability from one scoring system to another. Standardized coefficients are an integral part of path analysis and are sometimes used as the path coefficients. Interpretations of path coefficients (standardized), according to the findings given above, do not appear to require modification by the assignment of numbers to ordinal categories. A wide range of values would indicate unreliable coefficients and would negate any meaningful solution of the identification problem (that is, estimating the unknown parameters in a model from available empirical data). However, as Blalock (1967 and 1968) has pointed out, standardized as compared to unstandardized coefficients may not be as adequate for problems where the comparison of populations is necessary to determine whether or not the underlying causal processes are basically similar. Standardized coefficients appear to be more adequate for problems of generalizing to a specific population, because they can be used to assess the direct and joint contributions of the several independent variables.

Conclusions

The results of the tests based on assigning interval scores to ordinal categories suggest: (1) certain interval statistics can be used interchangeably with ordinal statistics and interpreted as ordinal, (2) certain interval statistics (e.g., variance) can be computed where no ordinal equivalent exists and can be interpreted with accuracy, (3) certain interval statistics can be given their interval interpretation with only negligible error if the variable is "nearly" interval, and (4) certain interval statistics can be given their interval interpretations with caution (even if the variable is "purely" ordinal), because the "true" scoring system and the assigned scoring system, especially the equidistant system, are almost always close as measured by r and r^2.

Consequently, treating ordinal variables as if they are interval has these advantages: (1) the use of more powerful, sensitive, better developed and interpretable statistics with known sampling error, (2) the retention of more knowledge about the characteristics of the data, and (3) greater versatility in statistical manipulation, e.g., partial and multiple correlation and regression, analysis of variance and covariance, and most pictorial presentations.

The study suggests two research strategies when analyzing ordinal variables. First, assign a linear scoring system according to the available evidence on the distances between ranks. Second, use all available rank order categories, rather than collapsing them into a smaller number, because the greater the number of ranks the greater the stability and confidence in the assigned scoring system (unless the dichotomization of ranks is suspected). The all-too-frequent strategy of dichotomizing or trichotomizing variables should be avoided if possible.[7]

A final word of caution is necessary. The researcher should know and report the actual scales of his data, and any interval statistics selected should be interpreted with care. Further exploration and tests are necessary for added confidence in treating ordinal data as if they are interval. The more conservative procedure, of course, is to treat ordinal data as strictly ordinal, and thereby avoid the possibility of attributing a property to a given scale which it does not possess.

References

ABELSON, ROBERT P. and JOHN W. TUKEY, 1959, "Efficient conversation of nonmetric information to metric information." *Proceedings of the Social Statistics Section,* American Statistical Association: 226–230.
ANDERSON, NORMAN A., 1961, "Scales and statistics: Parametric and nonparametric." *Psychological Bulletin.* 58 (July):305–316.
BLALOCK, HUBERT M., JR., 1967, Letter. *American Journal of Sociology* 72 (May): 675–677. 1968 "Theory building and causal inferences." Pp. 155–196 in Hubert M. Blalock, Jr. and Ann B. Blalock (eds.), *Methodology in Social Research.* New York: McGraw-Hill.

7. Another reason against the use of dichotomies or trichotomies is that often a large amount of information is lost by such drastic collapsing.

BONEAU, C. A., 1960, "The effects of violations of assumptions underlying the 't' test." *Psychological Bulletin*, 57 (January):49–64.

BREED, WARREN, 1963, "Occupational mobility and suicide among white males." *American Sociological Review*, 28 (April):179–189.

DUBLIN, L. I., 1963, *Suicide: A Sociological and Statistical Study*. New York: Ronald Press.

HIRSH, JOSEPH, 1959, Suicide." *Mental Hygiene*, 43 (October):516–526.

LABOVITZ, SANFORD, 1967, "Some observations on measurement and statistics." *Social Forces*, 46 (December):151–160. 1968a. "Reply to Champion and Morris." *Social Forces*, 46 (June):543–545. 1968b "Variation in suicide rates." Pp. 57–74 in Jack P. Gibbs (ed.) , *Suicide*. New York: Harper and Row.

LINDQUIST, E. F., 1953, *Design and Analysis of Experiments*. Pp. 78–90. New York: Houghton Mifflin.

MARIS, RONALD, 1967, "Suicide, status, and mobility in Chicago." *Social Forces* 46 (December):246–256.

MORRIS, RAYMOND N., 1968, "Commentary." *Social Forces*, 46 (June):541–543.

POWELL, ELWIN M., 1958, "Occupation, status, and suicide: Toward a redefinition of anomie." *American Sociological Review*, 23 (April):131–140.

REISS, ALBERT J., JR., with O. D. DUNCAN, PAUL K. HATT, and CECIL C. NORTH, 1961, *Occupations and Social Status*. New York: Free Press.

SOMERS, ROBERT H., 1962, "A new asymmetric measure of association for ordinal variables," *American Sociological Review*, 27 (December):799–811.

Chapter 31

ON THE STATISTICAL TREATMENT OF FOOTBALL NUMBERS

Frederic M. Lord
Educational Testing Group

Professor X sold "football numbers." The television audience had to have some way to tell which player it was who caught the forward pass. So each player had to wear a number on his football uniform. It didn't matter what number, just so long as it wasn't more than a two-digit number.

Professor X loved numbers. Before retiring from teaching, Professor X had been chairman of the Department of Psychometrics. He would administer tests to all his students at every possible opportunity. He could hardly wait until the tests were scored. He would quickly stuff the scores in his pockets and hurry back to his office where he would lock the door, take the scores out again, add them up, and then calculate means and standard deviations for hours on end.

Professor X locked his door so that none of his students would catch him in his folly. He taught his students very carefully: "Test scores are ordinal numbers, not cardinal numbers. Ordinal numbers cannot be added. *A fortiori*, test scores cannot be multiplied or squared." The pro-

Reprinted with permission of the author and the publisher from the *American Psychologist*, Vol. 8, 1953, pp. 750–51. Copyright 1953 by the American Psychological Association.

fessor required his students to read the most up-to-date references on the theory of measurement (e.g., 1, 2, 3). Even the poorest student would quickly explain that it was wrong to compute means or standard deviations of test scores.

When the continual reproaches of conscience finally brought about a nervous breakdown, Professor X retired. In appreciation of his careful teaching, the university gave him the "football numbers" concession, together with a large supply of cloth numbers and a vending machine to sell them.

The first thing the professor did was to make a list of all the numbers given to him. The University had been generous and he found that he had exactly 100,000,000,000,000,000 two-digit cloth numbers to start out with. When he had listed them all on sheets of tabulating paper, he shuffled the pices of cloth for two whole weeks. Then he put them in the vending machine.

If the numbers had been ordinal numbers, the Professor would have been sorely tempted to add them up, to square them, and to compute means and standard deviations. But these were not even serial numbers; they were only "football numbers"—they might as well have been letters of the alphabet. For instance, there were 2,681,793,401,686,191 pieces of cloth bearing the numbers "69," but there were only six pieces of cloth bearing the number "68," etc., etc. The numbers were for designation purposes only; there was no sense to them.

The first week, while the sophomore team brought its numbers, everything went fine. The second week the freshman team bought its numbers. By the end of the week there was trouble. Information secretly reached the professor that the numbers in the machine had been tampered with in some unspecified fashion.

The professor had barely had time to decide to investigate when the freshman team appeared in a body to complain. They said they had bought 1,600 numbers from the machine, and they complained that the numbers were too low. The sophomore team was laughing at them because they had such low numbers. The freshmen were all for routing the sophomores out of their beds one by one and throwing them in the river.

Alarmed at this possibility, the professor temporized and persuaded the freshmen to wait while he consulted the statistician who lived across the street. Perhaps, after all, the freshmen had gotten low numbers just by chance. Hastily he put on his bowler hat, took his tabulating sheets, and knocked on the door of the statistician.

Now the statistician knew the story of the poor professor's resignation from his teaching. So, when the problem had been explained to him the statistician chose not to use the elegant nonparametric methods of modern statistical analysis. Instead he took the professor's list of the 100 quadrillion "football numbers" that had been put into the machine. He added them all together and divided by 100 quadrillion.

"The population mean," he said, "is 54.3."

"But these numbers are not cardinal numbers," the professor expostulated. "You can't add them."

"Oh, can't I?" said the statistician. "I just did. Furthermore, after squaring each number, adding the squares, and proceeding in the usual fashion, I find the population standard deviation to be exactly 16.0."

"But you can't multiply 'football numbers,'" the professor wailed. "Why, they aren't even ordinal numbers, like test scores."

"The numbers don't know that," said the statistician. "Since the numbers don't remember where they came from, they always behave just the same way, regardless."

The professor gasped.

"Now the 1,600 'football numbers' the freshmen bought have a mean of 50.3," the statistician continued. "When I divide the difference between population and sample means by the population standard deviation. . . ."

"Divide!" moaned the professor.

". . . And then multiply by $\sqrt{1,600}$, I find a critical ratio of 10," the statistician went on, ignoring the interruption. "Now, if your population of 'football numbers' had happened to have a normal frequency distribution, I would be able rigorously to assure you that the sample of 1,600 obtained by the freshmen could have arisen from random sampling only once in 65,618,050,000,000,000,000,000 times; for in this case these numbers obviously would obey all the rules that apply to sampling from any normal population."

"You cannot . . ." began the professor.

"Since the population is obviously not normal, it will in this case suffice to use Tchebycheff's inequality,"[1] the statistician continued calmly. "The probability of obtaining a value of 10 for such a critical ratio in random sampling from any population whatsoever is always less than .01. It is therefore highly implausible that the numbers obtained by the freshmen were actually a random sample on all numbers put into the machine."

"You cannot add and multiply any numbers except cardinal numbers," said the professor.

---

1. Tchebycheff's inequality, in a convenient variant, states that in random sampling the probability that a critical ratio of the type calculated here will exceed any chosen constant, c, is always less than $1/c^2$, irrespective of the shape of the population distribution. It is impossible to devise a set of numbers for which this inequality will not hold.

"If you doubt my conclusions," the statistician said coldly as he showed the professor to the door, "I suggest you try and see how often you can get a sample of 1,600 numbers from your machine with a mean below 50.3 or above 58.3. Good night."

To date, after reshuffling the numbers, the professor has drawn (with replacement) a little over 1,000,000,000 samples of 1,600 from his machine. Of these, only two samples have had means below 50.3 or above 58.3. He is continuing his sampling, since he enjoys the computations. But he has put a lock on his machine so that the sophomores cannot tamper with the numbers again. He is happy because, when he has added together a sample of 1,600 "football numbers," he finds that the resulting sum obeys the same laws of sampling as they would if they were real honest-to-God cardinal numbers.

Next year, he thinks, he will arrange things so that the population distribution of his "football numbers" is approximately normal. Then the means and standard deviations that he calculates from these numbers will obey the usual mathematical relations that have been proven to be applicable to random samples from any normal population.

The following year, recovering from his nervous breakdown, Professor X will give up the "football numbers" concession and resume his teaching. He will no longer lock his door when he computes the means and standard deviations of test scores.

References

1. COOMBS, C. H. Mathematical models in psychological scaling *J. Amer. stat. Ass.,* 1951, 46, 480–489.
2. STEVENS, S. S. Mathematics, measurement, and psychophysics. In S. S. Stevens (Ed.), *Handbook of experimental psychology.* New York: Wiley, 1951, Pp. 1–49.
3. WEITZENHOFFER, A. M. Mathematical structures and psychological measurements. *Psychometrika,* 1951, 16, 387–406.

Selected
Bibliography

American Psychological Association (1954). Technical Recommendations for Psychological Tests and Diagnostic Techniques. *Psychol. Bull., Suppl.,* 51, No. 2, part 2 (38 pp.).

Anderson, N. A. (1961). Scales and Statistics: Parametric and Non-Parametric. *Psychol. Bull.,* 58 (July): 305–16.

Attneave, F. (1949). A Method of Graded Dichotomies for the Scaling of Judgments. *Psychol. Rev.,* 56, 334–40.

Bain, R. (1930). Theory and Measurement of Attitudes and Opinions. *Psychol. Bull.,* 27, 357–79.

Bartlett, C. J., L. C. Quay, and L. S. J. Wrightsman (1960). A Comparison of Two Methods of Attitude Measurement: Likert-Type and Forced Choice. *Educ. psychol. Measmt.,* 20, 699–704.

Bechtel G. (1971). A Dual Scaling Analysis for Paired Compositions. *Psychometrika,* 36, 2, 135–54.

Behan, F. L., and R. A. Behan (1954). Football Numbers (continued). *Amer. Psychologist* 9, 262–63.

Bendig, A. W. (1954). Reliability and the Number of Rating Scale Categories. *J. Appl. Psychol.,* 38, 38–40.

Berg, I. A. (1967). *Response Set in Personality Assessment.* Chicago: Aldine.

Bergmann, G., and K. W. Spence (1944). The Logic of Psychophysical Measurement. *Psychol. Rev.,* 51, 1–24.

Bock, R. D., and L. V. Jones (1968). *The Measurement and Prediction of Judgment and Choice.* San Francisco: Holden-Day.

Bohrnstedt, G. W. (1969). A Quick Method for Determining the Reliability and Validity of Multiple Item Scales. *Amer. Soc. Rev.* 34, 4, 542–48.

Boneau, C. A. (1960). The Effects of Violations of Assumptions Underlying the 't' Test. *Psychol. Bull.* 57 (January): 49–64.

Bonjean, C., R. Hill, and S. D. McLemore. (1968). *Sociological Measurement.* San Francisco: Chandler.

Borgatta, E. F. (1955). An Error Ratio for Scalogram Analysis, *Pub. Opin. Quart.* 19, (Spring) 86–100.

———— (1968). My Student, the Purist: A Lament. *Soc. Quart.* 9 (Winter), 29–34.

———— and D. G. Hays (1952). Some Limitations on the Arbitrary Classification of Non-scale Response Patterns in a Guttman Scale. *Pub. Opin. Quart.,* 16, 410–16.

Bradley, R. A. (1954). Incomplete Block Rank Analysis: on the Appropriateness of the Model for a Method of Paired Comparisons. *Biometrics,* 10, 375–90.

Brotemarkle, R. A., and S. W. Fernberger (1934). A Method for Investigating the Validity of the Categories of a Judgment Test. *J. Educ. Psychol.,* 25, 579–84.

Burke, C. J. (1953). Additive Scales and Statistics. *Psychol. Rev.,* 60, 73–75.

Burros, R. H. (1955). The Estimation of the Discriminal Dispersion in the Method of Successive Intervals. *Psychometrika,* 20, 299–305.

————— and W. A. Gibson. (1954). A Solution for Case III of the Law of Comparative Judgment. *Psychometrika,* 19, 57–64.

Burt, C. (1951). Test Construction and the Scaling of Items. *Brit. J. Psychol., Stat. Sect.,* 4, 95–129.

————— (1953). Scale Analysis and Factor Analysis. *Brit. J. Stat. Psychol.,* 6, 5–23.

Campbell, D. P., and W. W. Sorenson (1963). Response Set in Interest Inventory Triads. *Educ. Psychol. Measmt.,* 23, 145–52.

Campbell, D. T. (1950). The Indirect Assessment of Social Attitudes. *Psychol. Bull.,* 47, 15–38.

————— and D. W. Fiske (1959). Convergent and Discriminant Validation by the Multitrait-Multimethod Matrix. *Psychol. Bull.,* 56, 81–105.

Campbell, N. R. (1938). Symposium: Measurement and its Importance for Philosophy. *Proc. Arist. Soc. Suppl.,* 17, 121–42. London: Harrison.

Canter, R. R., and J. Hirsch (1955). An Experimental Comparison of Several Psychological Scales of Weight. *Amer. J. Psychol.,* 68, 645–49.

Castle, P. F. C. (1953). A note on the Scale-Product Technique of Attitude Scale Construction. *Occup. Psychol.* 27, 104–09.

Coombs, C. H. (1948). A Rationale for the Measurement of Traits in Individuals. *Psychometrika,* 13, 59–68.

Coombs, C. H. (1951). Mathematical Models in Psychological Scaling. *J. Amer. Stat. Ass.,* 46, 480–89.

————— (1952). *A Theory of Psychological Scaling.* Engng. Res. Inst. Bull. No. 34. Ann Arbor: Univ. Michigan Press.

————— (1953). Theory and Methods of Social Measurement, in L. Festinger and D. Katz, (ed.), *Research Methods in the Behavioral Sciences.* New York: Dryden.

————— H. Raiffa, and R. M. Thrall (1954). Some Views on Mathematical Models and Measurement Theory, in C. H. Coombs, R. M. Thrall, and R. L. Davis (eds.), *Decision Processes.* New York: Wiley, 19–37. Wiley, 19–37.

————— (1956). The Scale Grid: Some Interrelations of Data Models. *Psychometrika,* 21, 313–30.

————— and Kao, R. C. (1960). On a Connection Between Factor Analysis and Multidimensional Unfolding, in H. Gulliksen and S. Messick (eds.), *Psychological Scaling.* New York: Wiley, pp. 145–54.

————— (1964). *A Theory of Data.* New York: Wiley.

Comrey, A. L. (1950). A Proposed Method for Absolute Ratio Scaling. *Psychometrika,* 15, 317–25.

Couch, A., and K. Keniston (1960). Yeasayers and Naysayers: Agreeing Response Set as a Personality Variable. *J. Abnorm. Soc. Psychol.,* 60, 151–74.

Cozan, L. W. (1959). Forced-Choice: Better than Other Rating Methods? *Personnel,* 36, 80–83.

Cronbach, L. J. (1950). Further Evidence on Response Sets and Test Design. *Educ. Psychol. Measmt.*, 10, 3–31.

———— (1951). Coefficient Alpha and the Internal Structure of Tests. *Psychometrika*, 16, 297–334.

———— and P. E. Meehl. (1955). Construct Validity in Psychological Tests. *Psychol. Bull.*, 52, 177–93.

Crowne, D. P., and D. Marlowe, (1960). A New Scale of Social Desirability Independent of Psychopathology. *J. Consult. Psychol.*, 24, 349–54.

Davis, F. B. (1951). Item Selection Techniques, in E. F. Lindquist (ed.), *Educational Measurement*. Washington, D.C.: Amer. Council on Education.

———— (1952). Item Analysis in Relation to Educational and Psychological Testing. *Psychol. Bull.*, 49, 97–119.

Davis, J. A. (1958). On Criteria for Scale Relationships. *Amer. Jour. Soc.* LXIII, No. 4, (Jan.), 371–80.

Diederich, G. W., S. J. Messick, and L. R. Tucker (1955). A General Least Squares Solution for Successive Intervals. *Res. Bull.* 55–24. Princeton: Educational Testing Service.

Edwards, A. L. (1946). A Critique of Neutral Items in Attitude Scales Constructed by the Method of Equal Appearing Intervals. *Psychol. Rev.*, 53, 159–69.

———— and K. C. Kenny (1946). A Comparison of the Thurstone and Likert Techniques of Attitude Scale Construction. *J. Appl. Psychol.*, 30, 72–83.

———— (1948). On Guttman's Scale Analysis. *Educ. Psychol. Measmt.*, VIII, 313–18.

———— and F. P. Kilpatrick (1948a). Scale Analysis and the Measurement of Social Attitudes. *Psychometrika*, 13, No. 2, (June) 99–114.

———— (1948b). A Technique for the Construction of Attitude Scales. *J. Appl. Psychol.*, 32. 374–84.

———— (1951). *Psychological Scaling by Means of Successive Intervals*. Chicago: University of Chicago Psychometric Laboratory.

———— (1952). The Scaling of Stimuli by the Method of Successive Intervals. *J. Appl. Psychol.*, 36, 118–22.

———— and L. L. Thurstone (1952). An Internal Consistency Check for Scale Values by the Method of Successive Intervals. *Psychometrika*, 17, 169–80.

———— (1956). A Technique for Increasing the Reproducibility of Cumulative Attitude Scales. *J. Appl. Psychol.*, 40, 263–65.

———— (1957a). *Techniques of Attitude Scale Construction*. New York: Appleton-Century-Crofts.

———— (1957b). *The Social Desirability Variable in Personality Assessment and Research*. New York: Dryden.

Edwards, W. (1960). Measurement of Utility and Subjective Probability, in H. Gulliksen and S. Messick (ed.) *Psychological Scaling*. New York: Wiley 109–27.

Ehrenberg, A. S. C. (1955). Mathematics and Measurement in Psychology. *Brit. J. Psych.*, XLVI, 20–29.

Ekman, G., and L. Sjoberg (1965). Scaling, in *Annu. Rev. Psychol.*, 16, 451–74.

Engen, T. (1956). An Evaluation of a Method for Developing Ratio-Scales. *Amer. J. Psychol.*, 69, 92–95.

Ferguson, L. W. (1939). The Requirements of an Adequate Attitude Scale. *Psychol. Bull.*, 26, 665–73.

Ford, R. N. (1950) . A Rapid Scoring Procedure for Scaling Attitude Questions. *Pub. Opin. Quart.*, 14, 507–32.

Forehand, G. A. (1962) . Relationships Among Response Sets and Cognitive Behaviors. *Educ. Psychol. Measmt.*, 22, 287–302.

Francis, R. G. and R. C. Stone (1956) . Measurement and Attitude Analysis. *Midwest Soc.* XVIII, No. 1, 16–27.

Ghiselli, E. E. (1964) . *Theory of Psychological Measurement.* New York: Mc-Graw-Hill.

Gibson, W. A. (1953) . A Least Squares Solution for Case IV of the Law of Comparative Judgment. *Psychometrika*, 18, 15–21.

Goodenough, W. H. (1944) . A Technique for Scale Analysis, *Educ. and Psychol. Measmt.*, 4, No 3 (Autumn) : 179–90.

Goodman, L. A. (1959a) . Simple Statistical Methods for Scalogram Analysis *Psychometrika*, 24, No. 1, 29–43.

————— (1959b) . A Coefficient of Reproducibility. *Psychometrika*, 24 (March) .

Gordon, C. (1963) . A Note on Computer Programs for Guttman Scaling. *Sociometry*, 26, pp. 129–30.

Green, B. F. (1954) . Attitude Measurement, in G. Lindzey (ed.) , *Handbook of Social Psychology.* Cambridge: Addison-Wesley.

————— (1956) . A Method of Scalogram Analysis Using Summary Statistics. *Psychometrika*, 21, 79–88.

————— (1960) . A Technical Note on the Method of Successive Catagories Using Category Means," in H. Gullikson and S. Messick, (ed.) , *Psychological Scaling.* New York: Wiley.

Greenberg. Marshall G. (1965) . A Method of Successive Cumulations for the Scaling of Pair-Comparison Preference Judgments. *Psychometrika*, 30, No. 4 (December) .

Grossnickel, Louise T. (1942) . The Scaling of Test Scores by the Method of Paired Comparisons. *Psychometrika*, 7, 43–64.

Guilford, J. P. (1928) . The Method of Paired Comparisons as a Psychometric Method. *Psychol. Rev.*, 35, 494–506.

————— (1931) . Some Empirical Tests of the Method of Paired Comparisons. *J. Gen. Psychol.*, 5, 64–76.

————— (1932) . A Generalized Psychophysical Law. *Psychol. Rev.*, 39, 73–85.

————— (1938) . The Computation of Psychological Values from Judgments in Absolute Categories. *J. Exp. Psychol.*, 22, 32–42.

————— (1941) The Phi Coefficient and Chi Square as Indices of Item Validity. *Psychometrika*, 6, 11–19.

————— (1954) . *Psychometric Methods,* 2nd ed. London: McGraw-Hill.

Gullahorn, J. E., and J. T. Gullahorn (1968) . The Utility of Applying both Guttman and Factor Analysis to Survey Data. *Sociometry*, 31, No. 2, 213–18.

Gulliksen, H. (1946) . Paired Comparisons and the Logic of Measurement. *Psychol. Rev.*, 53, 199–213.

————— (1950) . *Theory of Mental Tests.* New York: Wiley.

————— (1952) . A Least-Squares Solution for Successive Intervals. *Amer. Psychol.*, 7, 408. (Abstract.)

————— (1954) . A Least Squares Solution for Successive Intervals Assuming Unequal Standard Deviations. *Psychometrika*, 19, 117–39.

————— (1956) A Least Squares Solution for Paired Comparisons with Incomplete Data. *Psychometrika*, 21, 125–34.

———— and Messick, S. (1960). *Psychological Scaling: Theory and Applications*, New York: Wiley.

Guttman, L. (1945). A Basis for Analyzing Test-Retest Reliability. *Psychometrika*, 10, 255–82.

———— (1946). The Test-Retest Reliability of Qualitative Data." *Psychometrika*, 11, No. 2, 81–95.

———— (1946). An Approach to Quantifying Paired Comparisons and Rank Order. *Annals of Mathematical Statistics*, 17, 144–63.

———— (1947). The Cornell Technique for Scale and Intensity Analysis. *Educ. Psychol. Measmt.*, 7, (Summer) 247–79.

———— (1950). Problems of Reliability, in S. A. Stouffer et al., *Measurement and Prediction*. Princeton: Princeton Univ. Press, pp. 277–311.

———— (1950a). Relation of Scalogram Analysis to Other Techniques, in S. A. Stouffer et al., *Measurement and Prediction*. Princeton: Princteon Univ. Press, pp. 172–212.

———— (1950b). The Principal Components of Scale Analysis, in S. A. Stouffer, et. al., *Measurement and Prediction*. Princeton: Princeton Univ. Press. 312–61.

———— (1950c). The Problem of Attitude and Opinion Measurement, in S. A. Stouffer, et al., *Measurement and Prediction*. Princeton: Princeton Univ. Press, pp. 46–59.

———— (1950d). The Third Component of Scalable Attitudes. *Int. J. Opin. and Att. Res.*, 4, 285–87. (Abstract.)

———— (1954). The Principal Components of Scalable Attitudes in P. F. Lazarsfeld, (ed.), *Mathematical Thinking in the Social Sciences*. Glencoe, Ill.: Free Press.

Harman, H. H. (1967). *Modern Factor Analysis*. 2nd Ed. Chicago: Univ. of Chicago Press.

Harris, W. P. (1957). A Revised Law of Comparative Judgment. *Psychometrika*, 22, 189–98.

Hawkes, Roland K. (1971). The Multivariate Analysis of Ordinal Measures. *Amer. J. Soc.*, 76, 5, 908–26.

Hays, W. L. (1954). An Extension of the Unfolding Technique to r-Dimensions. Ph.D. Thesis, University of Michigan.

Hempel, C. G. (1952). Fundamentals of Concept Formation in Empirical Science, *International Encyclopedia of Unified Science Foundations of the Unity of Science*, Chicago: Univ. of Chic. Press, Vol. I-II.

Henry, A. F. (1952). A Method of Classifying Non-Scale Response Patterns in a Guttman Scale. *Pub. Opin. Quart.*, 16, 94–106.

Hevner, K. (1930). An Empirical Study of Three Psychophysical Methods, *J. Genetic Psychol.*, 4, 191–212.

Hicks, L. E. (1970). Some Properties of Ipsative, Normative, and Forced Choice Normative Measures. *Psychol. Bull.*, 74, No. 3, 167–84.

Hovland, C. I. (1938). A Note on Guilford's Generalized Psychophysical Law. *Psychol. Rev.*, 45, 430–34.

———— and M. Sherif (1952). Judgmental Phenomena and Scales of Attitude Measurement: Item Displacement in Thurstone Scales, *J. Abnorm. Soc. Psychol.*, 47, 822–32.

Humphreys, L. G. (1949). Test Homogeneity and its Measurement. *Amer. Psychol.*, 4, 245. (Abstract.)

Jackson, J. M. (1949). A Simple and More Rigorous Technique for Scale Analysis, in *A Manual of Scale Analysis*, Part II. Montreal: McGill Univ., Mimeo.

Jahn, J. A. (1951). Some Further Contributions to Guttman's Scale Analysis. *Amer. Soc. Rev.*, 16, 233–39.

Jardine, R. (1958). Ranking Methods and the Measurement of Attitudes. *J. Amer. Stat. Ass.*, 53, 720–28.

Jones, L. V., and W. W. Rozenboom (1956). The Validity of the Successive Intervals Method of Psychometric Scaling. *Psychometrika*, 21, 165–83.

————— (1960). Some Invariant Findings under the Method of Successive Intervals, in H. Gulliksen, and S. Messick (eds.), *Psychological Scaling*. New York: Wiley, pp. 7–20.

Kahn, L. A., and A. J. Bodine (1951). Guttman Scale Analysis by Means of IBM Equipment. *Educ. Psychol. Measmt.*, 11, 298–314.

Katz, D. (1944). The Measurement of Intensity, in H. Cantril (ed.) *Ganging Public Opinion*. Princeton, N.J.: Princeton University Press.

Kelley, H. H., C. I. Hovland, M. Schwartz, and R. P. Abelson. The Influence of Judges' Attitudes in Three Methods of Attitude Scaling (mimeo). 1953.

Kellog, W. N. (1929). An Experimental Comparison of Psychophysical Methods. *Arch. Psychol.*, No. 106.

————— (1930). An Experimental Evaluation of Equality Judgments in Psychophysics. *Arch. Psychol.*, No. 112.

Kendall, M. G., and B. B. Smith (1940). On the Method of Paired Comparisons. *Biometrika*, 31, 324–45.

Krantz, D. H., and A. Tversky (1971). Conjoint-Measurement Analysis of Composition Rules in Psychology. *Psychol. Rev.*, 78, No. 2, 151–69.

Kriedt, P. H., and K. E. Clark (1949). "Item Analysis" versus "Scale Analysis." *J. Appl. Psychol.*, 33, 114–21.

Labovitz, S. (1967). Some Observations on Measurement and Statistics." *Social Forces*, 46, No. 2 (December) 151–60.

————— (1968). Reply to Champion and Morris. *Social Forces*, 46 (June) 543–45.

Lazarsfeld, P. F. (ed.) (1954). *Mathematical Thinking in the Social Sciences*. Glencoe, Ill.: Free Press.

————— (1959). Latent Structure Analysis, in S. Koch (ed.), *Psychology: A Study of a Science*, Vol. 3. New York: McGraw-Hill, pp. 476–543.

————— (1960). Latent Structure Analysis and Test Theory, in H. Gulliksen and S. Messick (eds.), *Psychological Scaling*. New York: Wiley, pp. 83–95.

————— and N. W. Henry (1968). *Latent Structure Analysis*. New York: Houghton Mifflin.

Likert, R., S. Roslow and G. Murphy, (1934). A Simple and Reliable Method of Scoring the Thurstone Attitude Scales. *J. Soc. Psychol.*, 5, 228–38.

Linder, R. E. (1933). A Statistical Comparison of Psychophysical Methods. *Psychol. Monog.*, 44, no. 199, 1–20.

Loevinger, J. (1947). A Systematic Approach to the Construction and Evaluation of Tests of Ability. *Psychol. Monogr.*, 61, No. 4.

————— (1948). The Technique of Homogeneous Tests Compared With Some Aspects of "Scale Analysis" and "Factor Analysis." *Psychol. Bull.*, 45, 507–29.

Lord, F. M. (1954). Scaling. *Rev. Educ. Research*, 24, 375–93.

Lorge, I. (1951). The Fundamental Nature of Measurement, in E. F. Lindquist (ed.), *Educational Measurement*. Washington, D.C.: Amer. Council on Education.

Marder, E. (1952). Linear Segments: A Technique for Scalogram Analysis. *Pub. Opin. Quar.*, XVI, 417–31.

Manning, S. A., and E. H. Rosenstock (1968). *Classical Psychophysics and Scaling* New York: McGraw-Hill.

Maxwell, A. E. (1959). A Statistical Approach to Scalogram Analysis. *Educ. Psychol. Measmt.*, 19, 3, 337–49.

McNemar, Q. (1946). Opinion-Attitude Methodology, *Psychol. Bull.*, 43, No. 4 (July), 289–374.

Menzel, H. (1953). A New Coefficient for Scalogram Analysis. *Pub. Opin. Quart.*, 17, 268–80.

Messick, S. J. (1956). An Empirical Evaluation of Multidimensional Successive Intervals. *Psychometrika*, 21, 367–76.

Micklin, M., and M. Durbin (1969). Syntactic Dimensions of Attitude Scaling Techniques: Sources of Variation and Bias, *Sociometry*, 32, No. 2, 194–206.

Milholland, J. E. (1953). Dimensionality of Response Patterns for the Method of Single Stimuli. Ph.D. Thesis, University of Michigan.

———— (1955). Four Kinds of Reproducibility in Scale Analysis. *Educ. Psychol. Measmt.*, 15, 478–82.

———— (1964). Theory and Techniques of Assessment. *Annu. Rev. Psychol.*, 15, 311–46.

Miller, D. (1964). *Handbook of Research Design and Social Measurement*. New York: David McKay.

Mosier, C. I. (1940). A Modification of the Method of Successive Intervals. *Psychometrika*, 5, 101–07.

Mosteller, F. (1949). A Theory of Scalogram Analysis, Using Noncumulative Types of Items: A New Approach to Thurstone's Method of Scaling Attitudes. Rep. no. 9, Lab. of Soc. Relations, Harvard University.

———— (1951a). Remarks on the Method of Paired Comparisons. I. The Least Squares Solution Assuming Equal Standard Deviations and Equal Correlations. *Psychometrika*, 16, 3–11.

———— (1951b). Remarks on the Method of Paired Comparisons: II. The Effect of an Abberant Standard Deviation When Equal Standard Deviations and Equal Correlations are Assumed. *Psychometrika*, 16, 203–06.

———— (1951c). Remarks on the Method of Paired Comparisons: III. A Test of Significance for Paired Comparisons when Equal Standard Deviations and Equal Correlations are Assumed, *Psychometrika*, 16, 207–18.

Nevin, J. R. (1953). A Comparison of Two Attitude Scaling Techniques. *Educ. Psychol. Measmt.*, 13, 65–76.

Newman, E. B. (1933). The Validity of the Just Noticable Difference as a Unit of Psychological Magnitude. *Trans, Kans. Acad. Sci.*, 36, 172–75.

Ofshe, R., and L. Ofshe (1970). A Comparative Study of Two Scaling Models: Paired Comparison and Scalogram Analysis. *Sociometry*, 33, 4, 409–26.

———— and R. Anderson (1968). Testing a Measurement Model, in E. Borgatta (ed.), 1969. *Sociological Methodology*. San Francisco: Jossey-Bass.

Osgood, C. E., G. J. Suci, and P. H. Tannenbaum (1957). *The Measurement of Meaning*. Urbana, Ill.: Univ. of Illinois Press.

Peabody, D. (1966). Authoritarianism Scales and Response Bias. *Psychol. Bull.*, 65, 11–23.

Remmers, H. H. (1954). *Introduction to Opinion and Attitude Measurement.* New York: Harper.

Riley, M. W., J. W. Riley, Jr., J. Toby, and M. L. Toby (1954). *Sociological Studies in Scale Analysis.* New Brunswick, N.J.: Rutgers Univ. Press.

Ramsay, J. O. and B. Case (1970). Attitude Measurement and the Linear Model. *Psychol. Bull.,* 74, No. 3 185–92.

Ross, R. T. (1934). Optimum Orders for the Presentation of Pairs in the Method of Paired Comparisons. *J. Educ. Psychol.,* 25, 375–82.

Saffir, M. (1937). A Comparative Study of Scales Constructed by Three Psychophysical Methods. *Psychometrika,* 2, 179–98.

Sagi, P. C. (1959). A Statistical Test for the Significance of the Coefficient of Reproducibility. *Psychometrika,* 24, 19–27.

Saltz, E., M. Reece, and J. Ager (1962). Studies of Forced-Choice Methodology: Individual Differences in Social Desirability. *Educ. Psychol. Measmt.,* 22, 365–70.

Scates, D. E. (1937). The Essential Conditions of Measurement. *Psychometrika,* 2, 27–34.

Schuessler, K. R. (1952). Item Selection in Scale Analysis. *Amer. Soc. Rev.,* 17, 183–92.

Schulman, G. I., and C. R. Tittle (1968). Assimilation-Contrast Effects and Item Selection in Thurstone Scaling. *Social Forces,* 46, No. 4, 484–91.

Schutz, R. E., and R. J. Foster (1963). A Factor Analytic Study of Acquiescent and Extreme Response Set. *Educ. psychol. Measmt.,* 23, 435–47.

Scott, William A. (1960). Measures of Test Homogeneity. *Educ. psychol. Measmt.,* 20, 751–57.

———— (1968). Attitude Measurement, in G. Lindzey and E. Aronson, *Handbook of Social Psychology,* 2nd ed. Reading: Addison-Wesley.

Senders, V. L. (1953). A Comment on Burke's Additive Scales and Statistics. *Psychol. Rev.,* 60, 423–24.

Shaw, M. and J. Wright (1967). *Scales for the Measurement of Attitudes.* New York: McGraw-Hill.

Shepard, R. (1960). Similarity of Stimuli and Metric Properties of Behavioral Data, in H. Gulliksen and S. Messick, (eds.), *Psychological Scaling.* New York: Wiley.

Shuford, E. H., L. V. Jones, and R. D. Bock (1960). A Rational Origin Obtained by the Method of Contingent Paired Comparisons, *Psychometrika* 25, 343–56.

Siegel, S. (1956). *Non-Parametric Statistics.* New York: McGraw-Hill, p. 76.

———— (1956). A Method for Obtaining an Ordered Metric Scale. *Psychometrika,* 21, 207–16.

Sjoberg, L. (1967). Successive Intervals Scaling of Paired Comparisons, *Psychometrika,* 32, 3, 297–308.

Slater, P. (1956). Weighting Responses to Items in Attitude Scales. *Brit. J. Stat. Psychol.,* X, 41–48.

Smith, D. (1967). Correcting for Social Desirability Response Sets in Opinion-Attitude Survey Research. *Pub. Opin. Quart.,* 31, 87–94.

Smith, R. G., Jr. (1950). Reproducible Scales and the Assumption of Normality. *Educ. Psychol. Measmt.,* 10, 395–99.

———— (1951). "Randomness of Error" in Reproducible Scales. *Educ. and Psychol. Measmt.,* XI, 587–96.

Snider, J., and C. Osgood (1968). *The Semantic Differential: A Sourcebook*. Chicago: Aldine.

Somers, R. H. (1968). An Approach to the Multivariate Analysis of Ordinal Data. *Amer. Soc. Rev.*, 33, 6, 971–77.

Stevens, S. S. (1939). On the Problem of Scales for the Measurement of Psychological Magnitudes. *J. Unif. Sci.*, 9, 94–99.

———— (1946). On the Theory of Scales of Measurement, *Science*, 103, 677–80.

———— (1951). Mathematics, Measurement, and Psychophysics, in S. S. Stevens ed.), *Handbook of Experimental Psychology*. New York: Wiley, pp. 1–49.

———— (1956). On the Psychophysical Law. Harvard Univ., Psycho-Acoustic Lab. Rep. PNR–188.

———— (1957). On the Psychophysical Law. *Psychol. Rev.*, 64, 153–81.

———— (1958). Problems and Methods of Psychophysics. *Psychol. Bull.*, 54, 177–96.

———— (1960). Ratio Scales, Partition Scales, and Confusion Scales, in H. Gulliksen, and S. Messick, (eds.), *Psychological Scaling*. New York: Wiley, pp. 49–66.

———— (1971). Issues in Psychophysical Measurement. *Psychol. Rev.*, 78, 5, 451–56.

Stouffer, S. A., L. Guttman, et al. (1950). *Measurement and Prediction*. Princeton: Princeton Univ. Press.

Suchman, E. A., and L. Guttman (1947). A Solution to the Problem of Question "Bias." *Pub. Opin. Quart.*, 11, No. 3, (Fall) 445–55.

———— (1950a). The Logic of Scale Construction. *Educ. Psychol. Measmt.*, 10, 79–93.

———— (1950b). The Utility of Scalogram Analysis, in S. A. Stouffer et al., *Measurement and Prediction*. Princeton: Princeton Univ. Press., pp. 122–171.

———— (1950c). The Intensity Component in Attitude and Opinion Research, in S. A. Stouffer, et al., *Measurement and Prediction*. Princeton: Princeton Univ. Press, 213–76.

TenHouten, W. D. (1969). Scale Gradient Analysis: A Statistical Method For Constructing and Evaluating Guttman Scales. *Sociometry*, 32, No. 1. (March) 80–98.

Thomson, G. H. (1912). A Comparison of the Psychophysical Methods. *Brit. J. Psychol.*, 5, 203–41.

Thurstone, L. L. (1925). A Method of Scaling Psychological and Educational Tests. *J. Educ. Psychol.*, 16, 433–51.

———— (1927). Three Psychophysical Laws. *Psychol. Rev.*, XXXIV, 424–32.

———— (1927a). The Method of Paired Comparisons for Social Values. *J. Abnor. Soc. Psychol.*, XXI, 384–400.

———— (1927b). Equally Often Noticed Differences. *J. Educ. Psychol.*, XVIII, 289–93.

———— (1927c). A Mental Unit of Measurement, *Psychol. Rev.*, XXXIV, 415–23.

———— (1928a). Attitudes Can Be Measured. *Am. Jour. Soc.*, 33, 529–54.

———— (1928b). Scale Construction with Weighted Observations. *J. Educ. Psychol.*, 19, 441–53.

———— (1929a). Theory of Attitude Measurement. *Psychol. Rev.*, XXXVI, 222–41.

————— (1929b). Fechner's Law and the Method of Equal-Appearing Intervals. *Jour. of Experi. Psychol.*, XII, 214–24.

————— and E. J. Chave (1929c). *The Measurement of Attitude.* Chicago: Univ. of Chicago Press.

————— (1931a). The Measurement of Change in Social Attitudes. *J. Soc. Psychol.*, II, 230–35.

————— (1931b). Rank Order as a Psychophysical Method. *J. Experi. Psychol.*, XVI, 187–201.

————— (1932). Stimulus Dispersions in the Method of Constant Stimuli. *J. Experi. Psychol.*, XV, 284–97.

————— (1948). Psychophysical Methods, in T. G. Andrews (ed.), *Methods of Psychology.* New York: Wiley.

————— (1954). The Measurement of Values. *Psychol. Rev.*, LXI, 47–58.

————— (1959). *The Measurement of Values.* Chicago: Univ. of Chicago Press.

————— and L. V. Jones (1959). The Rational Origin for Measuring Subjective Values (completed in 1956), in L. L. Thurstone, *The Measurement of Value.* Chicago: Univ. of Chicago Press.

Torgerson, W. S. (1954). A Law of Categorical Judgment, *Amer. Psychologist,* 9, 483. (Abstract.)

————— (1958). *Theory and Methods of Scaling.* New York: Wiley.

————— (1960). Quantitative Judgment Scales, in H. Gulliksen and S. Messick (eds.), *Psychological Scaling.* New York: Wiley. pp. 21–31.

Uhrbrock, R. S., and M. W. Richardson (1933). Item Analysis. *Person. J.,* 12, 141–54.

Upshaw, H. S. (1962). Own Attitude as an Anchor in Equal-Appearing Intervals. *J. Abnorm. Soc. Psychol.*, 64, 85–96.

————— (1964). A Linear Alternative to Assimilation and Contrast: A Reply to Manis, *J. Abnorm. and Soc. Psychol.*, 68, 691–93.

————— (1965). The Effect of Variable Perspectives on Judgments of Opinion Statements for Thurstone Scales: Equal-appearing Intervals, *J. Pers. Soc. Psychol.*, 2, 60–69.

————— (1968). Attitude Measurement, in H. M. Blalock, Jr. and A. B. Blalock, *Methodology in Social Research.* New York: McGraw-Hill.

Urban, F. M. (1939). The Method of Equal Appearing Intervals. *Psychometrika,* 4, 117–31.

Volkmann, J. (1937). The Natural Number of Categories in Absolute Judgments. *Psychol. Bull.,* 34, 543.

————— (1938). The Compression of an Absolute Scale. *Psychol. Bull.,* 35, 676.

—————, W. A. Hunt, and M. McGourty (1940). Variability of judgment as a function of stimulus density. *Amer. J. Psychol.,* 53, 277–284.

————— (1951). Scales of judgment and their implications for social psychology. In J. H. Rohrer and M. Sherif (Eds.) *Social Psychology at the Crossroads.* New York: Harper.

Weitzenhoffer, A. M. (1951). Mathematical structures and psychological measurements. *Psychometrika,* 16, 387–406.

Werner, R. (1966). A Fortran Program for Guttman and Other Scalogram Analysis. Compter Data Analysis Working Paper Number 1. Syracuse, New York: Systems Research Committee.

Werts, C. E. & R. L. Linn (1970). Cautions in Applying Various Procedures for

Determining the Reliability and Validity of Multiple Item Scales, *Amer. Soc. Rev.*, 34, 4, 757–59.

Wherry, R. J. (1938). Orders for the presentation of pairs in the method of paired comparisons. *J. Exp. Psychol.*, 23, 651–60.

————— & Gaylord, R. H. (1943). The concept of test and item reliability in relation to factor pattern. *Psychometrika*, 8, 247–69.

—————, and R. H. Gaylord (1944). Factor pattern of test items and tests as a function of the correlation coefficient, content, difficulty and constant error factors. *Psychometrika*, 9, 237–44.

Wiener, N. (1920). A New Theory of Measurement: A Study in the Logic of Mathematics. *Proc. London Math. Soc.*, Ser., 2, 19, 181–205.

Willis, R. (1954). Estimating the Scalability of a Series of Items: An Application of Information Theory. *Psychol. Bull.*, 51, 511–16.

————— (1960). Manipulation of Item Marginal Frequencies by Means of Multiple Response Items. *Psychol. Rev.*, 67, 32–50.

Wilson, T. P. (1971). Critique of Ordinal Values. *Social Forces*, 49, 3, 432–44.

Witryol, S. L. (1954). Scaling Procedures Based on the Method of Paired Comparisons. *J. Appl. Psychol.*, 38, 31–37.

Wrightsman, L. (1966). *Characteristics of Positively-Scored and Negatively-Scored Items from Attitude Scales.* Nashville: Peabody Teachers' College.

Zavala, A. (1965). Development of the Forced-Choice Rating Scale Technique. *Psychol. Bull.*, 62, 117–24.

Zavalloni, Marisa, and S. W. Cook (1965). Influence of Judges' Attitudes on Ratings of Favorableness of Statements about a Social Group, *J. Pers. Soc. Psychol.*, 1, 43–54.

Appendix: Tables

Table I. Angles Corresponding to Percentages,
Angle = Arc sin √ Percentage

%	0	1	2	3	4	5	6	7	8	9
0.0	0	0.57	0.81	0.99	1.15	1.28	1.40	1.25	1.62	1.72
0.1	1.81	1.90	1.99	2.07	2.14	2.22	2.29	2.36	2.43	2.50
0.2	2.56	2.63	2.69	2.75	2.81	2.87	2.92	2.98	3.03	3.09
0.3	3.14	3.19	3.24	3.29	3.34	3.39	3.44	3.49	3.53	3.58
0.4	3.63	3.67	3.72	3.76	3.80	3.85	3.89	3.93	3.97	4.01
0.5	4.05	4.09	4.13	4.17	4.21	4.25	4.29	4.33	4.37	4.40
0.6	4.44	4.48	4.52	4.55	4.59	4.62	4.66	4.69	4.73	4.76
0.7	4.80	4.83	4.87	4.90	4.93	4.97	5.00	5.03	5.07	5.10
0.8	5.13	5.16	5.20	5.23	5.26	5.29	5.32	5.35	5.38	5.41
0.9	5.44	5.47	5.50	5.53	5.56	5.59	5.62	5.65	5.68	5.71
1	5.74	6.02	6.29	6.55	6.80	7.04	7.27	7.49	7.71	7.92
2	8.13	8.33	8.53	8.72	8.91	9.10	9.28	9.46	9.63	9.81
3	9.98	10.14	10.31	10.47	10.63	10.78	10.94	11.09	11.24	11.39
4	11.54	11.68	11.83	11.97	12.11	12.25	12.39	12.52	12.66	12.79
5	12.92	13.05	13.18	13.31	13.44	13.56	13.69	13.81	13.94	14.06
6	14.18	14.30	14.42	14.54	14.65	14.77	14.89	15.00	15.12	15.23
7	15.34	15.45	15.56	15.68	15.79	15.89	16.00	16.11	16.22	16.32
8	16.43	16.54	16.64	16.74	16.85	16.95	17.05	17.16	17.26	17.36
9	17.46	17.56	17.66	17.76	17.85	17.95	18.05	18.15	18.24	18.34

Table I.—(Continued)

%	0	1	2	3	4	5	6	7	8	9
10	18.44	18.53	18.63	18.72	18.81	18.91	19.00	19.09	19.19	19.28
11	19.37	19.46	19.55	19.64	19.73	19.82	19.91	20.00	20.09	20.18
12	20.27	20.36	20.44	20.53	20.62	20.70	20.79	20.88	20.96	21.05
13	21.13	21.22	21.30	21.39	21.47	21.56	21.64	21.72	21.81	21.89
14	21.97	22.06	22.14	22.22	22.30	22.38	22.46	22.55	22.63	22.71
15	22.79	22.87	22.95	23.03	23.11	23.19	23.26	23.34	23.42	23.50
16	23.58	23.66	23.73	23.81	23.89	23.97	24.04	24.12	24.20	24.27
17	24.35	24.43	24.50	24.58	24.65	24.73	24.80	24.88	24.95	25.03
18	25.10	25.18	25.25	25.33	25.40	25.48	25.55	25.62	25.70	25.77
19	25.84	25.92	25.99	26.06	26.13	26.21	26.28	26.35	26.42	26.49
20	26.56	26.64	26.71	26.78	26.85	26.92	26.99	27.06	27.13	27.20
21	27.28	27.35	27.42	27.49	27.56	27.63	27.69	27.76	27.83	27.90
22	27.97	28.04	28.11	28.18	28.25	28.32	28.38	28.45	28.52	28.59
23	28.66	28.73	28.79	28.86	28.93	29.00	29.06	29.13	29.20	29.27
24	29.33	29.40	29.47	29.53	29.60	29.67	29.73	29.80	29.87	29.93
25	30.00	30.07	30.13	30.20	30.26	30.33	30.40	30.46	30.53	30.59
26	30.66	30.72	30.79	30.85	30.92	30.98	31.05	31.11	31.18	31.24
27	31.31	31.37	31.44	31.50	31.56	31.63	31.69	31.76	31.82	31.88
28	31.95	32.01	32.08	32.14	32.20	32.27	32.33	32.39	32.46	32.52
29	32.58	32.65	32.71	32.77	32.83	32.90	32.96	33.02	33.09	33.15
30	33.21	33.27	33.34	33.40	33.46	33.52	33.58	33.65	33.71	33.77
31	33.83	33.89	33.96	34.02	34.08	34.14	34.20	34.27	34.33	34.39
32	34.45	34.51	34.57	34.63	34.70	34.76	34.82	34.88	34.94	35.00
33	35.06	35.12	35.18	35.24	35.30	35.37	35.43	35.49	35.55	35.61
34	35.67	35.73	35.79	35.85	35.91	35.97	36.03	36.09	36.15	36.21
35	36.27	36.33	36.39	36.45	36.51	36.57	36.63	36.69	36.75	36.81
36	36.87	36.93	36.99	37.05	37.11	37.17	37.23	37.29	37.35	37.41
37	37.47	37.52	37.58	37.64	37.70	37.76	37.82	37.88	37.94	38.00
38	38.06	38.12	38.17	38.23	38.29	38.35	38.41	38.47	38.53	38.59
39	38.65	38.70	38.76	38.82	38.88	38.94	39.00	39.06	39.11	39.17
40	39.23	39.29	39.35	39.41	39.47	39.52	39.58	39.64	39.70	39.76
41	39.82	39.87	39.93	39.99	40.05	40.11	40.16	40.22	40.28	40.34
42	40.40	40.46	40.51	40.57	40.63	40.69	40.74	40.80	40.86	40.92
43	40.98	41.03	41.09	41.15	41.21	41.27	41.32	41.38	41.44	41.50
44	41.55	41.61	41.67	41.73	41.78	41.84	41.90	41.96	42.02	42.07
45	42.13	42.19	42.25	42.30	42.36	42.42	42.48	42.53	42.59	42.65
46	42.71	42.76	42.82	42.88	42.94	42.99	43.05	43.11	43.17	43.22
47	43.28	43.34	43.39	43.45	43.51	43.57	43.62	43.68	43.74	43.80
48	43.85	43.91	43.97	44.03	44.08	44.14	44.20	44.25	44.31	44.37
49	44.43	44.48	44.54	44.60	44.66	44.71	44.77	44.83	44.89	44.94

Table I.—(Continued)

%	0	1	2	3	4	5	6	7	8	9
50	45.00	45.06	45.11	45.17	45.23	45.29	45.34	45.40	45.46	45.52
51	45.57	45.63	45.69	45.75	45.80	45.86	45.92	45.97	46.03	46.09
52	46.15	46.20	46.26	46.32	46.38	46.43	46.49	46.55	46.61	46.66
53	46.72	46.78	46.83	46.89	46.95	47.01	47.06	47.12	47.18	47.24
54	47.29	47.35	47.41	47.47	47.52	47.58	47.64	47.70	47.75	47.81
55	47.87	47.93	47.98	48.04	48.10	48.16	48.22	48.27	48.33	48.39
56	48.45	48.50	48.56	48.62	48.68	48.73	48.79	48.85	48.91	48.97
57	49.02	49.08	49.14	49.20	49.26	49.31	49.37	49.43	49.49	49.54
58	49.60	49.66	49.72	49.78	49.84	49.89	49.95	50.01	50.07	50.13
59	50.18	50.24	50.30	50.36	50.42	50.48	50.53	50.59	50.65	50.71
60	50.77	50.83	50.89	50.94	51.00	51.06	51.12	51.18	51.24	51.30
61	51.35	51.41	51.47	51.53	51.59	51.65	51.71	51.77	51.83	51.88
62	51.94	52.00	52.06	52.12	52.18	52.24	52.30	52.36	52.42	52.48
63	52.53	52.59	52.65	52.71	52.77	52.83	52.89	52.95	53.01	53.07
64	53.13	53.19	53.25	53.31	53.37	53.43	53.49	53.55	53.61	53.67
65	53.73	53.79	53.85	53.91	53.97	54.03	54.09	54.15	54.21	54.27
66	54.33	54.39	54.45	54.51	54.57	54.63	54.70	54.76	54.82	54.88
67	54.94	55.00	55.06	55.12	55.18	55.24	55.30	55.37	55.43	55.49
68	55.55	55.61	55.67	55.73	55.80	55.86	55.92	55.98	56.04	56.11
69	56.17	56.23	56.29	56.35	56.42	56.48	56.54	56.60	56.66	56.73
70	56.79	56.85	56.91	56.98	57.04	57.10	57.17	57.23	57.29	57.35
71	57.42	57.48	57.54	57.61	57.67	57.73	57.80	57.86	57.92	57.99
72	58.05	58.12	58.18	58.24	58.31	58.37	58.44	58.50	58.56	58.63
73	58.69	58.76	58.82	58.89	58.95	59.02	59.08	59.15	59.21	59.28
74	59.34	59.41	59.47	59.54	59.60	59.67	59.74	59.80	59.87	59.93
75	60.00	60.07	60.13	60.20	60.27	60.33	60.40	60.47	60.53	60.60
76	60.67	60.73	60.80	60.87	60.94	61.00	61.07	61.14	61.21	61.27
77	61.34	61.41	61.48	61.55	61.62	61.68	61.75	61.82	61.89	61.96
78	62.03	62.10	62.17	62.24	62.31	62.37	62.44	62.51	62.58	62.65
79	62.72	62.80	62.87	62.94	63.01	63.08	63.15	63.22	63.29	63.36
80	63.44	63.51	63.58	63.65	63.72	63.79	63.87	63.94	64.01	64.08
81	64.16	64.23	64.30	64.38	64.45	64.52	64.60	64.67	64.75	64.82
82	64.90	64.97	65.05	65.12	65.20	65.27	65.35	65.42	65.50	65.57
83	65.65	65.73	65.80	65.88	65.96	66.03	66.11	66.19	66.27	66.34
84	66.42	66.50	66.58	66.66	66.74	66.81	66.89	66.97	67.05	67.13
85	67.21	67.29	67.37	67.45	67.54	67.62	67.70	67.78	67.86	67.94
86	68.03	68.11	68.19	68.28	68.36	68.44	68.53	68.61	68.70	68.78
87	68.87	68.95	69.04	69.12	69.21	69.30	69.38	69.47	69.56	69.64
88	69.73	69.82	69.91	70.00	70.09	70.18	70.27	70.36	70.45	70.54
89	70.63	70.72	70.81	70.91	71.00	71.09	71.19	71.28	71.37	71.47

Table I.—(Continued)

%	0	1	2	3	4	5	6	7	8	9
90	71.56	71.66	71.76	71.85	71.95	72.05	72.15	72.24	72.34	72.44
91	72.54	72.64	72.74	72.84	72.95	73.05	73.15	73.26	73.36	73.46
92	73.57	73.68	73.78	73.89	74.00	74.11	74.21	74.32	74.44	74.55
93	74.66	74.77	74.88	75.00	75.11	75.23	75.35	75.46	75.58	75.70
94	75.82	75.94	76.06	76.19	76.31	76.44	76.56	76.69	76.82	76.95
95	77.08	77.21	77.34	77.48	77.61	77.75	77.89	78.03	78.17	78.32
96	78.46	78.61	78.76	78.91	79.06	79.22	79.37	79.53	79.69	79.86
97	80.02	80.19	80.37	80.54	80.72	80.90	81.09	81.28	81.47	81.67
98	81.87	82.08	82.29	82.51	82.73	82.96	83.20	83.45	83.71	83.98
99.0	84.26	84.29	84.32	84.35	84.38	84.41	84.44	84.47	84.50	84.53
99.1	84.56	84.59	84.62	84.65	84.68	84.71	84.74	84.77	84.80	84.84
99.2	84.87	84.90	84.93	84.97	85.00	85.03	85.07	85.10	85.13	85.17
99.3	85.20	85.24	85.27	85.31	85.34	85.38	85.41	85.45	85.48	85.52
99.4	85.56	85.60	85.63	85.67	85.71	85.75	85.79	85.83	85.87	85.91
99.5	85.95	85.99	86.03	86.07	86.11	86.15	86.20	86.24	86.28	86.33
99.6	86.37	86.42	86.47	86.51	86.56	86.61	86.66	86.71	86.76	86.81
99.7	86.86	86.91	86.97	87.02	87.08	87.13	87.19	87.25	87.31	87.37
99.8	87.44	87.50	87.57	87.64	87.71	87.78	87.86	87.93	88.01	88.10
99.9	88.19	88.28	88.38	88.48	88.60	88.72	88.85	89.01	89.19	89.43
100.0	90.00	..	..	..	..	..	..	..	..	..

Note: Reprinted by permission from *Statistical Methods,* 4th Edition, by George W. Snedecor, © 1946 by The Iowa State University Press, AMES, Iowa.

Table II. Table of normal deviates z corresponding to proportions p of a dichotomized unit normal distribution

p	0	1	2	3	4	5	6	7	8	9
.99	2.326	2.366	2.409	2.457	2.512	2.576	2.652	2.748	2.878	3.090
.98	2.054	2.075	2.097	2.120	2.144	2.170	2.197	2.226	2.257	2.290
.97	1.881	1.896	1.911	1.927	1.943	1.960	1.977	1.995	2.014	2.034
.96	1.751	1.762	1.774	1.787	1.799	1.812	1.825	1.838	1.852	1.866
.95	1.645	1.655	1.665	1.675	1.685	1.695	1.706	1.717	1.728	1.739
.94	1.555	1.563	1.572	1.580	1.589	1.598	1.607	1.616	1.626	1.635
.93	1.476	1.483	1.491	1.499	1.506	1.514	1.522	1.530	1.538	1.546
.92	1.405	1.412	1.419	1.426	1.433	1.440	1.447	1.454	1.461	1.468
.91	1.341	1.347	1.353	1.359	1.366	1.372	1.379	1.385	1.392	1.398
.90	1.282	1.287	1.293	1.299	1.305	1.311	1.317	1.323	1.329	1.335
.89	1.227	1.232	1.237	1.243	1.248	1.254	1.259	1.265	1.270	1.276
.88	1.175	1.180	1.185	1.190	1.195	1.200	1.206	1.211	1.216	1.221
.87	1.126	1.131	1.136	1.141	1.146	1.150	1.155	1.160	1.165	1.170
.86	1.080	1.085	1.089	1.094	1.098	1.103	1.108	1.112	1.117	1.122
.85	1.036	1.041	1.045	1.049	1.054	1.058	1.063	1.067	1.071	1.076
.84	.994	.999	1.003	1.007	1.011	1.015	1.019	1.024	1.028	1.032
.83	.954	.958	.962	.966	.970	.974	.978	.982	.986	.990
.82	.915	.919	.923	.927	.931	.935	.938	.942	.946	.950
.81	.878	.882	.885	.889	.893	.896	.900	.904	.908	.912
.80	.842	.845	.849	.852	.856	.860	.863	.867	.871	.874
.79	.806	.810	.813	.817	.820	.824	.827	.831	.834	.838
.78	.772	.776	.779	.782	.786	.789	.793	.796	.800	.803
.77	.739	.742	.745	.749	.752	.755	.759	.762	.765	.769
.76	.706	.710	.713	.716	.719	.722	.726	.729	.732	.736
.75	.674	.678	.681	.684	.687	.690	.693	.697	.700	.703
.74	.643	.646	.650	.653	.656	.659	.662	.665	.668	.671
.73	.613	.616	.619	.622	.625	.628	.631	.634	.637	.640
.72	.583	.586	.589	.592	.595	.598	.601	.604	.607	.610
.71	.553	.556	.559	.562	.565	.568	.571	.574	.577	.580
.70	.524	.527	.530	.533	.536	.539	.542	.545	.548	.550
.69	.496	.499	.502	.504	.507	.510	.513	.516	.519	.522
.68	.468	.470	.473	.476	.479	.482	.485	.487	.490	.493
.67	.440	.443	.445	.448	.451	.454	.457	.459	.462	.465
.66	.412	.415	.418	.421	.423	.426	.429	.432	.434	.437
.65	.385	.388	.391	.393	.396	.399	.402	.404	.407	.410
.64	.358	.361	.364	.366	.369	.372	.375	.377	.380	.383
.63	.332	.335	.337	.340	.342	.345	.348	.350	.353	.356
.62	.305	.308	.311	.313	.316	.319	.321	.324	.327	.329
.61	.279	.282	.285	.287	.290	.292	.295	.298	.300	.303
.60	.253	.256	.259	.261	.264	.266	.269	.272	.274	.277

Table II.—(Continued)

p	0	1	2	3	4	5	6	7	8	9
.59	.228	.230	.233	.235	.238	.240	.243	.246	.248	.251
.58	.202	.204	.207	.210	.212	.215	.217	.220	.222	.225
.57	.176	.179	.181	.184	.187	.189	.192	.194	.197	.199
.56	.151	.154	.156	.159	.161	.164	.166	.169	.171	.174
.55	.126	.128	.131	.133	.136	.138	.141	.143	.146	.148
.54	.100	.103	.105	.108	.111	.113	.116	.118	.121	.123
.53	.075	.078	.080	.083	.085	.088	.090	.093	.095	.098
.52	.050	.053	.055	.058	.060	.063	.065	.068	.070	.073
.51	.025	.028	.030	.033	.035	.038	.040	.043	.045	.048
.50	.000	.003	.005	.008	.010	.013	.015	.018	.020	.023
.49	− .025	− .023	− .020	− .018	− .015	− .013	− .010	− .008	− .005	− .003
.48	− .050	− .048	− .045	− .043	− .040	− .038	− .035	− .033	− .030	− .028
.47	− .075	− .073	− .070	− .068	− .065	− .063	− .060	− .058	− .055	− .053
.46	− .100	− .098	− .095	− .093	− .090	− .088	− .085	− .083	− .080	− .078
.45	− .126	− .123	− .121	− .118	− .116	− .113	− .111	− .108	− .105	− .103
.44	− .151	− .148	− .146	− .143	− .141	− .138	− .136	− .133	− .131	− .128
.43	− .176	− .174	− .171	− .169	− .166	− .164	− .161	− .159	− .156	− .154
.42	− .202	− .199	− .197	− .194	− .192	− .189	− .187	− .184	− .181	− .179
.41	− .228	− .225	− .222	− .220	− .217	− .215	− .212	− .210	− .207	− .204
.40	− .253	− .251	− .248	− .246	− .243	− .240	− .238	− .235	− .233	− .230
.39	− .279	− .277	− .274	− .272	− .269	− .266	− .264	− .261	− .259	− .256
.38	− .305	− .303	− .300	− .298	− .295	− .292	− .290	− .287	− .285	− .282
.37	− .332	− .329	− .327	− .324	− .321	− .319	− .316	− .313	− .311	− .308
.36	− .358	− .356	− .353	− .350	− .348	− .345	− .342	− .340	− .337	− .335
.35	− .385	− .383	− .380	− .377	− .375	− .372	− .369	− .366	− .364	− .361
.34	− .412	− .410	− .407	− .404	− .402	− .399	− .396	− .393	− .391	− .388
.33	− .440	− .437	− .434	− .432	− .429	− .426	− .423	− .421	− .418	− .415
.32	− .468	− .465	− .462	− .459	− .457	− .454	− .451	− .448	− .445	− .443
.31	− .496	− .493	− .490	− .487	− .485	− .482	− .479	− .476	− .473	− .470
.30	− .524	− .522	− .519	− .516	− .513	− .510	− .507	− .504	− .502	− .499
.29	− .553	− .550	− .548	− .545	− .542	− .539	− .536	− .533	− .530	− .527
.28	− .583	− .580	− .577	− .574	− .571	− .568	− .565	− .562	− .559	− .556
.27	− .613	− .610	− .607	− .604	− .601	− .598	− .595	− .592	− .589	− .586
.26	− .643	− .640	− .637	− .634	− .631	− .628	− .625	− .622	− .619	− .616
.25	− .674	− .671	− .668	− .665	− .662	− .659	− .656	− .653	− .650	− .646
.24	− .706	− .703	− .700	− .697	− .693	− .690	− .687	− .684	− .681	− .678
.23	− .739	− .736	− .732	− .729	− .726	− .722	− .719	− .716	− .713	− .710
.22	− .772	− .769	− .765	− .762	− .759	− .755	− .752	− .749	− .745	− .742
.21	− .806	− .803	− .800	− .796	− .793	− .789	− .786	− .782	− .779	− .776
.20	− .842	− .838	− .834	− .831	− .827	− .824	− .820	− .817	− .813	− .810

Table II.—(Continued)

p	0	1	2	3	4	5	6	7	8	9
.19	− .878	− .874	− .871	− .867	− .863	− .860	− .856	− .852	− .849	− .845
.18	− .915	− .912	− .908	− .904	− .900	− .896	− .893	− .889	− .885	− .882
.17	− .954	− .950	− .946	− .942	− .938	− .935	− .931	− .927	− .923	− .919
.16	− .994	− .990	− .986	− .982	− .978	− .974	− .970	− .966	− .962	− .958
.15	−1.036	−1.032	−1.028	−1.024	−1.019	−1.015	−1.011	−1.007	−1.003	− .999
.14	−1.080	−1.076	−1.071	−1.067	−1.063	−1.058	−1.054	−1.049	−1.045	−1.041
.13	−1.126	−1.122	−1.117	−1.112	−1.108	−1.103	−1.098	−1.094	−1.089	−1.085
.12	−1.175	−1.170	−1.165	−1.160	−1.155	−1.150	−1.146	−1.141	−1.136	−1.131
.11	−1.227	−1.221	−1.216	−1.211	−1.206	−1.200	−1.195	−1.190	−1.185	−1.180
.10	−1.282	−1.276	−1.270	−1.265	−1.259	−1.254	−1.248	−1.243	−1.237	−1.232
.09	−1.341	−1.335	−1.329	−1.323	−1.317	−1.311	−1.305	−1.299	−1.293	−1.287
.08	−1.405	−1.398	−1.392	−1.385	−1.379	−1.372	−1.366	−1.359	−1.353	−1.347
.07	−1.476	−1.468	−1.461	−1.454	−1.447	−1.440	−1.433	−1.426	−1.419	−1.412
.06	−1.555	−1.546	−1.538	−1.530	−1.522	−1.514	−1.506	−1.499	−1.491	−1.483
.05	−1.645	−1.635	−1.626	−1.616	−1.607	−1.598	−1.589	−1.580	−1.572	−1.563
.04	−1.751	−1.739	−1.728	−1.717	−1.706	−1.695	−1.685	−1.675	−1.665	−1.655
.03	−1.881	−1.866	−1.852	−1.838	−1.825	−1.812	−1.799	−1.787	−1.774	−1.762
.02	−2.054	−2.034	−2.014	−1.995	−1.977	−1.960	−1.943	−1.927	−1.911	−1.896
.01	−2.326	−2.290	−2.257	−2.226	−2.197	−2.170	−2.144	−2.120	−2.097	−2.075
.00		−3.090	−2.878	−2.748	−2.652	−2.576	−2.512	−2.457	−2.409	−2.366

NOTE: This table is reprinted from A. L. Edward, *Techniques of Attitude Scale Construction*, pp. 246–247. Copyright © 1957.

Table III. Table of X²*

Degrees of Freedom	P = .99	.98	.95	.90	.80	.70	.50	.30	.20	.10	.05	.02	.01
1	.000157	.000628	.00393	.0158	.0642	.148	.455	1.074	1.642	2.706	3.841	5.412	6.635
2	.0201	.0404	.103	.211	.446	.713	1.386	2.408	3.219	4.605	5.991	7.824	9.210
3	.115	.185	.352	.584	1.005	1.424	2.366	3.665	4.642	6.251	7.815	9.837	11.341
4	.297	.429	.711	1.064	1.649	2.195	3.357	4.878	5.989	7.779	9.488	11.668	13.277
5	.554	.752	1.145	1.610	2.343	3.000	4.351	6.064	7.289	9.236	11.070	13.388	15.086
6	.872	1.134	1.635	2.204	3.070	3.828	5.348	7.231	8.558	10.645	12.592	15.033	16.812
7	1.239	1.564	2.167	2.833	3.822	4.671	6.346	8.383	9.803	12.017	14.067	16.622	18.475
8	1.646	2.032	2.733	3.490	4.594	5.527	7.344	9.524	11.030	13.362	15.507	18.168	20.090
9	2.088	2.532	3.325	4.168	5.380	6.393	8.343	10.656	12.242	14.684	16.919	19.679	21.666
10	2.558	3.059	3.940	4.865	6.179	7.267	9.342	11.781	13.442	15.987	18.307	21.161	23.209
11	3.053	3.609	4.575	5.578	6.989	8.148	10.341	12.899	14.631	17.275	19.675	22.618	24.725
12	3.571	4.178	5.226	6.304	7.807	9.034	11.340	14.011	15.812	18.549	21.026	24.054	26.217
13	4.107	4.765	5.892	7.042	8.634	9.926	12.340	15.119	16.985	19.812	22.362	25.472	27.688
14	4.660	5.368	6.571	7.790	9.467	10.821	13.339	16.222	18.151	21.064	23.685	26.873	29.141
15	5.229	5.985	7.261	8.547	10.307	11.721	14.339	17.322	19.311	22.307	24.996	28.259	30.578
16	5.812	6.614	7.962	9.312	11.152	12.624	15.338	18.418	20.465	23.542	26.296	29.633	32.000
17	6.408	7.255	8.672	10.085	12.002	13.531	16.338	19.511	21.615	24.769	27.587	30.995	33.409
18	7.015	7.906	9.390	10.865	12.857	14.440	17.338	20.601	22.760	25.989	28.869	32.346	34.805
19	7.633	8.567	10.117	11.651	13.716	15.352	18.338	21.689	23.900	27.204	30.144	33.687	36.191
20	8.260	9.237	10.851	12.443	14.578	16.266	19.337	22.775	25.038	28.412	31.410	35.020	37.566

df													
21	8.897	9.915	11.591	13.240	15.445	17.182	20.337	23.858	26.171	29.615	32.671	36.343	38.992
22	9.542	10.600	12.338	14.041	16.314	18.101	21.337	24.939	27.301	30.813	33.924	37.659	40.289
23	10.196	11.293	13.091	14.848	17.187	19.021	22.337	26.018	28.429	32.007	35.172	38.968	41.638
24	10.856	11.992	13.848	15.659	18.062	19.943	23.337	27.096	29.553	33.196	36.415	40.270	42.980
25	11.524	12.697	14.611	16.473	18.940	20.867	24.337	28.172	30.675	34.382	37.652	41.566	44.314
26	12.198	13.409	15.379	17.292	19.820	21.792	25.336	29.246	31.795	35.563	38.885	42.856	45.642
27	12.879	14.125	16.151	18.114	20.703	22.719	26.336	30.319	32.912	36.741	40.113	44.140	46.963
28	13.565	14.847	16.928	18.939	21.588	23.647	27.336	31.391	34.027	37.916	41.337	45.419	48.278
29	14.256	15.574	17.708	19.768	22.475	24.577	28.336	32.461	35.139	39.087	42.557	46.693	49.588
30	14.953	16.306	18.493	20.599	23.364	25.508	29.336	33.530	36.250	40.256	43.773	47.962	50.892

*Table III is reprinted from Table III of Fisher: *Statistical Methods for Research Workers*, Oliver & Boyd Ltd., Edinburgh, by permission of the author and his publishers.

For larger values of df, the expression $\sqrt{2\chi^2} - \sqrt{2(df) - 1}$ may be used as a normal deviate with unit standard error.

Index

Abelson, R. P., 309, 310n, 394, 400n, 412n
Abelson-Tukey method, 309-310, 394, 400n
Absolute theories, 14
Acquiesence, 247-248
Addition, 9, 11, 12, 15, 16, 18, 20
Affine group, 35
Ager, J., 414n
Agreement
  coefficient of, 102-105
    formula, 103-104
Almqvist, 366n
Ambiguity, 234
Analysis of matter, 15
Anderson, Norman A., 398n, 400n, 407n
Anderson, R., 413n
Arc sin transformation, 94
Assenters, 254
Attitude
  continuum, 122, 235-237
  scalable, 129
  scale, equal-appearing intervals, 119-121
  score, 111
Attneave, F., 123, 128n, 407n
Attributes, universe of, 161, 162
Austin, G. A., 39n
Axioms, 276-278
  of quantity, 1, 5, 27

Bain, R., 407n
Balance errors, 58
Bales, Robert F., 365n
Ballin, M., 340n, 345n
Bar chart representation, 149
Bartlett, C. J., 407n
Bechtel, G., 407n

Behan, F. L. and R. A., 39n
Bendig, A. W., 407n
Berg, I. A., 407n
Bergmann, G., 27, 48, 54n
Binomial distribution, 286
Birkhoff, G. D., 27
Blalock, Ann B., 400n, 416n
Blalock, H. M. Jr., 399, 400n, 416n
Blood, Robert O., 382n
Bock, R. D., 407n, 414n
Bodine, A. J., 196n, 412n
Bogoslovsky, 9
Bohrnstedt, G. W., 407n
Boneau, C. A., 401n
Bonjean, C., 256n
Borgatta, Edgar F., 347-364, 413n
Boring, E. G., 54n
Bradley, R. A., 408n
Bredemeier, Harry C., 227n
Breed, Warren, 398, 401n
Brotemarkle, R. A., 408n
Bruner, J. S., 39n
Buchanan, 19n
Burke, Cletus J., 2, 42-55, 334, 335
Butcher, H. J., 258-266, 269, 271n

Campbell, Arthur, 313n
Campbell, D., 253, 254, 256n, 408n
Campbell-Fiske method, 254, 256n
Campbell, John D., 229
Campbell, J. T., 191, 196n
Campbell, N. R., 7n, 10n, 23n, 25-27, 36, 39n, 54n, 299n, 408n
Canter, R. R., 408n
Carnap, R., 27

Case, B., 414n
Castle, P. F. C., 259, 263-265, 267, 271n, 408n
Categories, 145
Chance reproducibility. *See* reproducibility, chance
Chave, E. J., 113-117, 121n, 127, 128n, 167, 243n, 346n, 415n
Chein, Isidore, 227n
Chilton, Roland J., 130, 205-222, 224n, 225, 227
Chi-square, 94, 96, 215
  significance test for coefficient of agreement, 104
  significance test for coefficient of consistence, 101, 102
Christie, R., 248, 256n
Circular triad, 98-102
  formula, 100
Clark, K. E., 170n, 337n, 338, 345n, 412n
Classical measurement theory, 25, 26
Clausen, John A., 224n
Coefficient of
  agreement. *See* agreement, coefficient of
  consistence. *See* consistence, coefficient of
  dissimilarity. *See* dissimilarity, coefficient of
  reproducibility. *See* reproducibility, coefficient of
  scalability. *See* scalability, coefficient of
Cohn, Richard, 227n
Comparative judgment, 81-92
Computer programs, 129
Comrey, A. L., 54n, 408n
Consistence, 135
  coefficient of, 98-102
    formula, 99, 100
  significance test, 101
Consistency, index of, 183, 185
  internal, 238, 242
Cook, S. W., 417n
Coombs, C. H., 2, 37, 39n, 41n, 136n, 273-332, 374-375, 405n
Coombs-Pruitt study, 303, 310n
Cornell technique, 217, 343, 359
Costner, Herbert L., 380
Couch, A., 408n
Cozan, L. W., 408n
Cournot, 7n
Couturat, 14n
Criterion, 138
  for scalability, 159-161, 203
Cronbach, L. J., 186n, 188, 192, 193, 195, 196n, 255, 256n, 409n
Cross-validation, 253
Crown, S., 259, 260-266, 269, 271n
Crowne, D. P., 409n

Crowne-Marlowe social desirability scale, 249
Cumulative, 134-136
  scales. *See* scalogram analysis

Davidson, D., 40n, 54n
Davis, F. B., 409n
Davis, H., 41n
Davis, J. A., 409n
Davis, R. L., 41n, 408n
Degrees of freedom, 95
  for chi-square test of coefficient of agreement, 104, 105
  for chi-square test of coefficient of consistence, 101, 102
Density, measurement of, 7-9
Derived measurement, 26, 27
Descartes, R., 3
Desirability, social. *See* social desirability
Detection theory, 30
Developmental process scaling, 333, 365-387
Diederich, G. W., 409n
Differences, 8
Dimensionality, 247
Dingler, 7n
Discriminal
  continuum, 61
  deviation, 83, 84, 86, 90
  difference, 69, 70, 72, 83, 85, 89, 90
  limen, 70, 87
  dispersion, 63, 65, 67, 68, 70, 72, 73, 77-80, 83, 84, 87, 88, 90
  normality of, 72
  processes, 60-62, 64, 65, 67, 68, 70, 73, 82, 90, 91
  Gaussian distribution of, 91
Dissenters, 254
Dissimilarity, coefficient of, 324
Distribution-free statistics. *See* nonparametric statistics
Dixon, Wilfred J., 213, 221n
Dodd, Stuart C., 167n
Double-barreled statements, 234, 245
Dublin, L. I., 398, 401n
Dudding, 10n
Duhem, 6n, 7n, 16n
Duncan, O. D., 389
Durbin, M., 413n

Edwards, Allen L., 57, 58, 98-105, 110-112, 113-121, 123, 124, 128n, 170n, 217, 221n, 248, 256n, 333, 336-346, 409n, 425n
Edwards, W., 40n, 409n
Ehrenberg, A. S. C., 409n

Ekman, G., 409n
Engen, T., 409n
Equal-appearing interval scaling, 37, 58, 113-121, 336
  calculations, 115-119
  graphical method, 117-119
Equal delta solution, 307, 320, 321, 323
Equal discriminability scale, 123
Equality, 7, 9, 13, 14, 29, 43
Equally often noticed differences, 78-79.
Equal-ratio scales, 37
Error-counting procedures, 216-221
Error, expected, 376
Error of measurement, 157
Error, pattern of, 160
Euclidian geometry, 44
Expected error, 376
  formula, 376
Extended measurement. See measurement, extended
Eysenck, H. J., 259-265, 266n, 269, 270, 271n

Factor analysis, 191, 252
Faris, Robert E. L., 366n
Farmerization, 246
Farnsworth, P. R., 340n, 345n
Fechner, G. T., 40n
Fechner's Law, 73-78, 81, 84, 91
Feigl, H., 27
Ferguson, L. W., 258, 266n, 409n
Fernberger, S. W., 408n
Festinger, L., 170n, 196n, 337n, 338, 346n, 408n
Fine, Paul, 227n
Fisher, R. A., 427n
Fisher exact test, 187
Fiske, D. W., 253, 254, 256n, 408n
Flanagan correlation coefficient, 189
Ford, R. N., 167n, 196n, 227n, 410n
  criteria, 229
Forehand, G. A., 410n
Foster, R. J., 414n
Francis, R. G., 410n
Freedman, Ronald, 312n, 313n
Fundamental measurement, 26, 27, 36

Galanter, E. H., 41n, 55n
Gamma, 256, 397
  partial, 397
Garner, W. R., 40n, 123, 128n
Gaylord, R. H., 190, 196n, 417n
Ghiselli, E. E., 410n
Gibbs, Jack, 401n
Gibson, W. A., 410n
Gleser, G., 255, 256n
Gödel, 24

Goldberg, David, 274, 311-332
Goode, Frank, 304, 309, 310, 319, 320
Goodenough, W. H., 170n, 381, 410n
Goodman, J. S., 39n
Goodman, L. A., 40n, 130, 206, 212-215, 217-220, 221n, 256n, 377, 378, 410n
Gordon, C., 410n
Graded dichotomies, method of, 123
Gradgram, 129
Graham, 44
Greenberg, Marshall G., 410n
Green, Bert F., 57, 58, 93-97, 122-128, 130, 183-185, 189, 195, 196n, 212-215, 217-220, 221n, 410n
Grossnickel, Louise T., 410n
Guilford, J. P., 93, 97n, 268, 270, 271n, 342, 346n, 410n
Gullahorn, J. E., 410n
Gullahorn, J. T., 410n
Gulliksen, H., 40n, 54n, 123, 128n, 270, 271n, 280, 310n, 408n, 410n, 412n, 414n, 415n
Guttman, Louis, 129, 130-133, 142-171, 171n, 172, 174-176, 178, 180, 182, 184, 188, 189, 192, 195, 196n, 197, 203, 204n, 205, 206, 216, 221n, 222n, 224n, 244, 251, 280, 293, 333, 336-338, 343, 344, 346n, 348n, 375, 415n
Guttman scaling, 129-230, 333, 334, 336, 344, 345, 347, 348, 350, 351, 363, 367, 368, 370n, 371, 372, 374, 375, 380, 381, 383, 384
Hagedorn, Robert, 388n
Hake, H. W., 123, 128n
Hall, Calvin S., 365n, 366n
Hardy, G. N., 24, 40n
Hare, A. Paul, 365n
Harman, H. H., 411n
Harris, W. P., 411n
Harvard Laboratory of Social Relations, 135, 136n
Hatt, Paul K., 401n
Hawks, Roland K., 411n
Hays, David G., 347-364, 407n
Hays, W. L., 411n
Head, Kendra B., 244-257
Hegel, 17n, 21n
Helmholtz, 7n, 12
Hempel, C. G., 27, 411n
Henry, Andrew F., 347-364, 411n
Henry, Neil W., 379n, 412n
Hertz, 7n
Heterochromatic light, 11
Hevner, K., 340n, 346n, 411n
Hicks, L. E., 411n
Hilbert, 5n
Hillegas, 88

Hill, Reuben, 365n, 366n
Hill, Richard, 256n, 407n
Hirsh, Joseph, 398, 401n, 408n
Hoelder, 5
Hoffman, Lois W., 382n
Homogeneity, 172, 247, 249, 251
  index of, 130, 180-183, 187, 189, 251
Homogeneous test, 180, 194
Horst, P., 132, 192, 195, 196n
Hovland, C. I., 411n, 412n
H scale, 347-364
H technique, 333, 347-364
Hull, Clark L., 54n
Hullian theory, 51-52
Humphre s, L. G., 190, 196n, 411n
Hunt, W. A., 416n

Illumination, 10
Inconsistency in judgments, 98
Information theory, 30
Intensity, 132
Inter-item correlations, 246, 251
Internal consistency, 238, 242
Interval scales and scaling, 2, 28, 31, 35,
  38, 43, 278
I scale, 283-292, 295, 296, 297, 298, 300-303,
  314, 315, 328, 329
Isotonic group, 35
Item selection, 117, 119, 120, 126

Jackson, J. M., 130, 176, 178-180, 181, 182,
  183, 185, 187, 189, 195, 196n, 412n
Jahn, J. A., 412n
Jardine, R., 412n
Joint distribution, 282, 293, 295, 296
Joint frequency matrix, 3ᵀ0, 371n
Jones, L. V., 407n, 412n, 414n, 416n
J scale, 282, 284, 287-292, 295, 296, 298,
  301-304, 313-315, 327-331
Jurgensen, C. E., 118n, 121n, 346n
Just noticeable differences, 59, 82
Just noticeable increase, 75

Kahn, L. A., 412n
Kahn, L. H., 196n
Kantor, 43
Kao, R. C., 408n
Katz, D., 408n, 412n
Kelley, E. Lowell, 312n
Kelley, H. H., 412n
Kelley, T. L., 89n
Kellog, W. N., 412n
Kendall, M. G., 99, 101, 102, 104, 105n,
  412n
  coefficient of agreement. See agreement,
  coefficient of

Kendall (Continued)
  coefficient of consistence. See consistence,
  coefficient of
  significance test for agreement. See chi-
  square test of agreement
  significance test for consistence. See chi-
  square test of consistence
Keniston, K., 408n
Kenney, K. C., 345n
Kilpatrick, F. P., 115n, 118, 170n, 333, 336-
  346
Kiser, Clyde V., 311n
Known groups, 253
Known group validity. See validity, known
  group
Koch, S., 412n
Krantz, D. H., 412n
Kriedt, P. H., 170n, 337n, 338, 345n, 412n
Kruskal, W. H., 32n, 40n, 256
Kuder, G. F., 112n
Kuder-Richardson estimates of reliability,
  112, 192, 193
Kurtosis, 209

Labovitz, Sanford, 334, 335, 388-401, 412n
Lambert, 10n
Latent continuum, 132, 136
Latent distance scale, 347, 351
Latent structure analysis, 129
Law of Comparative Judgment. See com-
  parative judgment, law of
Lazarsfeld, P. F., 129, 131n, 193, 196n,
  224n, 281n, 347, 348n, 351, 375, 412n
Leik, Robert K., 333, 334, 365-387
Levels of measurement, 2, 335
Likert, Rensis, 231, 232, 233-243, 258-260,
  263-266, 269, 333, 336, 344, 346n,
  412n
Likert scales. See summated rating scales
Lindauer, F., 248, 256n
Linder, R. E., 412n
Lindquist, E. F., 401n
Lindzey, Gardner, 365n, 366n
Linear group, 35
Linear interval scale, 35
Linear separations, 72
Linn, R. L., 416n
Loevinger, Jane, 130, 180-183, 185, 187, 189,
  190, 192, 195, 196n, 215n, 216, 222,
  251, 412n
Logarithmic interval scale, 28, 34-37
Long, J. A., 183, 196n
Long's index, 183
Loomis, Charles P., 365n, 366n, 367n
Loomis, Zona K., 365n, 366n, 367n
Lord, F. M., 40n, 196n, 334, 335, 402-405,
  412n

Lorge, I., 412n

Magnitude, 6, 7, 14, 15
Magnitudes, stimulus. See stimulus magnitudes
Manning, S. A., 413n
Marder, E., 196n, 413n
Marginal distributions, range of. See range of marginal distributions
Marginals, 155
Maris, Ronald N., 398, 401n
Marlowe, D., 408n
Marschak, J., 276
Marshall, Harvey, 388n
Mathematics, preceding discovery, 4
Matthews, Merlyn, 333, 334, 365-387
Maxwell, A. E., 413n
McGill, W. J., 40n
McGourty, M., 416n
McLemore, S. D., 256n, 407n
McNemar, Q., 413n
Mean
    arithmetic, 30, 120, 126
    geometric, 30
    harmonic, 30
Mean errors per respondent, 373
Mean method of scoring, 126
Measurement, 52
    definition of, 22, 24, 34, 45
    directed position, 42-48, 334, 335
    derived. See derived measurement
    extended, 53
    fundamental. See fundamental measurement
    independent position, 42-48, 334, 335
    levels of, 2, 335
    model, 52
    of error, 157
    operation, 52, 53
    properties, 45
    restricted, 53
    result, 52
    stochastic view of, 53
Median, 30, 126
    method of scoring, 11, 112, 126
    scale value, 111, 114, 116, 120
    formula, 116-117
Meinong, 6, 21n
Mela, Donald, 279
Menzel, Herbert, 224n, 380n, 413n
Messick, S., 310n, 408n, 412n, 413n, 414n, 415n
Method
    of constant stimuli, 85
    of equal-appearing intervals. See equal-appearing intervals

Method (Continued)
    of graded dichotomies. See graded dichotomies
    of paired-comparisons. See paired-comparisons
    of successive intervals. See successive intervals
Micklin, M., 413n
Midpoints, 287-289, 291, 315
Milholland, J. E., 413n
Miller, D., 256n, 413n
Mishler, Elliot, 312n
Modal discriminal process, 63, 66, 67, 71, 78, 82, 83, 90
Moore, Mary, 227n
Morgenstern, O., 40n, 276
Morris, Raymond N., 396, 401n
Mosier, C. I., 123, 128n, 268, 271n, 413n
Mosteller, C. Frederick, 94, 97n, 281n, 413n
Murphy, G., 259, 260, 263, 266n, 412n
Murphy, G. and L. B., 144, 145, 234, 243n

Nagel, Ernest, 1, 3-21, 276
Naturally ordered items, 386
Nay-sayers, 254
Neumann, J. von, 40n, 276
Nevin, J. R., 413n
Newcomb, T. M., 144, 145
Newman, E. B., 413n
Noland, E. W., 171n, 196n
Nominal scales and scaling, 2, 28, 30, 31, 35, 38, 43, 278
Nonparametric statistics, 34, 47, 48, 50
Nonscale types, 133, 160
Norms, 249, 250
North, Cecil C., 401n
Northwood, L. K., 383n
Number, 8
    of items, 160
    of possible pairs, 71
    of response categories, 161
    in science, 46
Nye, F. Ivan, 382n

Occam's razor, 14
Ofshe, L., 413n
Ofshe, R., 413n
Oliver, R. A. C., 258n, 264, 266n
Open-ended questions, 245
Oppenheimer, R., 40n
Order, 8
Ordered
    items, naturally, 386
    metric scale, 37, 38, 278, 304, 319
Ordering of response types, 373-374
Ordinal scales and scaling, 2, 28, 31, 35, 38, 43, 278

Osgood, C., 257n, 413n, 415n

Pace, C. R., 167n
Paired-comparison scaling, 37, 58, 93-97, 98-105, 115, 125
  computations, 93-97, 98-105
Parallelogram, 149, 151
Partially ordered scale, 278
Participant observation, 167
Pattern of error, 160
Peabody, D., 413n
Pearson, 355n
Perloff, R., 191, 196n
Permutation group, 35
Phi coefficient, 130, 186-188, 189, 342
Phi-gamma hypothesis, 73
Pilkington, G. W., 267-271, 271n
Plateau, J. A. F., 40n
Platonic origins of measurement-directed-ness, 43
Plus percentage ratio, 130, 178-180, 181, 187, 189
  formula, 179
Poisson distribution, 286
Poppleton, Pamela K., 267-271, 271n
Possible pairs. See number of possible pairs
Potter, Robert G. Jr., 312n
Powell, Elwin M., 398, 401n
Power group, 35
Prime intervals, 318, 319, 320, 321, 325
Principal components of scalable attitudes, 132
Pruitt, D., 303, 310n
Psychological
  continuum, 59, 62, 65-69, 72-75, 78, 79, 82, 85, 110, 111, 114, 120, 121, 282
  measurement, 66
Psychophysical
  analysis, 62
  relations, 63
Purdy, Ross, 388n

Quantity, axioms of. See axioms
Quasi scale, 132, 138, 141, 160
Quay, L. C., 407n
Q values. See method of equal-appearing interval scaling

Rackham, Horace H., 281n
Raiffa, Howard, 39n, 276, 408n
Ramsay, J. O., 414n
Rand Corporation, 135, 136n
Range of marginal distributions, 159
Ratio scales and scaling, 2, 24, 27, 28, 31, 35, 43, 278
Reece, M., 414n

Reed, Robert, 383n
Reiss, Albert J., 389, 390n, 401n
Relative theories, 14
Relativity of scales, 163
Reliability, 112, 121, 139, 172, 192, 232, 242, 249, 250, 356
  coefficients, 121, 192, 232, 342
  parallel forms, 251
  split-half, 237, 251
  test-retest, 250, 251, 356
Remmers, H. H., 414n
Reproducibility, 140, 160, 161, 172-196, 206, 218, 224-226, 333, 337, 338, 343, 349, 352, 356, 358, 363
  chance, 130, 207, 208, 213, 218, 224, 225, 226, 375
  coefficients of, 130, 137, 146, 157, 159, 170, 172-196, 208, 210, 214, 224, 251, 377, 378, 381, 384, 385
  formula, 178, 183-185, 378
  error of, 138, 139n, 184, 378, 385
  marginal, 379
  minimum, 179
Response categories, 138
  number of, 161
Response set, 247
Restricted measurement. See measurement, restricted
Rho, partial, 397
Rice, S. A., 234, 243n
Richardson, M. W., 112, 416n
Riley, John W. Jr., 227n, 414n
Riley, Matilda White, 207, 214n, 222n, 227n, 414n
Robinson, John P., 232, 244-257
Rodgers, Roy H., 365n, 366
Rohrer, J. H., 416n
Rorer, L., 247, 256n
Rosenstock, E. H., 413n
Rosenthal, I., 256n
Roslow, S., 259, 260, 263, 266n, 412n
Ross, Robert T., 58, 106-109, 414n
Rozenboom, W. W., 412n
Runge, 7n, 8n, 21n
Rush, Jerrold G., 244-257
Russell, B., 12-15, 24, 40n

Saffir, M. A., 122, 123, 128n, 414n
Sagi, P. C., 130, 206, 208, 212, 215n, 216, 222n, 224n, 312n, 414n
Saltz, Eli, 130, 172-196, 414n
Sampling, 249
Savage, L. J., 32n, 40n
Scalable attitudes, 129
  principal components of, 132
Scalable universe, 162, 165

Scalability, 224, 225, 247, 380, 384, 385
 coefficient of, 381
  chance, 225
  significance test, 334, 385
 index, 334, 380
Scale
 definition of, 144, 145
 discrimination technique, 336-346
 pattern, 158
 product weights, 264, 269, 270
 score, 146
Scales of measurement, 27, 43
Scalogram
 analysis, 129-230
 board, 129, 175
  technique, 169
 criteria, 140
Scates, D. E., 414n
Schlick, 7n
Schooler, Carmi, 130, 223-230
Schuessler, Karl F., 130, 197-204, 206, 207,
  209, 212, 215, 217, 218, 222n, 224n,
  414n
Schulman, G. I., 414n
Schutz, R. E., 414n
Schwartz, M., 412n
Scott, W. A., 251, 256n, 414n
Seashore, R. H., 340n, 346n
Senders, Virginia L., 40n, 47, 54n, 414n
Sense distance. See discriminal difference
Sensory stimulus intensities, 75
Shapiro, G., 171n
Sharp, Harry, 312n
Shaw, M., 257n, 414n
Shepard, R. N., 310, 414n
Sherif, M., 411n, 416n
Shuford, E. H., 414n
Siegel, S., 40n, 47, 54n, 414n
Sigerfoos, C. C., 120, 121n
Similarity group, 35
Simultaneous contrast, 86
Sjoberg, Gideon, 366n
Sjoberg, L., 409n, 414n
Skew, measure of, 209
Skinner, B. F., 44
Slater, P., 414n
Sletto, R. F., 166n
Smith, B. B., 412n
Smith, D., 249, 257n, 414n
Smith, R. G. Jr., 414n
Snedecor, George W., 422n
Snider, J., 257n, 415n
Social desirability, 248
Somers, Robert H., 389, 401n, 415n
Sorenson, W. W., 408n

Space
 criterion, 108
 errors, 58, 108, 109
Spaier, 3, 16
Spence, K., 44, 54n, 407n
Split-half reliability, 237, 251
S-scale, 313, 314, 315, 318, 326, 330, 331
Standard error, 159, 208, 219
Standardization, 399
Star, Shirley, 224n
Statistical practice, 48
Statistics, 46
Stefan-Boltzmann law of energy radiation,
  21
Steinburn, Thomas, 379n
Stevens, S. Smith, 2, 22-41, 43, 46, 47, 54n,
  55n, 274, 276, 277, 282, 290, 299n,
  405n, 415n
Stigler, G. J., 41n
Stimulus
 continuum, 61, 74
 differences, 78, 81
 magnitudes, 59, 74, 75, 77
Stone, R. C., 410n
Stouffer, Samuel A., 130, 131-141, 142n,
  205n, 224n, 281n, 347-364, 415n
Subjective dispersion. See discriminal dis-
  persion
Subliminal comparisons, 86
Successive contrast, 86
Successive-interval scaling, 58, 122-128
 computations, 124-126
Suchman, E., 129, 131n, 171n, 224n, 348n,
  415n
Suci, G. J., 413n
Summary statistics method, 130
Summated rating scales, 231-272, 333, 336,
  341, 344
Suppes, P., 40n, 54n
Supraliminal comparisons, 86
Surrogative measurement, 16, 20

Tannenbaum, P. H., 413n
Tau, partial, 397
Tchebycheff's inequality, 404n
TenHouten, W. D., 415n
Test-retest reliability. See reliability, test-
  retest
Tetrachoric coefficient of correlation, 152n
Time
 criterion, 108
 errors, 58, 109
Thomasson, Richard, 313n
Thomson, G. H., 415n
Thorndike, 88, 92
Thrall, R. M., 39n, 41n, 408n

Thurstone, L. L., 41n, 57, 59-80, 81-92, 93, 94, 97n, 113-117, 121n, 122, 124, 127, 128n, 136n, 167n, 191, 232, 234, 237, 243n, 258, 299n, 336, 339, 346n, 415n
Thurstone
  case I, 84, 91
  case II, 85, 91
  case III, 85, 91
  case IV, 87, 92
  case V, 57, 88, 92, 94
  scales, 57-128, 234, 258, 267, 268, 336, 342, 344, 345
  score, 268
  weighted, 259, 264, 265
  weighted scoring, 232, 269
Tittle, C. R., 414n
Toby, Jackson, 227n, 414n
Toby, Marcia L., 227n, 414n
Tolman, E. C., 44
Torgerson, W. S., 197, 200, 204n, 374-375, 416n
Tukey, J. W., 32n, 309, 310, 194, 400n
Tversky, A., 412n
Type I error, 48
Type II error, 48

U. See coefficient of agreement
Udy, S. H., 366n
Uhrbrock, R. S., 416n
Unfolding
  methods, 273-332
  theory, 273-332
Unidimensionality, 343, 344
Universe
  of attributes, 161, 162
  of content, 165, 166, 244
  scalable, 162, 165
Upshaw, H. S., 416n
Urban, F. M., 416n

Validation, 249
Validity, 250
  known group, 253
Vernon, P. E., 258n, 265, 266n
Volkmann, J., 41n, 416n
Von Neumann, J., 40n, 276

Walsh, 11n
Warrain, 6n
Watson, J., 43
Weber's law, 73-78, 81, 84, 91
Weitzenhoffer, A. M., 405n, 416n
Werner, R., 213, 222n, 416n
Werts, C. E., 416n
Westoff, Charles, 312n
Whelpton, P. K., 311n, 313n
Wherry, J. J., 190, 196n, 417n
White, Benjamin W., 130, 172-196, 281n
Whitehead, A. N., 28, 41n
Wiener, N., 41n, 417n
Wiksell, 366n
Wilcoxon, 34
Williams, Robin, 366n
Willis, R., 417n
Wilson, Thomas P., 334n, 417n
Witryol, S. L., 417n
Woodger, J. H., 275
Wright, J., 257n, 414n
Wrightsman, L., 248, 257n, 407n, 417n
Wronski, 6n

Yea-sayers, 247, 254

Zavala, A., 417n
Zavalloni, Marisa, 417n
Zeta. See coefficient of consistence
Zetterberg, Hans L., 366n